THE
SOUTHERN
ARK

THE SOUTHERN ARK

Zoological Discovery in
New Zealand
1769 – 1900

J. R. H. ANDREWS

UNIVERSITY OF HAWAII PRESS
HONOLULU • HAWAII

First published by Century Hutchinson New Zealand Limited
© J. R. H. Andrews 1986
Published in the United States of America by
University of Hawaii Press
2840 Kolowalu Street
Honolulu, Hawaii 96822
Printed in Hong Kong

Library of Congress Cataloging-in-Publication Data
Andrews, J. R. H.
    The southern ark.

    Bibliography: p.
    Includes index.
    1. Zoology — New Zealand — History. I. Title.
QL340.A53 1988   591.9931   88-10701
ISBN 0-8248-1192-5

CONTENTS

PREFACE

IN THE 130 years covered by this book, the study of botany and zoology underwent enormous changes. More than any other branches of science they were tied — together with geology — to trade and commerce and the development of empires: an inextricable link that advanced both scientific and colonialist causes. It was then fortunate that the discovery and development of New Zealand occurred during this period, and became so intimately part of it. It was fortunate too that this country's limited fauna, both fossil and extant, should attract such widespread attention; should reveal itself so gradually decade after decade to maintain a prominent place in the scientific arenas of the northern hemisphere. The kiwis, the giant Moas, the rare Takahe or *Notornis* and the Tuatara, were the more obvious examples. But even among the lesser faunal lights were specimens sought after, collected or described by the great and famous of the civilized northern world. There was a time when a shell from New Zealand might be cradled in the hand of a collector from those parts and viewed with the same wonder that would be accorded a rock from a distant planet.

It comes as something of a surprise, therefore, to learn that detailed historical accounts of the biological sciences in New Zealand are relatively few.[1] For example: among biographies of naturalists from the period of our interest, those of Colenso, Haast and Reischek stand virtually alone and only recently has the life of Sir Walter Buller begun to receive attention beyond the existing résumés.[2] Specialized histories covering particular periods or groups of animals are also scarce, although there is David Medway's meticulous work on the ornithology of Cook's voyages; the New Zealand material in the works of Dr Peter Whitehead on Cook's naturalists; and Sir Charles Fleming's article on the history of the New Zealand Institute and the Royal Society is in the process of being expanded into a considerably larger work.[3] Much of scientific interest can also be taken from accounts of the major voyages of exploration, beginning with J. C. Beaglehole's works on the Cook voyages (see sources below) as well as the useful synopsis of the voyages by Fell and his colleagues.[4] Short articles on the evolution debate, and the history of entomology and marine biology have also been published and so has, at greater length, the story of New Zealand's animal and plant importations.[5]

In the more general categories of New Zealand science history, the work of Callaghan, Dick and Hoare must be mentioned; the latter's booklets based on his Cook lecture series and his contributions, with others, to the proceedings of the first conference on the History of Science in New Zealand are a particularly valuable introduction to the subject.[6]

This very brief review has not taken account of those innumerable scientific publications whose introductions and reviews of literature are invaluable sources for the student of science history — an early example is J. E. Gray's preface to his catalogue of the New Zealand fauna in Dieffenbach's *Travels....* published in 1843. Many such writers are taxonomists; biologists who specialize in the classification and naming of animals and plants, and who inevitably become historians of science as they probe into the early literature to validate the scientific names of their specimens, or grub round in libraries and museums for old correspondence, shipping lists or street directories and so secure the provenance of this or that type species. Most prominent in the periodical literature

1. Hoare, 1984, p.1.

2. Bagnall and Petersen, 1948; von Haast, 1948; King, 1981; Bagnall, 1966; Turbott, 1967; Galbreath, 1984.

3. Medway, 1976, 1979; Whitehead, 1968, 1969, 1978a, 1978b; Fleming, 1968.

4. Fell, et al., 1953.

5. Stenhouse, 1984; Miller, 1953; Putnam, 1977; Thomson, 1922; Druett, 1983; Miller, 1984.

6. Callaghan, 1957; Dick, 1951; Hoare, 1977a, 1977b, 1984; Hoare, 1977a, p.9, finds medical history somewhat better placed than the other life sciences.

were the *Transactions and Proceedings of the New Zealand Institute*. From 1869 these provide a continuing documentary record of the progress of New Zealand science, and the leisurely writing style of the period demonstrates something of the wit and fire that characterized the deliberations and exchanges of the contributors. Further back this record can be traced in the scientific journals of England and Europe.

In combing through the manuscripts and published works one finds that an attempt at a history of zoological discovery in New Zealand is feasible, but the subject is at once both sprawling and intricate. How then to go about it? The possibility of a history of 'natural history' incorporating botany and geology as well as zoology was quickly put aside. Many of the distinguished figures referred to in the later pages delved into more than one field, often throwing in astronomy, meteorology or hydrography as well. But such an approach would result in a work that was either superficial or intolerably long.

The scope of zoology alone is still immense, prompting the question of what to put in and what to leave out. The work of Bagnall and Petersen, and Hoare and others gave me a feeling for the major periods and influences in New Zealand's scientific history: the voyages of exploration, the missionaries and so on. I became aware that particular animals, or groups of animals — the flightless birds for example — helped turn the flywheel of New Zealand science, and would need special mention or even chapters to themselves. The logical answer, it seemed, was to incorporate these within an examination of the major periods, thus placing in starker relief the individuals, the times and the circumstances in which they worked, and the animals they studied. There was also the advantage here of maintaining some sort of chronological order which would be easier to follow. The choice to proceed by way of a narrative, with appropriate pauses for explanation or analysis was a difficult one. Too many pauses and the narrative falls apart, too few and the work loses some of its purpose. I hope I have achieved a reasonable balance. The choice of time at which to conclude the work — 1900 — was initially rather arbitrarily selected; the end of a century seemed a convenient place at which to stop, and I had no wish to enter the complexities and length of a work on the twentieth century. Events proved however that the date chosen was in several respects a good one.

For whom was this work to be written? It was soon clear that natural history and the sciences that emerged from it have long held a constituency of enthusiastic amateurs — indeed, the very development of these sciences for a time depended on them. Even today, the natural sciences have maintained an amateur following that continues to earn the respect and cooperation of its professional counterpart. Amateur naturalists are therefore entitled to expect, along with other lay followers of science, a work that can be understood by the non-specialist. In attempting to appeal to a general as well as a specialized readership I have tried to leave the text relatively unencumbered by scientific detail and technical terms, reserving such matter for the footnotes.

For scientific colleagues I hope the work will provide a useful background and an analysis that for the time being will be unavailable elsewhere in easily accessible form. Perhaps it may help them understand how the institutions in which they work, and even the roles they themselves now play, have evolved. More specifically, it is hoped that those who labour in the various branches of zoology might see here the tentative beginnings of a history of their own specializations, and that some will be tempted to extend the study beyond the limits set. Finally, scientists need constant reminding that theirs is a human endeavour, inseparable from the personalities and motivations of its practitioners and the social, cultural and economic milieu in which it is found. Such influences will buffet and change the direction of science as much as those factors that come from within science itself. Within this web of conflicting forces we can detect the issue of the balance between science's leading and its reacting within the culture. This is not a question answered within this book, but it is at least exposed.

In conclusion I hope that historians will overlook my inadvertent infringements of their discipline and find that the treatment of the subject from a different perspective is useful and stimulating enough to encourage further interest in this relatively neglected field.

Textual Notes

Animals have generally been referred to in the text by their standard common name (where that exists) and in the footnotes by their scientific name. The common names for birds and insects have been taken from the *Annotated Checklist of the Birds of New Zealand* by Kinsky *et al.*, and the *Standard Names for Common Insects of New Zealand* by Ferro *et al.*, respectively. For fish and shells I used *Collins' Guide to the Sea Fishes of New Zealand* by Ayling and Cox, *New Zealand Freshwater Fishes* by McDowell and *Common Seashells* by Penniket. Reptile names are those suggested in *New Zealand Herpetology* (D. G. Newman, Ed.). There will be some inconsistencies where a species exists for which there is no common name and in instances such as the Moa where the accepted common name covers a number of species. Scientific names used are, as far as can be ascertained, those currently favoured by authorities in the various groups. The names are latinised and binomial, the first (generic) name (e.g. *Apteryx*) being followed by the species name (e.g. *australis*). A third might be added in the case of a subspecies, thus: *Apteryx australis mantelli* – the North Island Brown Kiwi.[7] Listings of species collected on the earlier voyages are as complete as I can make them, although given the complexities of their recording, and the name changes that have taken place since some may have been overlooked. Post 1840, such listings become impracticable and only records of particular significance are noted. Dates used in the citation of literature are the dates of issue, rather than the nominal date of publication. For example early numbers of *Transactions and Proceedings of the New Zealand Institute* were almost invariably issued the year following the nominal publication date. Illustrations were an important part of the zoological literature, and in many instances they dominated the text. Their value, particularly in earlier times, sometimes went beyond showing the reader what the specimen actually looked like. This happened when the 'type' or original specimen from which the description was made became lost or destroyed. Greater scientific value is attached to such illustrations, which are referred to by some zoologists as iconotypes. Through most of the period covered by this book, very high artistic and technical standards were set by those who prepared and published the illustrations, and the period also saw considerable changes in the style and technique of reproduction. The stylistic changes even came to reflect developments in the discipline of zoology itself. The choice of illustrations made for this book has attempted to reflect the foregoing, with many of those appearing being unpublished (beyond their original appearance); in some cases the artist's originals have been reproduced for the first time.

7. The standard common names are prefixed with capital letters e.g. the North Island Brown Kiwi. Less specifically lower case initial letters are used e.g. 'kiwis', 'kakas'.

Sources

Primary sources (including original illustrations) were drawn on from the following institutions or locations: Alexander Turnbull Library (Buller, Hutton, Gideon and Walter Mantell, Taylor, Yate); National Museum Archives (Buller, Hudson, McLachlan); National Archives (Buller, Pollack); Gibbs family (Hudson); Royal College of Surgeons,

London (Buckland, Clift, Earl, Home, Owen); Linnean Society of London (Forster, Lesson, Swainson); Merseyside County Museum (Derby, Accessions Records); Knowsley Hall Library (Gould); British Museum (Book of Presents); British Library (Williams, Broderip); British Museum (Natural History) (Owen, Keulemans, Accessions Records); Hope Entomological Collections Library, Oxford University (Hudson, Jones, McLachlan); National Maritime Museum, Greenwich (Lloyds Register); Muséum National d'Histoire Naturelle, Bibliothèque Centrale (Werner, Cuvier, Dumont d'Urville, Duperrey, Gaimard, Lesson, Quoy); Muséum National d'Histoire Naturelle, Ornithologie, Entomologie (Accessions Records). In addition, many original printed works written by the early naturalists, in book or periodical form, were consulted at the above libraries and in the Beaglehole Room, Victoria University of Wellington Library.

I have also used a number of secondary sources, mainly those relating to voyage journals and accounts. In particular I have made frequent reference to the works of J. C. Beaglehole on the Cook voyages and the lives of Cook and Banks, M. E. Hoare on the Forsters, and Olive Wright and John Dunmore on the French voyages. Without the thorough scholarship of these authors, the task of writing this book would have been much harder. Several publications have been found useful sources of reference without always being directly quoted in the text, e.g. the *Bibliography of New Zealand Marine Zoology 1769-1899* by Freed, the *Annotated Index to Some Early New Zealand Bird Literature* by Oliver, and Fildes *Selective Indexes to Certain Books Relating to Early New Zealand*.

ACKNOWLEDGEMENTS

A NUMBER OF PEOPLE made the task of writing this book both pleasant and instructive. Pleasant, because of their unflagging encouragement and interest, and instructive because my knowledge of the zoological groups was measurably greater at the end of the project than it was at the beginning, and friendly historians patiently introduced me to some of their conventions, as well as sharing their knowledge of New Zealand and European history. Among those who assisted me the following deserve special mention: Ken Pounder, Charles Clark and Lance Earney of Hutchinson for their enthusiastic support throughout; Michael Gifkins, their editor, and Lindsay Missen, the designer; Dr Peter Whitehead of the British Museum (Natural History) and David Medway of New Plymouth, whose knowledge of science history, particularly of Cook's period, was generously shared; Miss Clem Fisher, of the Merseyside County Museum for assistance with kiwi specimens and accession records; M. Jean Roche of the Muséum National d'Histoire Naturelle, Ornithologie, for hospitality and help in searching records; M. Yves Laissus of the Bibliothèque Centrale, Muséum National d'Histoire Naturelle, for arranging examinations of manuscripts; Professor Jack Garrick, Drs Peter Castle and George Gibbs of the Victoria University Zoology Department who patiently answered many questions on matters of fish and insect taxonomy, and the last again for assistance in arranging the loan of G. V. Hudson's diaries; Sir Charles Fleming for helpful discussion and the loan of books; Dr Bruce Sampson of the Victoria University Botany Department, concurrently writing a book on New Zealand botanical art and able to offer much helpful advice; Dr John Yaldwyn and Dr Alan Baker of the National Museum for assistance with Moa and marine mammal literature; Ray Grover, National Archivist, for help in arranging materials and many useful suggestions regarding the manuscript; Katherine Coleridge and Nan Taylor of the Beaglehole Room, Victoria University Library, for generous assistance with the literature; Dr David McKay of the Victoria University History Department, for help with Banks and Cook materials; Dr Michael Hoare, late of the Manuscripts Section, Alexander Turnbull Library, for enthusiastic support and helpful discussion and Mrs Marion Minson and Miss Moira Long of the same library for help with illustrations; Dr Ross Galbreath, Entomology Division, DSIR, for help in checking the Buller portion of the manuscript; the late Colin Roderick for several useful discussions on kiwis and their history; Miss Nesta Black, who with unfailing good humour typed the manuscript; Ms W. Baker, Ms R. Yeoman and Mr J. Gellen for assistance with the indexes; and members of my family, particularly Pamela Andrews, who provided helpful criticism and put up with considerable inconvenience over the term of the book's preparation.

I am also indebted to a number of others for assistance as follows: from the departments of the Victoria University of Wellington; Professor J. Wells, Drs B. Bell, D. Burton, C. Daugherty, R. Wear, Mrs R. Ainsworth, Mrs D. Fox, Messrs T. Worthy, A. Hoverd, C. Roberts (Zoology); Dr J. Dawson, Mrs O. Vincent (Botany); Professors P. Munz, D. Hamer, Drs T. Beaglehole and J. Phillips (History); Professor G. Hawke (Economics); Professor H. Delbrük (Modern Languages); Dr C. Parkin (Philosophy), Mrs D. Freed, Mrs C. Dunlop and Mrs B. Harris (Library); Professor R. Cave (Librarianship); Mr H.

Orsman (English); Mr J. Casey and Miss I. Anderson (Photographers).

Professor J. Robb (Auckland University, Zoology Department); Drs J. Dugdale, G. Ramsay, R. Craw and W. Kuschel (Entomology Division, DSIR); Mr L. Paul (Fisheries Research Division, MAF); Mr E. Dawson (Oceanographic Institute, DSIR); Dr M. Williams and Mr B. Reid (Wildlife Service, Department of Internal Affairs); Drs F. Climo, G. Hardy, Messrs S. Bartle, R. Ordish, R. Palma, A. Marshall, F. Wiley and Mrs M. Angelo (National Museum); Mr A. Bagnall; Mr P. Parkinson, Mr D. Retter (Alexander Turnbull Library); Dr W. Czernohorsky (Auckland Institute and Museum); Mrs S. Natusch (Wellington); Dr A. Wheeler, Mr A. Harvey, Mrs A. Datta, Miss J. Jeffrey, Miss A. Jackson (British Museum — Natural History); Miss J. Wallace (British Museum); Mrs I. Young (Knowsley Hall); Mrs A. Smith (Library, Hope Entomological Collections, Oxford); M. Quentin (Muséum National d'Histoire Naturelle — Entomologie); Mme C. Hustache (Bibliothèque Centrale, Muséum National d'Histoire Naturelle).

The cooperation of the British Museum (Natural History) Library, the Bibliothèque Centrale, Muséum National d'Histoire Naturelle, the Australian Museum, the Victoria University Library, the Alexander Turnbull Library, and the National Museum Library, in allowing the reproduction of illustrations and in all other facets of library assistance, is gratefully acknowledged. I am also indebted to the Earl of Derby for his permission to use the library at Knowsley Hall.

Finally I thank the following for financial assistance: The British Council, and the Council's representative in Wellington, Mr D. Howell; the Internal Research Committee of the Victoria University of Wellington, and the Hutton Fund of the Royal Society of New Zealand.

Chapter One

OF EMPIRES AND 'GENTLEMEN AMATEURS'

*In Ao-pouri, the world of darkness, Rangi-nui, the sky
father, dwelt with his wife, Papa-tua-nuku, the earth mother, and the
living things of the earth were created. First the plants, to cover Papa,
followed by the trees of the forest. After them, insects of every kind. The aitanga-
pekepeketua, the ancestors of the tuatara, the great lizard, and the lice of the
earth mother. Then the crabs, univalves, bivalves — all things that have shells,
and then animals, dogs of every species. Next came the birds, and the gods parted
the parents to create the 'World of Light' — the sun, moon and the stars. The
waters of the heavens were heated and species of fish descended to earth.
Finally, Tane made the first woman Hineahuone, and her daughter,
Hine-titama, from whom mankind was created.*[1]

I T WAS THE EIGHTEENTH CENTURY, the first tentative European contact with New Zealand had been made and the flora and fauna of that country were long since integrated into a culture. As part of a vocabulary, diet and mythology, the plants and animals had become subservient to the spiritual and bodily needs of the country's first human inhabitants, the Maori. The abundant fish, shellfish and crustaceans of the rivers and coasts, and the birds of the forests, were sources of food. Reptiles, eaten in earlier times, were the inspiration for symbolic carvings and for legends, amplified into monsters of enormous size and ferocity. The giant flightless birds were virtually a memory, the subject of stories told to grandchildren, but in this case the stories were true, as the bleached relics in the swamps, river valleys and caves would bear testimony. Insects were ever-present — the shrill, summer chirping of Kihikihi[2] the cicada, Namu the sandfly, Waeroa the mosquito and Kutu the louse — and at times too familiar. Mammals were few; the dog and the rat they had brought with them, and they knew Pekapeka the bat. Only the tiny amphibian fauna remained obscured from the Maori;[3] perhaps it too had been consigned to mythology. The hard and skeletal remains of living creatures were converted into hooks and lures to catch more of their kind. Some were turned into ornaments of dress and habitation. The first introductions had been made, habitats modified, forests burnt, and species already doomed by nature were being hastened to their end. All man's relationships with nature — confrontational, passive, yielding — had been observed by the Maori, including the identification and observation of animals for their own sake.[4] Even, by instinct or ritual, their conservation and management. But for the greater part the animals were recorded only in oral traditions, in stylised carvings, or in rock designs.

Now the time was approaching when another culture might attempt description of these animals — descriptions written and figured, an essential cataloguing that might contribute to the basis of new ideas and philosophies, as well as serving the requirements of the collector's cabinet or the coarser needs of the larder or commerce. This was a task that Europeans, with their written language and illustrative skills, could undertake. But for them the time was not yet ripe. They must wait for social and economic change at home — a change from economies based on agriculture and commerce; a change from the preoccupations of wars that were to do with trade and global supremacy. Needed

1. Derived from Miller, 1952, pp.1-5; Smith, 1913, pp.117, 136-137, 156, 157.

2. The Clapping Cicada.

3. Thomson, 1853, p.68. The Maoris seemed unaware of the New Zealand frogs' existence and had no name for them.

4. Miller, 1971, pp.147-148.

too were commercial concerns, an interest in things scientific, and to appeal to more romantic and adventurous instincts, the lure of an undiscovered and exotic southern continent. Previous European contact with New Zealand in 1642 had come too soon in the history of scientific development and was too transitory to be of use — we do not know of any animals collected or described from Tasman's voyage in his encounter with New Zealand in 1642. Thus it was not until the latter half of the 18th century that the necessary elements began to assemble themselves in the right proportions, in sufficient quantity and at the right time for New Zealand to be discovered in what we might call a scientific sense.

To identify all these elements, to position them correctly in the mesh of social, economic and scientific factors that led to the zoological discovery of New Zealand, would be a lengthy and difficult task. But as the 18th century moves into its second half, some of the more significant of them are at least able to be identified.[5]

Clive's victory at Plassey in 1756 had helped found the British Empire in India; in 1758-59, British naval and land forces cut off the French, first at Louisburg and then at Quebec, initiating the fall of French Canada. Present at the latter campaign as a ship's master, and learning the techniques of marine surveying, was James Cook, yet to become one of the principal figures of this story.[6] The victories in Canada, India and elsewhere ensured that Britain was predominant among the colonial powers when the Treaty of Paris was signed in 1763, and the seas for the time being were hers to command and explore.

In the brief period of peace that followed, a new energy and prosperity invested Britain as industry and agriculture were allowed to develop. The 1760s were to be the years of the Bridgewater Canal, the spinning-jenny, Watt's steam engine, Josiah Wedgwood and the Staffordshire potteries. The century had seen progress in the physical sciences in Britain; natural history had been given some slight momentum by Sir Hans Sloane's presidency of the Royal Society and, in 1759, the opening of the British Museum as a result of his bequest. In 1760, the King and Queen were interested in botany, and in the early years of the decade entomologists and other naturalists began to forgather.[7] Now perhaps the time was right for natural history to be removed from the stagnation in which it had been languishing for several decades, and a time for Britain to join more fully the revival of scientific learning that characterized the middle years of the 18th century in Europe.[8]

European progress in natural history had been seen in the works of the insect men, De Geer, Bonnet, Réaumur, Lyonnet; in the development of societies such as the influential and exclusive Gesellschaft der Naturforschender Freunde of Danzig; and in the person of Peter Simon Pallas from Berlin, an indefatigable field worker who would visit England and Holland before his journeys carried him through the Russian Empire — a counterweight to the Pacific exploration that took place at the same time. But in the compilation and popularization of natural history there was little to equal the vast work of the Frenchman Louis LeClerc, Compte de Buffon, Keeper of the King's Gardens and the Royal Museum. He ground out volume after volume of his *Histoire Naturelle. . .* while dressed in full court costume, complete with orders, day after day and for more than 40 years.[9] Although later said to be responsible for 'a vastly long look at nature, an immobility in the centre of the world . . . a monument of books built on demolished truths . . .'[10] he had compiled a major integration and organization of the known content of natural history and no work had greater impact at the time.

If Buffon had created a monument to natural history, then the path ahead was laid

5. Some of these have already been raised in great detail in the context of Cook's voyages and the lives of Cook and Joseph Banks: e.g. Beaglehole in Banks, 1962, pp.1-3; Beaglehole in Cook, 1955, pp.xxi-cxxii; Beaglehole, 1974, pp.99-127.

6. Beaglehole, 1974, pp.26-59.

7. Sloane's collection was the largest in private hands in Europe. He has another claim to fame as the inventor of milk chocolate. (Stearn, 1981, pp.4, 5); Allen, 1976, pp.43, 45.

8. Cole, 1975, p.452.

9. Allen, 1976, p.39.

10. Gaillard, *in* Mançeron, 1983, p.138.

by another man, the Swedish botanist Carl von Linné, better known as Linnaeus, who had included animals in the classical 10th edition of *Systema Naturae* in 1758 and thus revolutionized zoological nomenclature. Here at last was an appealing system — a sensible and easily understood classification of both plants and animals. The basic unit would be the species, the combination of the latinised generic and specific names being unique to a particular life-form. Surely such a system would be grasped at with relief by perplexed natural historians overwhelmed by floods of new species? But in a way the floods were necessary to encourage acceptance, leaving the system to infiltrate slowly but surely into the zoological and botanical writings of Britain and Europe, fostered by the talented Linnaean students — apostles was not too strong a word[11] — who travelled, collected, systematized, described, corresponded and extended his influence, and some of whom appeared in England. Yet the revival of natural history in that country was also to depend on a widening of the circle of intellectual debate by the inclusion of women, and on local naturalists with their respectable and sometimes laborious compilations, such as the *British Zoology* by Thomas Pennant. A country doctor whose prolific writings in natural history made him a leading British naturalist, Pennant's refusal to accept the Linnaean system was a stumbling-block to progress. Less well known but well thought of nonetheless was George Edwards and his *Natural History of Birds*, continued as *Gleanings in Natural History*. While Edwards, Pennant and others were providing basic reference materials for future British expeditions, a certain revival of spirit was occurring near Selbourne, a small village not far from London, where the observations and writings of Gilbert White, then curate of Faringdon, were being assembled and would eventually comprise a book that would have a profound effect on natural history in his own century and beyond.[12]

It was against this background and in such surroundings that an energetic and personable young man called Joseph Banks underwent his university instruction in botany, inherited his father's large fortune, and began to take an interest in scientific expeditions. It was Banks's energy and vitality, which were later to contribute so much to British science, which secured him an education in the first place, such were the shortcomings of the universities of the day. Then and for some years to come, the fate of natural history would be bound to individuals rather than institutions — universities were not yet the obvious places in which to carry out research. Indeed, as has been said in the case of Joseph Banks, they were sometimes not even the obvious places in which to be educated.[13] Museums were in their infancy and it would be years before the British Museum would be a reliable institution in which to deposit specimens. Naturalists of the day often followed another calling — particularly as clergymen or physicians — and, scattered as they were far and wide, communication among them while not impossible was slow and unreliable. There were relatively few societies for natural history or its branches, although natural history was an interest of society. It was an age that encouraged amateurs,[14] the heyday of the collector and the classifier; the voyager abroad and the compiler at home. But from time to time Banks, Forster and their contemporaries would show by thought or deed that they prefaced the growing disjunction between 'natural history' and 'biology' that would appear by the end of the century,[15] and prepared the way for a more utilitarian view of science.[16]

This all too briefly was the state of affairs before 1768. The observation of a rare astronomical event[17] was now to give scientific purpose to a voyage to the southern oceans, where the rumoured southern continent had become an irresistible attraction, and there were manifold discoveries in natural history to be made. It must be conceded

11. It was in fact the term used by Linnaeus himself of his students. Beaglehole, 1974, p.145.

12. The classical *Natural History and Antiquities of Selbourne* published in 1789.

13. Cameron, 1952, pp.2, 3. See also Banks, 1962; Smith, 1911 for detail of Banks' life.

14. Beaglehole, *in* Banks, 1962, p.123.

15. 'Natural History' would retain its amateur constituency but 'biology' would become the preserve of the more academic scientists. This is not to say that natural history ceased to be of interest to the 'professionals' on the contrary it attracted professional support throughout the 19th century (see Farber, 1982, pp.125, 126).

16. McKay, 1979, pp.25, 28.

17. The transit of Venus.

18. The identity of the birds cannot be established with certainty: the Fluttering Shearwater *(Puffinus gavia)* and possibly a Cook's Petrel *(Pterodroma cookii)* – but there are other possibilities. D. Medway, *pers. comm.*

19. Banks, 1962, p.399.

20. Sydney Parkinson mentions *'Beroe Coaretata'* (probably one of the so-called Comb-jellies) seen the previous day (5 October) and porpoises on 4 October: in Parkinson, 1773, pp.86, 87. A *Beroe* was described by Solander on 6 October, Sol. MS. Z4, Zoology Library, British Museum (Natural History).

that in Cook's first Pacific exploration, natural history rode in on the back of the physical sciences, but it was significant that it did so in the person of Joseph Banks. Whatever his later failings, and whatever the stature of those who followed him, he is crucial to the beginnings of zoological discovery in New Zealand. And yet standing behind Banks was the appropriateness of the times – economic, political and social – and the pendulum's swing in favour of developments in the natural sciences, particularly in Britain. The disciplinary setting had been provided by Linnaeus and his school, with an immediate representative in the form of Daniel Solander who would accompany the expedition to the Pacific. We are also fortunate in the appointment of a sympathetic and skilful captain – without Cook's tolerance of Banks and his entourage, his interest in their work, and his skill as a navigator, the natural history discoveries would have been far fewer.

It is this miscellany of people and events that eventually leads us to the observation of a small shark at sunset on 6 October, 1769, and then the afternoon of the next day with Joseph Banks in his small boat firing his musket at sea birds[18] and dipping a net into the still waters of Poverty Bay for salps, polyzoans and seaweed.[19] Here would be the physical beginnings of zoological discovery in New Zealand – the first field work (and that was all it was to be) carried out by a European naturalist in New Zealand waters.[20]

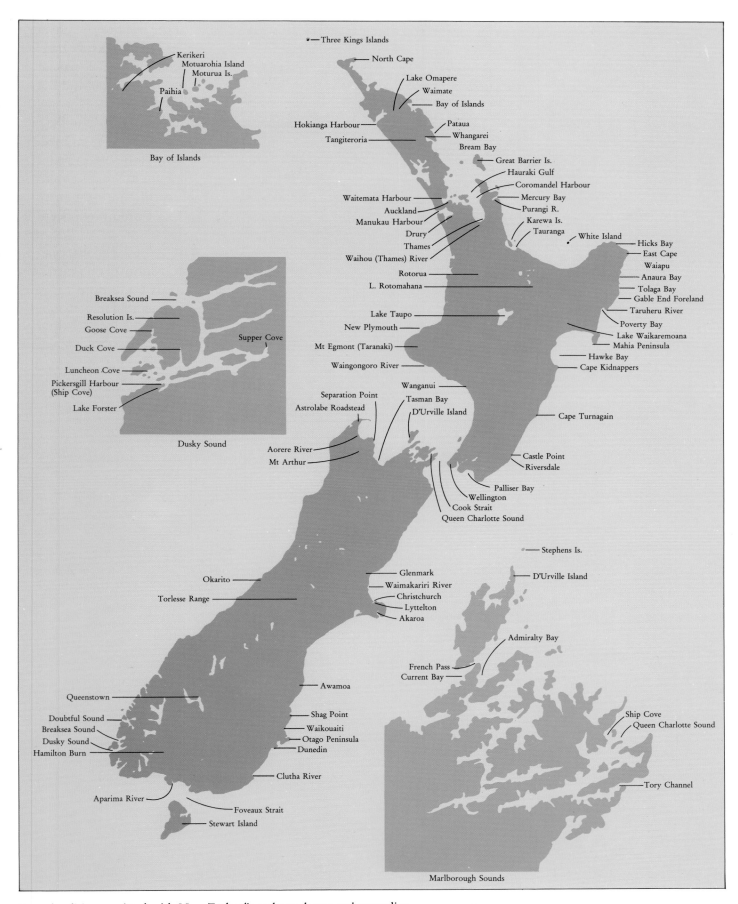

Some localities associated with New Zealand's early explorers and naturalists.

NEW ZEALAND REDISCOVERED: *ENDEAVOUR* AND *ST. JEAN BAPTISTE*

I T WAS SOME 13 MONTHS before those first collections that the *Endeavour* had left England, with Lieutenant James Cook in command and aboard her the largest scientific party ever assembled for a voyage of exploration. Civilian expedition scientists were not new, even in the Pacific.[1] On Buffon's recommendation the botanist Philibert Commerson sailed under F. Chesnard de la Giraudais on the *Étoile*, the store ship accompanying Bougainville and the *Boudeuse* on a circumnavigation that crossed the Pacific in 1767-68.[2] Banks himself had gone on the *Niger* to Newfoundland,[3] but the scientific corps of the *Endeavour* voyage was far greater than anything that preceded it and its potential was comparatively enormous. Apart from Banks himself, there was Daniel Solander, the Linnaean student who had come to England to spread knowledge of the new system of classification and stayed to become a pillar of the natural history establishment; Herman Spöring who went as Banks's secretary, but was to demonstrate a notable talent for zoological illustration; Sydney Parkinson, a first class natural history draughtsman; Alexander Buchan, employed for landscape and ethnological drawing; and Banks's four servants.

Then, as now, it was a matter of practical necessity for the navy or the military to accommodate civilian scientists — the former having the resources basic to major exploration, the latter having the appropriate scientific background. That the interests of natural history, and of Banks in particular, should be so well served on this present voyage is nonetheless a little surprising. After all, the naturalists and their party came to outnumber handsomely the astronomers who were to pursue the original scientific purpose of the expedition. It could be anticipated, as Cook and other commanders were to discover, that civilians would not always be cooperative shipmates, lacking the necessary responses to discipline or the physical toughness that such excursions required. Problems like these would in fact inevitably lead to a period of greater use of naval surgeons in the role of naturalists, with other officers doing the work of physical scientists.[4]

If sombre thoughts like these surfaced during the preparation for the voyage then they were not greatly evident as the men engaged instead in the more positive aspects of organising and equipping a major scientific expedition. Preparation was vital to the success of the scientific programme and just as the astronomers needed their telescopes, clocks and other instruments[5] so the naturalists needed all the paraphernalia of collecting and preservation — but what to take and how much of it? There was no going back for forgotten bits and pieces and the few ports of call were unlikely to provide them with much beyond the normal maintenance and provisioning of ships. Here Joseph Banks's previous experience on the *Niger* two years before must have helped greatly. Then he had worked alone, so that many of the requirements of biological field work, afloat and ashore, were by now familiar to him. Also, his colleague Daniel Solander was an experienced natural historian and no doubt able to be helpful in choosing the right equipment and supplies.[6]

1. Natural history and exploration had long been associated, and European colonies had their naturalist observers and collectors, e.g. Sloane in the West Indies and Marcgrave in Brazil in the 17th century. Also artists had accompanied explorers for some hundreds of years before the 18th century, (Blunt in Carr, 1983, p.14).

2. Dunmore, 1965, p.64; Brosse, 1983, pp.24-25.

3. Lysaght, 1971.

4. The problem did not confine itself to English expeditions. A disastrous attempt to mix civilian scientists and the Navy was Baudin's exploration of the Western Pacific; Dunmore, 1969, pp.11, 12.

5. Beaglehole, 1974 p.137.

6. They had additional advice from colleagues such as Fabricius — a student of Linnaeus; Beaglehole *in* Banks, 1962, v.1, p.70.

7. The seines were made of a soft, loose twine which decayed and lacked the strength to hold large fish; Cook to Navy Board, 20 January, 1769; Beaglehole, in Cook, 1961, p.911.

8. Letter from Ellis to Linnaeus, quoted by Beaglehole *in* Banks, 1962, v.1, p.30.

9. Cole, 1975, pp.445-448. The developing use of alcohol and glass preserving jars was significant in the improvement of natural history collections in the 18th century. Spirits had already been in use for 100 years or so with wine, brandy or even rum being distilled to give alcohol of a suitable strength.

10. A. Wheeler, *pers. comm.* Forster had recommended engraved lead 'tickets' attached to the specimen by wire; Stresemann, 1975, p.79.

11. Whitehead, 1969, p.179. Cole, 1975, p.449.

12. Beaglehole, *in* Banks, 1962, v.1, pp.33-34, 183; v.2, p.99.

13. Ellis to Linnaeus, quoted by Beaglehole in Banks, 1962, v.1, p.30.

14. Banks, 1962, v.1, p.158.

15. Lysaght, 1979, p.16.

16. Banks, 1962, v.1, p.153.

They took various sorts of nets for dipping, casting and seining, most of a coarser mesh for fish, but including a fine dipping net which would be needed for the smaller marine invertebrates[7] and a 'drudge' or dredge for taking bottom samples. Fishing lines and hooks were plentiful — some of these would be used by the crew to replenish their food supplies. There were 'all sorts of machines for catching and preserving insects'[8] which no doubt included a butterfly net or two, pins for laying them out, and boxes for storage. Birds and mammals would be collected with the aid of a musket and stored, as the larger fish would be, in casks of spirits.[9] Spirit-preserved animals needed tags for identification — these would have to be of soft metal (probably lead or zinc) easily engraved with a number or name.[10] Smaller vertebrates and invertebrates needed glass bottles, which came in all shapes and sizes including wide-mouthed bottles with ground glass stoppers and filled with spirits, provided by the surgeon-anatomist John Hunter. The bottles were probably part of a large stock made to order for Hunter.[11] Added to these essentials were numerous odds and ends — labels to fix to the bottles, ink, pens, paper and notebooks. And this was just for zoology; there were as well materials for pressing plants and other equipment for the botanical collections. Then there were the books — essential reference works which might provide some clue to species identification. Included of course were Linnaeus's *Systema Naturae* (in its 12th edition in the year of their departure), Pennant's *British Zoology*, Buffon's *Histoire Naturelle*. . . and the more recent *Elenchus Zoophytorum* and *Miscellanea Zoologica* of Simon Pallas; Edwards's *A Natural History of Birds* and Brisson's *Ornithologie*; and 'a suitable stock of books relating to the natural history of the Indies . . .'[12] together with others including, as might be expected, many botanical reference works. It was indeed as John Ellis, the English natural historian, wrote: 'No people ever went to sea better fitted out for the purpose of Natural History . . .'[13]

The sights and sounds of lading can be imagined — stacks of boxes and casks, the clinking of bottles, the shouting of directions as Banks's servants carried equipment aboard for stowage in the cabins and elsewhere below decks. The nets were placed in the hold, out of reach as we learn later on, but the ex-Whitby collier was not designed as a research vessel and problems of this sort were inevitable as its 97-foot length (30 metres) was already stretched to the limits to accommodate Banks and his retinue.

And what were the feelings of the members of the natural history party as the time of departure came closer? Banks, buoyed up by enthusiasm and no doubt too busy to dwell on future discomforts and dangers, was to some extent already expedition-hardened — although once underway the enormity of the venture struck him.[14] Solander, like Banks, had volunteered himself to the expedition: he was not in the normal sense an employee and to make some sacrifice in the cause of natural history would be an heroic gesture in the Linnaean tradition. But the others, essentially employees of Banks, must have entertained some of the foreboding expressed by Sydney Parkinson in his will, made only 12 days before he left, in which he described himself as going on a long and hazardous voyage from which 'God knows I may never return'.[15] These thoughts would have occurred repeatedly to all on board as the ship made its way from Galleons Reach down the Thames to Plymouth, where Banks and Solander finally joined them. Nonetheless they left England on August 26th, 1768, according to Banks 'all in excellent health and spirits perfectly prepard (in Mind at least) to undergo with Chearfullness any fatigues or dangers that may occur in our intended Voyage.'[16]

In recording their passage to Madeira and beyond, Banks provides us with one natural history observation after another. Finding much that was new so close to home, he entertained very high hopes for greater discoveries further on, and his delight and enthusiasm as plankton, large marine invertebrates, fish and birds were captured and hauled over the side are very evident in his writing. It was here, at sea, that zoology was to have its day — both now and for the rest of the voyage[17] — owing to the relative absence of plant specimens which would otherwise have preoccupied the enthusiastic botanist in Banks. At this stage of the journey Parkinson, not yet under pressure, produced some of his best and most finished work and was generally untroubled by the bothersome insects that would, when on land, attack his drawings and specimens.[18] It is a satisfying comment on the development of the science, and their application to it, that when circumstances permitted, Banks and his colleagues went beyond descriptive work to anatomical dissection, behavioural 'experiments' (albeit crude) and parasitological examinations.

They worked night and day using nets, lines, fish spears, cloths trailed astern to lure crabs, and muskets to shoot birds, even albatrosses. Banks was not at all perturbed by the folklore applying to these and shot them with impunity. Specimens were examined on the large fixed table in the relatively well-lit great cabin, which also functioned as a dining room, writing room and library. Parkinson sketched and painted under Banks's direction, working well into the night or finishing off the work the next day if he could not complete it at one sitting. Solander made descriptive notes which were later fair-copied by Spöring. Often they worked for long stretches, starting at 8.00 a.m. and breaking off at 2.00 p.m. when they would eat and wait for the cabin to air[19] for a couple of hours before recommencing and working until darkness fell. With the help of Banks's journal[20] we can recreate the scene in the great cabin. Banks at his writing desk, reference books at hand, attends to his journal or letters, rising now and then to talk to his colleagues. A bunch of seaweed hangs from a cabin beam, barnacle-encrusted wood with perhaps a fish or some other specimens lie on the table, taking the attention of Solander and Parkinson. Labelling materials, the plant press or a jar or two of preservative are nearby; paints and large folio sheets cover the remaining surface. The air is stuffy, the penetrating smell of seaweed or fish mixing with the vinous smell of spirits and the lingering smells of forgotten meals and of men long at sea. The sounds are the murmur of voices, sometimes excited, often muted, with the scratching of pens and the rustling of papers; the noises of the ship, the sea and the wind form a background. Movement is constant; apart from the four men (Spöring is in there somewhere) Cook and his officers are in and out of the cabin, going about the business of the ship, and now and then one of Banks's servants enters to take a specimen to the casks, fetch a book or to attend to some other need. These are the sights and sounds of natural historians at sea — a pattern to be repeated at other times and with other vessels.

One particular aspect of their activities would be experimental, but at the same time outside the bounds of their normal scientific investigation: this was the eating of their specimens. Many years later Banks was to assert, 'I believe I have eaten my way into the Animal Kingdom farther than any other man,'[21] and on this voyage there was an unrivalled opportunity for him and his colleagues to consolidate that claim. There was the justifiable need to '. . . discover what might be available as food, either fresh or salted, on emergency' and as well Banks was conscious of the need to conserve their supply of spirits.[22] But in this century and the next, straight-out curiosity, the search for ever more succulent flesh and piquant flavours, drove many a natural historian to sacrifice the juicier parts of his specimens for the pot and leave only the inedible for science and posterity. Thus only the head and legs of a bird might be saved, or the skin of a mammal,

17. Wheeler *in* Carr, 1983, p.203.

18. Banks, 1962, v.1, p.260.

19. Beaglehole, *in* Banks, 1962, v.1, p.33. Solander did not like the lingering smell of food in the cabin.

20. Beaglehole, *in* Banks, 1962, v.1, pp.33-34, 396.

21. Beaglehole, 1974, p.261.

22. Whitehead, 1969, p.167.

but very little that was useful would remain of a fish, and even invertebrates were not safe.

The *Endeavour* made its way south from Madeira to Rio de Janeiro and then to Tierra del Fuego, landings at these places enabling them to botanise. It was at Tierra del Fuego that the natural history party received its first major setback — a sharp reminder of the hazards of field work in an area of difficult terrain and capricious weather. As a result, two of Banks's servants died, Buchan had an epileptic fit and others, including Solander, were close to succumbing to the cold. But for some lucky navigation, warmth from a fire, and a vulture that was later cooked and shared by all, more might have died and the scientific effort of the expedition compromised.[23]

During the 13-week stay at Tahiti zoology took its place with botany and ethnology. The flora and fauna were diverse, the people friendly. Parkinson recalled the collecting efforts of Banks and Solander and their direction of him in the drawing of the colourful tropical plants, fish and birds.[24] His painting was sometimes carried out under unpleasant conditions;[25] it was at this time that new pressure was placed on the natural history draughtsmen when not long after their arrival the unfortunate Buchan, who was already ill, had another epileptic fit and died. The work was now shared by Parkinson and (to a lesser extent) Spöring; the unfinished nature of some of Parkinson's work dates from about this period.

Heading south from the Society Islands, Banks and the others turned their attention back to the birds and marine life — bunches of seaweed, barnacles, seals and porpoises — and then to the anticipation of land, and finally, on October 6th, 1769, to the small shark at sunset. New Zealand, or as it appeared to them, the southern continent, had been reached at last.

They went ashore on the 8th and on the two days following, but altercations with the Maoris prevented any serious collecting. It was an unpromising beginning: misunderstandings, warlike gestures, pilfering, and then the muskets and swift retribution. No one was happy with the outcome and the little field work that Banks and Solander were able to carry out was done under the protection of the marines. One or two birds were shot (the species unidentified) and numerous 'grey' ducks seen, but they left Poverty Bay with little else but 40 species of plants and the taste of a black bird with orange flesh and a flavour a bit like that of stewed shellfish.[26] A farewell visit by the Maoris allowed some fleeting human parasitology: '. . . more lice than ever I saw before!,' remarked Banks of his visitors' hair.[27] The *Endeavour* sailed south, around the Mahia Peninsula following the curve of Hawke Bay, the snow-covered ranges a backdrop to their continuing encounters with the Maoris. Near Cape Kidnappers, canoes laden with nets and fishing gear approached them and they traded with the occupants for fish, amongst which were the first specimens drawn and painted by Parkinson in New Zealand waters, including the Tarahiki, the Marblefish, Butterfish, and Scarlet Wrasse.[28] Already fish

23. At this distance it is not easy to be certain of the precise reasons for their predicament or of its severity. Physical fitness was obviously a problem and they were probably inadequately clad; rum hastened the end of Dorlton and Richmond. Although Beaglehole does not rank it among the great crises of the voyage, and suggests that Banks made too much of the lack of food and low temperatures, our knowledge of hypothermia today suggests that Banks was right in his concern. See Beaglehole *in* Banks, 1962, v.1, p.223.

24. Parkinson, 1773, p.36.

25. Banks, 1962, v.1, pp.260-261. The flies attacked the paint on Parkinson's work and also the fish he painted. Attempts were made to trap them with tar and molasses.

26. Parkinson, 1773, p.89. The orange-fleshed bird was possibly a Tui. Banks, 1962, v.1, pp.404, 406.

27. Banks, 1962, v.1, p.407.

28. It is questionable whether these fish were caught by those on the *Endeavour*, or were among those obtained from the Maoris. Cook said that the latter fish were 'stinking', causing Beaglehole to suggest that they were dried (in Cook, 1955, p.177). This seems unlikely, as Cook referred to the canoes as 'fishing boats', Banks noted that they had stone weights, 'netts', and other fishing implements with them (Banks, 1962, v.1, p.412), and Parkinson that 'none of their fish was quite fresh and some of it stank intolerably.' (Parkinson, 1773, p.94). The four fish painted were: Tarahiki *(Nemadactylus macropterus)*, Butterfish *(Odax pullus)*, Scarlet Wrasse *(Pseudolabrus miles)* and Marblefish *(Aplodactylus arctidens)*. These were described by Solander, with the Red Pigfish *(Bodianus vulpinus)* also from here; in *Pisces Australiae*, Sol. MS. Z2, Zoology Library, British Museum (Natural History). (A Scorpion fish and Leatherjacket incorrectly included here by Richardson, 1843, pp.18, 29.) The Butterfish and Marblefish are herbivorous (i.e. would not take a baited hook) and because of their habitat could be caught only by set nets, traps or spears — none of which were likely to be in use on the *Endeavour* at this time. Therefore the fish collected and drawn were probably acquired in exchange with the Maoris on 15 October off Cape Kidnappers. 'Mataruwhow', a name used in error by Parkinson and Banks for Cape Kidnappers, is their rendering of the Maori name for Scinde Island (Napier).

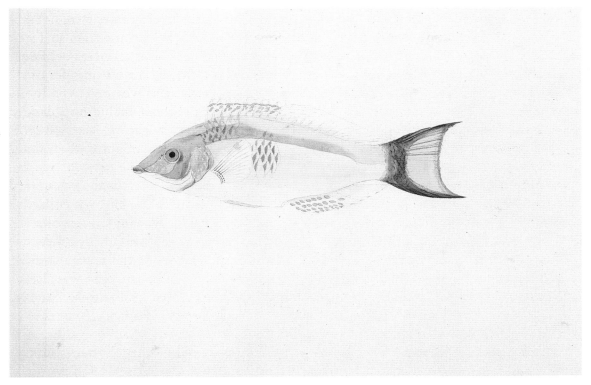

The Scarlet Wrasse (*Pseudolabrus miles*) from the vicinity of Cape Kidnappers, painted by Parkinson (vol. 2. pl. 38). Among the first of their New Zealand specimens illustrated. Most of Parkinson's New Zealand work was in this condition — partly painted with accompanying notes to guide its completion. Published by permission of the Trustees of the British Museum. *(see p.10)*

were predominant in Parkinson's folios — their transient colours, so easily lost in spirit preservation, and the fact that so little would remain after an exploratory tasting, meant that these, of all the vertebrates, had to have some priority in illustration.

On 16 October at Cape Turnagain the ship went about and headed north; it was not until they were anchored at Anaura Bay, some distance up the coast from their first landfall, that collecting of any account was done ashore. Banks and Solander landed shortly after some seamen sent to fill the water casks and finding the Maoris relatively friendly felt relaxed enough to attempt an exploration of the Bay. They shot some 'most beautiful birds', referred to as parrots by Parkinson, who noted the absence of 'ground fowl or domestic poultry' and remained optimistic about the presence of mammals of 'the larger kind'.[29] At the end of the day the ship's boats were still ferrying water, so to avoid delay in sorting and cataloguing the day's collections Banks persuaded the Maoris to launch a canoe for them. At the first attempt they capsized in the surf, but a second effort, with the group divided in two, reached the ship safely. If specimens were lost Banks makes no mention of it. The ship backtracked to Tolaga Bay to the south, where they took on wood and water while Banks and the others enjoyed some success at collecting. Red Gurnard, the Spotty, Sand Flounder, Red Cod, Kahawai and Stargazers were added to their list but not drawn.[30] Here eventuated one of the voyage's unsolved zoological mysteries when on Pourewa Island at the south point of the bay Spöring saw a large bird with an enormous tail fly overhead. His description fits that of no known New Zealand bird and it has been thought that he might have been mistaken in what he saw.[31] From Tolaga Bay the *Endeavour* headed north, around East Cape into the Bay of Plenty and up the coast to the Coromandel Peninsula. Off Cape Runaway they were sold some mussels and crayfish by the Maoris, and it might have been a crayfish from this source,

29. Parkinson, 1773, p.99. The bird was probably the North Island Kaka. Barnacles (*Lepas* sp.) were described from here by Solander; Sol. MS. Z4, Zoology Library, British Museum (Natural History).

30. Richardson, 1843, pp.17, 18, 27. Respectively: *Chelidonichthys kumu, Pseudolabrus celidotus, Rhombosolea plebeia, Pseudophycis bachus, Arripis trutta, ? Genyagnus novaezelandiae.*

31. Banks, 1962, v.1, p.421.

32. On *verso* of Pl.12, v.2 of Parkinson's drawings.

33. Cook, 1955, p.194 and Parkinson, 1773, p.164 both refer to 'Mullet' possibly Kahawhai, Grey Mullet, or Yellow-eyed Mullet.

34. Cook, 1955, p.194 notes cockles, clams, and mussels from this locality.

35. He drew very few birds from New Zealand and Australia.

36. Respectively: *Caranx georgianus, Arripis trutta, Chelidonichthys kumu,* and *Zeus japonicus.* The last in Parkinson, 1773, p.104. The Red Gurnard noted in Sol. MS. Z2, Zoology Library, British Museum (Natural History).

37. Trevally particularly form large schools on the North Island east coast (Ayling and Cox, 1982, p.215). Banks, 1962, p.428, refers to 'Macarel of 2 sorts' which he might have confused with Kahawhai and Trevally. Beaglehole's identification of Southern and New Zealand Mackerel (Banks, 1962, v.1, p.428) may not be correct in view of what was drawn here.

38. Banks, 1962, v.1, p.430.

39. Banks, 1962, v.1, p.431.

or one earlier from Anaura Bay, that was lightly sketched in pencil by Parkinson — one of his few invertebrate drawings from New Zealand.[32]

The morning of 4 November saw them anchored off the Purangi River to observe the transit of Mercury (the place was later called Mercury Bay) where Banks, thinking the area held promise, landed the following day at the mouth of the river. Here the seine was hauled by one of the smaller boats, but only a few 'mullet'[33] were caught; the trawl and the dredge did little better later, except for a few shells.[34] Several birds were shot, probably Variable Oystercatchers, but Parkinson, preoccupied with plants and fish, did not draw them.[35] Once again, their lack of success at fishing did not handicap them as on the fourth and fifth day of their stay the Maoris came alongside and traded the vast numbers of fish in their canoes — so many in fact that the bulk of the catch was salted down. Parkinson drew and partly painted the Trevally, Kahawai, and the Red Gurnard, and noted John Dorys, in his journal.[36] Trevally and Kahawai form large schools and probably constituted a major part of the Maori catch.[37] Other animals collected or seen at Mercury Bay found their way more easily to the table than to the collecting jar or into Parkinson's folios. A shag hunt resulted in 20 of the birds being shot and eaten — 'since our fowls and ducks have been gone we find ourselves able to eat any kind of Birds (for indeed we throw away none) . . .' said Banks,[38] and this might be part of the reason that so few were illustrated. Pipis were found to be delicious and he devoted a whole day's entry to praise of the oysters they ate there.[39] Crayfish too were also in good supply and well appreciated. If the zoology of Mercury Bay was not exactly fascinating to the naturalists it certainly fed them well, and they were very satisfied with their plant collecting. Before they left on 15 November Banks and Solander stocked up on botanical specimens so that the hardworking Parkinson could finish dealing with them at sea.

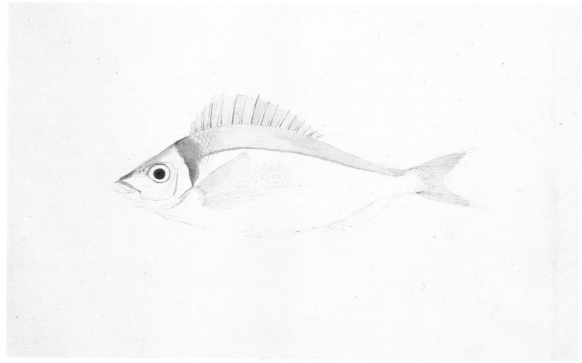

The Tarakihi (*Nemadactylus macropterus*) from the vicinity of Cape Kidnappers, painted by Parkinson (vol. 2. pl. 52). Also among the first of of their New Zealand specimens illustrated. Published by permission of the Trustees of the British Museum. *(see p.10)*

The *Endeavour* rounded the Coromandel Peninsula and sailed into the Hauraki Gulf, finally anchoring near the mouth of the Waihou (Thames) River on 20 November. The next day they set out in the small boats to examine their surroundings, and in the process 'caught a considerable quantity of fish, with hook and line, of the Scienna or bream kind'.[40] This was possibly their first encounter with New Zealand Snapper, caught in greater numbers further north when they put their lines down in what was to be called Bream Bay.[41] Beyond Bream Bay they continued northwards, stopping next at the Bay of Islands. Who would have guessed that this rain-soaked place with its sometimes hostile Maori population would prove such a popular destination for future generations of sailors and explorers? Though their difficulties with the local population were diverting they did find some time for natural history. Fish were caught on a line,[42] and the seine was dragged on 2 and 3 December. As Cook wrote later, '. . . we got only fish, some few we caught our selves with hook and line and in the Saine but by far the greatest part we purchass'd of the Natives and these of Various sorts, such as Shirks, sting-rays, Breams, Mullet, Mackarel and several other sorts; their way of catching them are the same as ours, (viz) with hooks and lines and with saines . . .'.[43] Parkinson must have painted the 'other sorts'; . . . the Goatfish, Hapuku, Blue Cod and the Sea Perch. From the Bay of Islands they also collected a slug-like marine mollusc, the Duck-bill Limpet — the only New Zealand invertebrate painted by Parkinson — and a crab.[44]

Leaving the Bay of Islands they tacked into unfavourable winds until eventually North Cape was passed. The weather had calmed by 24 December and taking advantage of the lull Banks launched a small boat in order to shoot some gannets.[45] The hapless birds were immortalized in two ways: Parkinson partly coloured a drawing of one of them, the only New Zealand bird to be illustrated on the voyage, and the others substituted for goose in a pie used to celebrate Christmas — a great success in spite of their unseasonal circumstances. Their passage down the west coast to the North Island was not marked by any unusual event or notable zoological discovery, although Banks shot some petrels and an albatross, and with a further albatross conducted an 'experiment' to see if the bird could take off in calm water, unaided by a wave. The unfortunate bird was, for a time, hotly pursued by the furiously rowing Banks; its attempts to lift off failed at first but finally it beat its way into the air with some encouragement from his musket.[46] North of Hokianga a turtle was seen and a large sunfish; more seabirds; and then the towering peak of Mount Egmont before crossing the entrance to Cook Strait and entering Queen Charlotte Sound with Cook observing a 'sea lyon' before anchoring at Ship Cove on 15 January.

They were to spend more than a fortnight here, in reasonably agreeable surroundings. Relations with the Maoris were more or less friendly, the fishing and shooting excellent, and the dawn chorus of bellbirds and others a delight to the ear. Only the plant collecting was a little disappointing to Banks and Solander. For day-to-day observation of natural history here we do better by turning to Parkinson's journal rather than that of Banks. For Banks, the shooting of shags (and eating them), and fishing off the rocks with Solander, were memorable pleasures but he doesn't bother to tell us what fish he caught. But Parkinson recalled '. . . cuttle-fish, large breams, (some of which weighed twelve pounds, and were very delicious food, having the taste of fine salmon) small grey breams, small and large barracootas, flying gurnards, horse-mackerel, dog-fish, soles, dabs, mullets, drums, scorpenas or rock-fish, cole-fish, the beautiful fish called chimera, and shaggs.' Here he painted the Rough Skate, Red Cod, and the Spotty.[47] The words, 'There is also a little sandfly which is very troublesome . . .' followed by a description of the bite,

40. Parkinson, 1773, p.106. Blue Cod *(Parapercis colias)* also taken from here; Sol. MS. Z2, Zoology Library, British Museum (Natural History).

41. They were painted by Parkinson and recorded from 'Ohoorage' (Hauraki Gulf). Beaglehole in Cook, 1955, p.210, thought the fish were Snapper but changed to Tarakihi in Banks, 1962, v.1, p.438. Solander described a Tarakihi that day in his *Pisces Australiae* (pp.29-30) but all the evidence suggests that Beaglehole was right the first time.

42. Parkinson, 1773, p.109.

43. Cook, 1955, p.219.

44. Respectively: *Upeneichthys lineatus (porosus), Polyprion oxygeneios Parapercis colias, Scorpaena papillosus,* (all drawn by Parkinson), also Scorpion Fish *(Scorpaena cardinalis)* Tarakihi *(Nemadactylus macropterus),* Trevally *(Caranx georgianus),* Leatherjacket, *(Parika scaber);* in Sol. MS. Z2, Zoology Library, British Museum (Natural History). *Scutus breviculus* (the only NZ invertebrate painted by Parkinson) and *Notomithrax peronii* (the last possibly from this voyage, Whitehead, 1969, p.177).

45. *Sula bassana serrator,* the Australian Gannet, in the neighbourhood of the Three Kings Islands. Also a Flesh-footed Shearwater *(Puffinus carneipes hullianus)* Solander description MS. Z4, Solander slip Aves 171-172, Zoology Library, British Museum (Natural History). See also Lysaght, 1959.

46. Banks, 1962, v.1, pp.450-451. Off Hokianga the Sooty and Little

continued overleaf

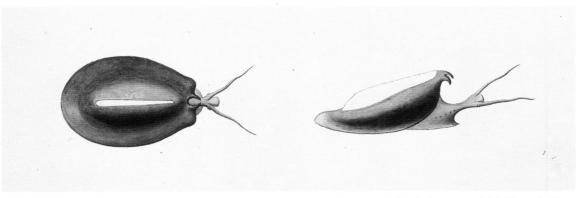

The Shield Shell or the Duck-bill Limpet *(Scutus breviculus)* from the Bay of Islands (Motu Arohia). The only New Zealand invertebrate animal known to have been painted by Parkinson (vol. 3, pl. 73). Published by permission of the Trustees of the British Museum. *(see p.13)*

The Australian Gannet *(Sula bassana serrator),* shot near Three Kings Islands, the only bird to have been drawn by Parkinson from the region of New Zealand (vol. 1, pl. 30). Published by permission of the Trustees of the British Museum. *(see p.13)*

were to become a familiar entry in the journals of those who explored New Zealand — particularly in southern parts. The birds of Queen Charlotte Sound he found to include 'parrots, wood-pigeons, water-hens; three sorts of birds having wattles; Hawks; with a great variety of birds that sing all night.'[48]

By 3 February, Banks and Solander were assembling their vast collections of seeds, shells, and other odds and ends before the ship made ready to leave Queen Charlotte Sound. Three days later they were out in Cook Strait heading for Cape Turnagain and the discovery that the land they had sailed around was indeed an island. When on 26 March they had virtually completed a circumnavigation of the South Island and anchored in Admiralty Bay, the dream of a southern continent, shared by an ever-decreasing number on the ship, had evaporated completely. They had not put foot ashore since leaving Queen Charlotte Sound, and a disappointed Banks was forced to watch the enticing entrances of the Fiordland sounds pass by, the time and the circumstances compelling

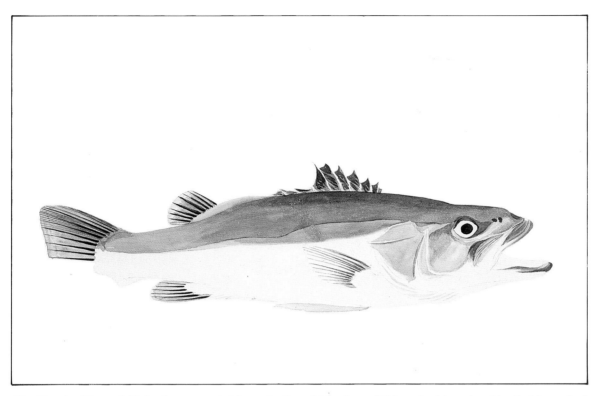

The Hapuku (Groper) (*Polyprion oxygeneios*) from the Bay of Islands — off Motu Arohia, painted by Parkinson (vol. 2, pl. 74). Published by permission of the Trustees of the British Museum. *(see p.13)*

Cook to reject the idea of making an entry. The naturalists had to be content with shooting a few seabirds, and sighting albatrosses, penguins and porpoises. They fished and botanised last on D'Urville Island before leaving New Zealand on 31 March, 1770.[49]

This was a rather uneventful and anticlimactic end to the first investigation of New Zealand natural history. The aura of the unexplored had been penetrated, and the flora and fauna found different, even novel, but they were not remarkable. If hidden within the country there were natural history secrets that would fire the European imagination, then they were not to be shared on this occasion. And so it was, on the day before he left New Zealand, that Banks sat down to his journal to '. . . give myself liberty of conjecturing and drawing conclusions from what I have observd . . .'.[50] In the zoological part of his account (he included also botany, geology and, in some detail, ethnology) he begins with the quadrupeds, thinking that apart from the seals and the sea-lion, there were no native mammals in New Zealand. He correctly assessed the dog and the rat as introductions of the Maori, but he was unaware of the two species of native bat.[51]

Banks might have been said to have been a little precipitate in his judgment on mammals; after all, the interior of both the North and South Islands and much of the coastal outline of the South Island remained unexamined. But here he was closer to the mark than he was with some of the other animal groups. 'Of Birds there are not many species . . .' he wrote, and to one whose ornithological experience was based on Europe, that statement would not be entirely unfair — New Zealand's bird fauna is that of an oceanic island and somewhat sparse. If the bird fauna was less various and less luxuriant in its appearance, then it made up these deficiencies with the quality of its song, the most melodious that Banks had ever heard. He and his colleagues had quickly recognized those birds for which there were analagous species in Europe — the ducks, shags, hawks, owls and quail — but the more unique or characteristic of the New Zealand birds — the Tui, fantail, kokako

49. Banks, 1962, v.1, p.465 A skua (possibly the Southern Skua), *(Stercorarius skua lonnbergi)* one of the few animals described from this period; Sol. MS. Z4 Zoology library, British Musuem (Natural History). See also Lysaght, 1959. The locality was mistakenly thought to be Murderers Bay, where some fish (e.g. Gemfish) were described in *Pisces Australiae* MS. Z2, Zoology Library, British Museum (Natural History).

50. Banks, 1962, v.2, p.1; in which he acknowledges he may often be mistaken. Cook also summarized the natural history, Cook, 1955, pp.276-277.

51. The native bats *Chalinobus tuberculatus* and *Mystacina tuberculata* also overlooked, together with the Kiore, the rat *(Rattus exulans)* introduced by the Maoris; Banks, 1962, v.2,

— of which a few were probably collected, were not singled out for comment in Banks's natural history summary.[52] The kiwi was missed altogether, not surprisingly as the bird is a secretive forest-dwelling one, but hidden in the statement about Maori dress (cloaks) made of 'the skins of divers birds' was probably the first unconscious European observation of a remnant of the kiwi. Their companions of the open sea — the albatrosses, shearwaters, and petrels — could also be seen on the coast, with the additional novelty of penguins. The reptiles and amphibians were obvious omissions, the latter being perhaps beyond their reach, but lizards were commonplace.

Last of the terrestrial animals, and apparently least in Banks's view, were the insects which he claimed rivalled birds in their paucity. He notes '. . . a few Butterflys and Beetles, flesh flies very like those in Europe, Mosquetos, and sandflies maybe exactly the same as those of North America, make up the whole list.' Weighed against the considerable lepidopteran fauna of Europe, the New Zealand butterflies must have been seen as sparse and unspectacular; other insect groups, apart perhaps from the beetles, were not notably well represented. Here the reaction of the 18th century naturalist had a lot in common with later observers who would similarly remark on New Zealand insects. Nevertheless it would take many more generations of naturalists to collect in their wake before the majority were accounted for. Above all else, the sandflies were grossly underestimated — that 'we never met with any great abundance; . . .' simply meant that they did not visit the right places. Although insects barely feature in Banks's journal, it is evident that the naturalists made some effort to collect them for unlike the shells on the beach they had to be pursued, shaken out of bushes and parted from rotting logs. Some may even have crawled from botanical specimens as Parkinson sketched them. The list of species tells us that they collected on the beach, beside river banks, above high water marks, and back into the fringing forest. They collected only during the daytime, thus missing nocturnal species; but the Red Admiral and the Yellow Admiral, 25-odd species of beetles, and sundry others were taken into the collection.[53]

But it was in the sea, or at the water's edge, that the New Zealand fauna redeemed itself. It made 'abundant recompense . . .' said Banks, although this success was measured in terms of well-filled stomachs before he briefly considered the diversity of the marine organisms. In the middle of his descriptions the succulent crayfish took his attention. Then there were 'excellent oysters and many sorts of shell fish and cockles, clams etc.,' a statement that almost describes the extent of their modest shell-collecting: a few relatively common species, the result of a casual pocketing while walking on the beach, or even the leavings of some feast with the Maoris. The Common Cockle, the Green Mussel, the Large Ostrich Foot and at least ten other species were known to have been taken.[54] It was the fish however that were predominant, faring rather better than the birds. Nineteen species were drawn by Parkinson and several more described by Solander.[55]

52. The Laughing Owl *(Sceloglaux albifacies)* and the Quail *(Coturnix novaezelandiae novaezelandiae)* are now extinct. No New Zealand land birds were drawn; apart from the Tui and South Island Kokako few were collected (Medway, 1976 pp.48, 133). Parkinson collected birds as well as insects (Medway, p.48).

53. See note 5, Chapter 4. Some of the insects collected would not be commonly encountered today (G. W. Gibbs, pers. comm.)

54. Wilkins, 1955, p.97. To his list of eight species (see Note 46, Chapter 4) can be added: The Opal Top Shell *(Cantharidus opalus opalus)* (Walch, 1774, Pl.3, Figs. 5 & 6); The Silver Paua *(Haliotus australis)* (Spengler, 1776 p.150-152) and ? the same in Dalston Catalogue, Anon, 1775, pp.6, 16, 21; The Periwinkle *(Littorina cincta)* (Zorn von Plobsheim, 1775 p.167) and the Duck-bill Limpet *(Scutus breviculus)* (Parkinson drawings). We can also add the brachiopod *(Terebratella sanguinea)* (Wilkins, 1955, p.85). See also Dance, 1972, p.357.

55. The 19 species painted by Parkinson: Carpet Shark *(Cephaloscyllium isabella)*, Spiny Dogfish *(Squalus acanthias)*, Rough Skate *(Raja nasuta)* Kahawai *(Arripis trutta)*, Red Cod *(Pseudophycis bachus)*, Snapper *(Chrysophrys auratus)*, Marblefish *(Aplodactylus arctidens)*, Tarakihi *(Nemadactylus macropterus)*, Gemfish *(Rexea solandri)*, Scarlet Wrasse *(Pseudolabrus miles)*, Spotty *(Pseudolabrus celidotus)*, Butterfish *(Odax pullus)*, Scorpion Fish *(Scorpaena cardinalis)*, Red Gurnard *(Chelidonichthys kumu)*, Sea Perch *(Scorpaena papillosus)*, Blue Cod *(Parapercis colias)*, Blue Moki *(Latridopsis ciliaris)*, Hapuku *(Polyprion oxygeneios)*, Goatfish *(Upeneichthys lineatus)*. A further 19 were described by Solander in *Pisces Australiae* Sol. MS, Z2, Zoology library, British Museum (Natural History) — some beyond accurate identification or synonyms, but the Barracouta *(Thyrsites atun)*, Leatherjacket *(Parika scaber)*, Trevally *(Caranx georgianus)*, Saury *(Scomberesox saurus)*, Red Pigfish *(Bodianus vulpinus)*, Elephant Fish *(Callorhynchus milii)* and Sand Flounder *(Rhombosolea plebeia)* are among them. Only a few (e.g. Marblefish and Scorpionfish) were to be the basis of types.

And so, Banks moves on to botany and ethnology, leaving us with a brief and rather selective summary even though there was no intention that his journal should be a complete scientific document. They would convert the output of Parkinson's brush and pencil and the notes of Solander into a distinctive and official published record – at this stage there can have been no doubt in their minds of that. The exploration itself was selective, as only limited sections of the coast were visited and the interior, not in their brief, remained simply a prospect. Under such circumstances a substantial floral and faunal survey was impossible and much of the real essence of New Zealand zoology would inevitably escape them. But in a very general sense their impressions were more or less right, and they could be certain that the specimens that they had gathered up would arouse great interest from eager students and collectors at home and abroad. Also, they would leave in the almost certain knowledge that this would not be the last time that New Zealand natural history would be subject to European scrutiny.

The journey back to England, via Australia, New Guinea, Indonesia and the Cape of Good Hope need not concern us in its detail, but two events are worth recording. The first was potentially disastrous and the second, while not fatal to the expedition as a whole, held significant implications for the development of the voyage's natural history. On 10 June, 1770, the *Endeavour* struck part of the Great Barrier Reef (the part became known as Endeavour Reef) just before 11 p.m. At one stage Banks gave up hope of the ship staying afloat and started to pack what he thought he could save of his possessions. Although self-preservation would have been uppermost in his mind, he must have spared at least a thought for the potentially doomed specimens, drawings and descriptions. But care and good seamanship, with everyone including Banks taking their share of the hard work, brought them through the crisis. Zoologists have since commented in their frustration at trying to untangle some of the twisted threads of this expedition's zoology, that it would have been better had the reef claimed her.[56] The damage however was to bring them problems of a more severe and lasting nature, compelling them to pay a lengthy call at Batavia (Djakarta) in Indonesia.

This was virtually the only port in this part of the world where ships could undergo major repairs and replenishment, and the *Endeavour* was in such a condition that a stay of some length seemed unavoidable. But apart from the undoubted quality of its marine yard, Batavia was a place to steer clear of. A miasma of uncleanliness and disease, with tertian and the even more virulent quartan malaria together with incapacitating and fatal dysenteries running rampant, it was a haven for ships but a graveyard for mariners.[57] Solander and Banks soon went down with malaria, Solander being the more badly affected. Both retreated to the country where the air was better and were joined by Spöring and some of the others. Banks had a supply of quinine bark which undoubtedly reduced the severity of his symptoms, but he did not appear to have enough to share with his sick colleagues, or perhaps he was not aware of the importance of quinine to malaria sufferers.

The death toll at Batavia reached seven, including Green, the astronomer, and Monkhouse, the surgeon. More succumbed to malaria and dysentery when they prolonged their visit to Indonesia at Princes Island.[58] On 24 January Spöring died, followed on the 26th by Parkinson. The terse entries in Banks's journal recording these tragic events probably reflect his own discomfort at the time rather than any lack of sympathy for his unfortunate assistants. Their party now decimated, Banks and Solander both recovered. Not for the first time the success of the natural history effort had hung in the balance. The loss of the one person who could quickly and accurately work up the sketches of

56. Whitehead, 1969, p.184.

57. Typhoid included. Watt, 1979, p.142.

58. Panaitan, western end of Sunda Strait.

animals and plants into publishable plates — one whose knowledge and skill had gone beyond that of an artist with ability in natural history drawing — was an omen that they were to be beset by problems of a quite different kind once they reached England, problems that in their way were almost as damaging as would have been the loss of the ship or the passing of one of the principal investigators. But such reflection, had it existed, was pushed into the background and disappeared in the general clamour and euphoria of their return on 12 July, 1771.

Taken from the *Endeavour* at Galleons Reach and transported up to London by sailing barge were the casks of animals in spirit, chests full of specimen bottles, animal skins, boxes of shells and other dried specimens, stacks of dried plants, curios and artifacts. The slightly putrescent air of imperfect preservation would surround the loads that made their way from the river in horse-drawn wagons, jolting over the cobbles, to Banks's residence in New Burlington Street. Here, as some of the chests were triumphantly broached to display their exotic contents or to give away a duplicate specimen, the chaos of unpacking, sorting and storing must have been considerable. Now or perhaps even earlier in the transit from ship to shore, the first jar may have been broken, a label here and there detached, a specimen lost in the litter of arrival. Some would remain for years unpacked.[59] Later, in 1776, all would be moved to 32 Soho Square, together with the books, notes and drawings, to form a kernel of natural history, a place where future collections would be received along with the curious and the interested, friends and scholars.

Re-immersed in the scientific and social world, and in all manner of other things including the prospect of a second voyage, Banks retained Solander as his librarian and assistant. Dividing his time between Banks's residence and the British Museum, where he was an under-librarian, Solander began the laborious task of writing up the species diagnoses and descriptions and supervising the artists and engravers employed to prepare the botanical illustrations. But the botany of the voyage was never to be published; nor, lagging somewhat further behind, was the zoology.[60]

Several reasons can be given for the failure of Banks and Solander to publish their scientific discoveries, quantified by Banks as '. . . about 1,000 species of plants that have not been described by any botanical author; 500 fishes; as many birds; with insects, sea and land, innumerable . . .'.[61] There was no conscious decision not to publish the zoology; instead, there appeared to be a steady weakening of resolve until events overtook the collections and the collectors alike. Viewed by the scientific standards of later generations it might seem inexcusable, or at best careless, but then the disciplines, the canons of zoology, were far less entrenched in those times. Particularly, the regularised describing, naming, and curation of animal specimens were still in their infancy, and the publication of the results of any scientific endeavour, while desirable, was not generally considered compulsory. Only Linnaeus, with the orderly mind of the systematizer and classifier, and with a keen sense of foreboding, was aware of the consequences of prolonged inaction.[62] The distractions of their continuing public acclaim, the loss of part of their technical and administrative infrastructure with the deaths of Parkinson and Spöring, the inadequacy of museums as institutions for curating large collections — one digs deep to excuse the likeable personalities of Banks and Solander, but it is within these personalities that further reasons can be found.

Banks, primarily a botanist — the plants came very close to being published — was generous in the disposal of animal specimens and had diverse and demanding interests. In the background was his large estate. Then he was a scientific impresario, an organiser

59. Whitehead, 1969, p.161.

60. John Cleveley, John and James Miller and Frederick Nodder were the artists. Some biological notes occur in Parkinson's Journal, published by his brother Stanfield in 1773, following a disputation with Banks and several others — see Banks, 1962, v.1, pp.59, 60. Some botany was published in the 20th C.

61. Cameron, 1952, p.319. 'Insects' includes all marine invertebrates.

62. Beaglehole, *in* Banks, 1962, pp.70-71.

of expeditions and people on a grand scale; but the excitement and the challenge were more in the preparation, chase, and collection than in the hard graft of writing up results, a problem that does not restrict itself to scientists of his period. He was indeed, as has been said of him, notwithstanding his education, Fellowship of the Royal Society and all, a 'Gentleman Amateur of Science'[63] in an age that predated much professional science. Solander on the other hand while undoubtedly a gentleman was certainly no amateur. His background of Linnaean teaching, his exact descriptions, his laborious cataloguing of specimens, his positions with the British Museum and with Banks, all bring him closer to one's idea of the professional scientist. Although a taker of copious natural history notes, he kept no daily journal and was a poor correspondent so he comes to life only through the writings of others.[64] But he was always there — enthusiastic, hardworking and sensible, but unfortunately, as far as the Cook voyages were concerned, virtually unpublished. A new edition of the *Systema Naturae* was intended: volumes of 'Solander slips' — small slips of paper bearing species diagnoses and other information — were the raw material for this. Manuscripts on the fishes collected on the voyage (such as the *Pisces Australes* which contains the New Zealand fish) and another manuscript containing the birds and other animals, give all the appearance of being intended for publication. But publication isn't everything — even now those valuable manuscripts are still available[65] — and at the time the specimens, the notes, the library, the houses at New Burlington Street and Soho Square, and especially Banks and Solander themselves constituted a resource in natural history that was unrivalled in England or anywhere else.

The slow exposure of parts of the collection to scientific scrutiny (only parts, because Linnaeus's worst fears were realized) is a story in itself, a sequel that merges with the accounts of the two other voyages undertaken by Cook in the South Pacific. However, before turning to these, there is another voyage driven by some of the same forces which brought Cook to New Zealand, although with results that barely register on any zoological scale. It nevertheless sounded a note of things to come in the exploration of New Zealand and its natural history.

Unlike others of its kind, the voyage of the *St Jean Baptiste* in the Pacific, under the command of Jean de Surville, was privately organized, their motivation being trade and the lure of rich southern lands following the collapse of the French East India Company after the Seven Years War.[66] This meant that the instructions were not detailed and they carried no natural historian. Their route from the Hooghly River,[67] through South-east Asia and the Solomon Islands, saw them in New Zealand waters little more than two months later than Cook. They sighted New Zealand off Hokianga Harbour, sailed around North Cape, narrowly missing an encounter with Cook, and anchored in Doubtless Bay where they landed on 18 December, 1769, with many of the crew sick with scurvy. Supplies were replenished — fish, a few vegetables and fresh water — and a pair of pigs were given to the Maoris, with the explanation that if kept they would be the source of more pigs. History does not tell us the fate of these animals or of a cock and a hen also presented, but it is doubtful that they survived to reach parenthood and were probably quickly dispatched to meet the needs of the day, leaving their claim to fame that they were the first attempt by Europeans to introduce animals to New Zealand soil. On 26 December the French fished and hunted, but the birds obtained were unidentified — possibly Tuis and Pukeko were among them.[68] Otherwise the only sign of vertebrate life was a few lizards, animals that apparently went unnoticed by Banks and his entourage. In his journal, the First Officer Guillaume Labé describes some of the fish given them by the Maoris as 50 centimetres long and shaped like a bonito. He also mentions the

63. *Ibid.* p.3. In an age that lacked science administrators in Government he was increasingly called upon to assist; D. L. McKay, *pers. comm.*

64. *Ibid.* pp.26, 53.

65. In the British Museum (Natural History) Zoology Library.

66. Dunmore, 1965, p.114.

67. A branch of the Ganges.

68. De Surville, 1981, p.148. Tui *(Prosthemadera novaeseelandiae)*, Pukeko *(Porphyrio porphyrio melanotus)*.

69. Labé, 1981, pp.196, 245-258. Dunmore's (as editor) identification of this as the Butterfly Tuna is unlikely (p.196); it was possibly one of the Kahawai *(Arripis trutta)* or the Jack or Horse Mackerel *(Trachurus spp.)*. The birds respectively: the Long-billed Curlew *(Numenius madagascariensis)*, Tui *(Prosthemadera novaeseelandiae)*, Parakeet *(Cyanoramphus novaezelandiae novaezelandiae or C. auriceps auriceps)* New Zealand Pigeon *(Hemiphaga novaeseelandiae novaeseelandiae)*, Brown Teal *(Anas aucklandica chlorotis)*, ? Grey Duck *(Anas superciliosa superciliosa)*. Sea birds not identifiable. Large shell probably the Large Trumpet *(Charonia lampas capax)*.

70. McNab, 1908-14, v.2, p.341. North Island Kokako *(Callaeas cinerea wilsoni)*. The lizard possibly the New Zealand Oviparous Skink *(Leiolopisma suteri)* or the Shore Skink *(L. smithi)*.

71. Hawkesworth, 1774.

72. Beaglehole *in* Banks, 1962, v.1, p.70.

73. Hoare, *in* Forster, 1982, pp.49-52.

74. Anderson, the surgeon's mate on this voyage, would also make collections.

native dog, Long-billed Curlews, Tuis, New Zealand Pigeons, and less specifically: teal, parakeets, small wild duck, sea birds, and shells — clams, mussels, and a large shell used as a trumpet by the Maoris.[69] Surville, in his brief summary of the animals encountered, thought the fish good; he mentions gurnard, plaice and mackerel. Another journal, that of the Second Officer Jean Pottier de l'Horme, describes the North Island Kokako, Tui, various wildfowl, seabirds, and a little black lizard 10 centimetres long.[70]

Some birds which they enjoyed but did not describe were killed for food on Sunday, 31 December, and on New Year's Eve they left, the country dismissed as a promising trading prospect and the crew still not completely fit. De Surville sailed on to South America, where at Chilca, Peru, he put off in a boat to go ashore. The conditions were rough, the boat capsized and encumbered by his heavy dress uniform, decorations and sword, de Surville drowned. After many trials and some considerable time the *St Jean Baptiste* made France in August 1773, under the command of Labé. It is just possible that a few animal specimens might have reached France with them, particularly shells, but no positive record exists. New Zealand natural history had not figured largely on this voyage, but it was a brief and early signal of French interest in the country, interest that was to be revived many years later in a far more substantial and scientific form. The journals of the Frenchmen did not become widely available as public records, and the French were soon to become better informed about New Zealand with the publication and translation into French of Cook's Voyage in 1774.[71]

Turning back now to the aftermath of that first singular voyage and to the events taking place in England: the air was still thick with excitement over their return when a second voyage, again with Cook in command and Banks and Solander as naturalists, was proposed. It was this news that reduced Linnaeus to such a miserable state[72] but the momentum was such that his concerns, if heeded at all, were eventually brushed aside. It might have been the old man's wish that some circumstance would intervene to prevent their departure and redirect their attention to the organization and publication of material from the previous expedition, but if so his wish was only partially granted. There is little reason to recall the circumstances that led to the replacement of Banks and Solander as naturalists on the second voyage;[73] there is even less reason to deplore them, since there is nothing to suggest that the ultimate outcome of the second voyage would have been any better had they gone and certainly nothing as far as zoology was concerned. Instead, their premature departure had the effect of introducing two new figures on to the scene — Johann Rheinhold Forster and his son Johann George Adam Forster, who would assist him and prepare the illustrations.[74]

Chapter Three

THE WANDERING SCHOLARS AND COOK'S LAST VOYAGES

APPOINTED ON 11 JUNE, little over a month was scant time for a scientist to equip himself and his small staff, in this case one son and one servant, for a journey to the South Seas.[1] Cook, ready to assist, had already seen to some of the nets, seeking a superior quality to that used on the previous voyage; he recommended those made of three-thread twine which 'can be got at the Shortest notice of James Davidson No. 27, Fish Street Hill'.[2] A bare week after their appointment the Forsters received £1700 and hastily went about buying equipment and supplies. Their request for four puncheons of double proof spirits of 80 gallons each (320 in all — a total of 1440 litres) for the preservation of specimens was swiftly actioned by the Victualling Board.[3] Ultimately given £4000 for his support on the expedition, Forster claimed that £1500 was spent on equipment and supplies. These were bought hastily and therefore expensively: never very good with money, he would inevitably have paid more than he needed to.[4] The relationship between Forster and the jilted naturalists of the earlier expedition was strained and Banks in particular was not very forthcoming with advice and assistance.[5] But if the Forsters had mixed feelings about the support for their part of the expedition they need not have worried about Linnaeus, who had already warmed to Forster's generosity and promise as a scientist and who had by now washed his hands of Banks and Solander. The burden of trust that he had placed on these two was now borne by Forster, virtually alone except for his son. It was a lot to ask.

Forster came to his task with a personality which would be his undoing but an advantageous background. The town of Danzig, contacts with Der Gesellschaft der Naturforschende Freunde,[6] his school friend Karl Gottfried Woide, the plight of the German universities — these will all surface again as surprising intersections in the zoological history of a country distant from Europe. Yet in spite of his credentials and his experience in natural history, the Forster who appeared with his son on the doorstep of English science in the autumn of 1766 was not an obvious future candidate for the job of naturalist on a British voyage of exploration. He could barely speak English, he had a family in Danzig to support, and he had hardly any money. At first sight, he was an altogether unattractive prospect for advancement. Digging deeper, however, one finds a man well capable of field work and of putting together and executing a scientific programme, and able to see at least some of this work through to publication.

It was these qualities which helped his rapid rise in both the social and the learned worlds — a rise which though turbulent and marked by altercation, by projects begun but not completed, was also distinguished by eventual recognition of his learning and his achievements, and his election to a Fellowship of the Royal Society. All the while he had been strengthening his association with natural history and its more influential practitioners;[7] now he could throw out stronger hints of his desire to join a voyage to the South Seas. When Banks began to fall out with the Admiralty in May of 1772, Forster with his supporters was ready to make his move. After bumping and lurching along

1. Ernst Scholient — he seems not to have had much part in the natural history effort.

2. Beaglehole, *in* Cook, 1961; v.2 pp.911 and 919.

3. *Ibid.*, p.941.

4. *Ibid.*, p.xlv. Beaglehole doubts that he spent this enormous sum.

5. Hoare, *in* Forster, 1982, p.52.

6. A society whose members studied the natural and physical sciences in Danzig. For detail of Forster's life see Hoare, 1976.

7. Including Anna Blackburne, Pennant, Banks, Solander, Tunstall, among others.

a hard road, he had arrived and the way ahead was clear. Surely, with his invaluable son, he would return from a hugely successful expedition, add substantially to the literature, and secure his reputation and future. But for the moment the expedition and its aftermath lay a long way ahead.[8] The *Resolution* left Plymouth Sound on 13 July, 1772, in company with the *Adventure* and Captain Furneaux.

It was a voyage in which New Zealand was to play an important and predetermined role. Queen Charlotte Sound was to be their springboard into high southern latitudes, where they would search again for the southern continent. There was no mention at this stage of Dusky Sound. Before they sailed into Antarctic waters they called into Cape Town and it was here that the scientific effort of the voyage received an unexpected bonus in the form of Anders Sparrman — another ex-student of Linnaeus, whom they found enjoying the study of the Cape's natural history. Forster spoke to Cook about the value of having such a man along with them, while Sparrman wrestled with the temptations of remaining at the Cape. In the end both were convinced and the new member came on board at Forster's expense.

Following their plunge into the icy latitudes after leaving the Cape, Sparrman must have wondered at the wisdom of his decision. It was a bleak period at times — fogs, the cold, snow, wind, damp bedding, ice in the rigging. The livestock gradually succumbed and some hapless sheep intended for release in New Zealand or Australia barely survived. But it wasn't always so — there were some fine days and observations of the fauna crowd Forster's journal: petrels, albatrosses, penguins, the occasional marine mammal (a seal or whale not closely identified).[9] As on the previous voyage, zoology predominated while at sea; every now and then Forster was able to put his formidable talents to describing a new species. But nearly four months of this sort of travel was beginning to take its toll and scurvy appeared among the crew. Tiring of his dreadful cabin conditions, Forster would have swapped the lot, £4000 and all, for a comfortable job at the British Museum.[10] Cook was aware of the need to make a landfall, and reasonably soon. When on 25 March, 1773, land was sighted and it proved to be southern New Zealand, Dusky Bay would do.

Forster's relief moved him to some anticipatory remarks on the plants and animals which were to be savoured both scientifically and gastronomically. He then embarked on lengthy and fulsome praise for the navigators who had brought them there, and the British in general. There was a sighting of four species of seabirds before they moved deeper into the Bay with its tree-covered islands, an observation of some Variable Oystercatchers, a safe anchorage close to the shore and, almost immediately, a boat dispatched for fishing.[11] Fish were landed almost as fast as they could put the lines down: Blue Cod, Scorpion Fish, Trumpeters, Scarlet Wrasse and Tarakihi, and then, after a hasty meal, they shot some birds: a South Island Robin and the Western Weka. George, with another party, took another robin and a South Island Fantail. He wrote later of '. . . a new store of animal and vegetable bodies and among them hardly any that were perfectly similar to the known species, and several not analagous even to the known genera'.[12] The Forsters introduction to New Zealand zoology had begun in earnest.

The entire period spent at Dusky Sound, nearly seven weeks, was one of almost continuous activity. For the naturalists it was a glorious mixture of hunting, shell-collecting, botanising, and some brief but uncomplicated encounters with the Maori. To some extent all three — the two Forsters and Sparrman — contributed to the tasks of collecting, drawing, and describing, but in the main it was the elder Forster who undertook the animal descriptions and generally supervised the others. George described the plants and did most of the drawings and paintings, while Sparrman helped with the

8. For a full account of the machinations needed to displace Banks see Hoare, 1982, pp.46-51. While all this was going on the *Marquis de Castries* (du Clesmeur) and *Mascarin* (Marion du Fresne) were already anchored in the Bay of Islands (4.5.1772). The French voyage, mounted for trade and strategic reasons, saw the deaths of Marion and 27 of his crew at the hands of the Maoris. There was little of zoological consequence. There are references to birds (e.g. quail; Dunmore, 1965, v.1, p.184); Marion's second-in-command, Julien Crozet, refers to Pukeko, New Zealand Pigeon, and North Island Kaka (Fleming, 1982a, pp.225, 290, 298).

9. Detailed identifications and annotations (particularly ichthyology and ornithology) are footnotes to Forster, 1982, v.1-4. These need not be repeated here in full.

10. *Ibid.*, v.2, p.233, 234.

11. Forster, 1982, pp.239, 240; The Variable Oystercatcher, *Haematopus unicolor*.

12. Forster, G., 1777, p.123. The fish were Blue Cod *(Parapercis colias)*, Scorpion Fish *(Scorpaena cardinalis)*, Common Trumpeter *(Latris lineata)*, Scarlet Wrasse *(Pseudolabrus miles)*, Tarakihi *(Nemadactylus macropterus)*; the birds — South Island Robin *Petroica australis australis)*, Western Weka *(Gallirallus australis australis)*, South Island Fantail *(Rhipidura fuliginosa fuliginosa)*.

The New Zealand Dotterel (*Charadrius obscurus*) painted by George Forster from Dusky Sound (vol. 1, pl. 122). One of the more lively of the Forster renditions. Published by permission of the Trustees of the British Museum. *(see p.25)*

plants, some of the animal descriptions, and collected both for the expedition and for himself.[13] Bad weather – and there was plenty of it – periodically enabled them to relieve the congestion of specimens and notes by confining them to the ship. On April 10 Forster was compelled to write: 'It rained still & was vastly bad foggy weather: We finished some Descriptions, cleaned Shells, brought a side plants etc: but few things would keep, on account of the moist weather: my Cabin was a Magazine of all the various kinds of plants, fish, birds, Shells, Seeds, etc. hitherto collected: which made it vastly damp, dirty, crammed, & caused very noxious vapours, & an offensive smell, . . . it was so dark, that I was obliged to light a candle during day, when I wanted to write something . . .'.[14] Given to peevish complaints he might have been, but his working and living conditions appeared rather less favourable than those enjoyed by the naturalists on the first voyage.

Even though he had to return to a dank and smelly workplace, it was not a disordered Forster who came in each evening from the field. He had arranged his work so that the correct zoological terminologies and conventions were ready to hand for the efficient and accurate description of specimens. Perishable or urgent specimens were immediately

13. Sparrman, 1953, p.27.

14. Forster, 1982, p.251

committed to his notebooks and long hours were often worked before retiring. George, whose output of animal drawings was to be substantial while in New Zealand, did well to produce them in quarters no better than his father's, although he had some use of the great cabin, with its rather better lighting.[15] Watching his progress was William Hodges, the official artist to the expedition and a competent painter of landscapes, ready to help and advise the younger man.[16] The overall picture is of hard work of almost dour intensity, with much of it done on board the ship when it was wet or dark, the heavily wooded edge of the Sound pressing in on them and a presiding air of melancholy at times providing an atmosphere altogether different from the one on the *Endeavour*. It was partly to do with their circumstances, but it was also a matter of style.

Things were much better in the field. Both the Forsters enjoyed hunting; so too did Sparrman, although there were moments of doubt.[17] The birds and the seals were easy targets and the naturalists were cheerfully joined in the sport by Cook and his officers. Cook himself makes references to animals seen and collected and on 1 April he describes a 'Duck of Blue grey Plumage with the end of its Bill as soft as the lips of any other animal, as it is altogether unknown I shall endeavour to preserve the Whole in spirits'.[18] One notable piece of collecting took place with a musket while on board the ship. Forster, using a stone on shore for target practice, allowed some shot to fall short, striking one of a school of fish. Cook recovered the unlucky animal which was drawn by George and described by his father.[19]

The crew had their sport too — good line fishing was possible from the ship and close to the shore, as well as from the smaller boats, and drop-nets (lobster pots) were used to catch crayfish.[20] The ease of catching fish meant that they had no need to rely on the small Maori population in the area for specimens and fresh food. The crew also collected shells, with the expectation that they would be able to sell them either to the naturalists or to collectors on their return home. We discover later that this activity did not have the approval of Forster, as they appeared to collect indiscriminately, neglected their specimens and, when they purchased shells from the local population, they competed with the naturalists and helped push the prices up.[21]

For Forster and his assistants a typical day's activity might read like the one described on 9 April: '. . . we fished in the boat & I shot a Gannet. We went to the very place where I & Capt. *Cook* had been the very first day of our arrival here. We saw vast Schools of small Fish, playing on the water. We landed & collected vast numbers of Ear-Shells of a very large Size on the Shore between the Stones. I shot a Waterhen, & collected besides Starfish, small Shells, Coats of Mail, & some Corals.'[22] Sometimes the collecting was more vigorous, such as at Duck Cove on 6 April: 'Here we saw two grey Ducks, with white, soft bills of the *whistling* kind, such as we had seen April ye 2^d in the Indian Cove: which we shot. We saw some smaller ducks of the *Gadwall* kind & we killed 12 of them & three Curlews or Oister catchers.'[23] Sometimes there was downright carnage when a 'Duck-shooting' party was formed: 'At our return we found that about 9 Shags, about 40 Waterhens, 27 Ducks, 1 Curlew, 1 Woodcock, 1 Sandpiper, 1 large Pigeon, several Pohebirds, 2 large Parrots, a Parrokeet, & several other small birds had been killed.'[24]

Now and again they would leave the water, the beach and the bush margins to discover the hazards of walking in the dense wet Fiordland forest: the rotten logs, the moss, the vines and entangling branches, all made worse by the persistent rain. Lack of quadrupeds and a sparse human population had left the bush all but untracked; it was easier to follow a stream, such as the one by their anchorage where at its source Forster found a small

15. *Ibid*, pp.81, 94.

16. *Ibid*, p.95.

17. Sparrman, 1953, p.31

18. Cook, 1961, p.114. This was the Blue Duck *(Hymenolaimus malacorhynchos)*.

19. Forster, 1982, pp.268-269; it was a Jack Mackerel *(Trachurus declivis)*.

20. And on one occasion a Hagfish.

21. Hoare, in Forster, 1982, v.1, p.117. This did not seem to be a problem in New Zealand.

22. Forster, 1982, p.250. Ear-shells = Paua, *Haliotis* sp.; Coats of Mail = a species of chiton; the corals are possibly one of *Flabellum rubrum* or *profunda*.

23. *Ibid.*, p.248. Birds respectively: Blue Duck, Brown Teal *(Anas aucklandica chlorotis)*, and Curlew = Variable Oystercatcher.

24. *Ibid.*, p.256. The Large Pigeon = New Zealand Pigeon, Pohebirds = Tui, the large Parrots = South Island Kaka *(Nestor meridionalis meridionalis)*, Parakeet = Red-crowned Parakeet.

lake, a place of quiet beauty.[25] Their expeditions to collect plants and animals sometimes compelled them to camp out overnight – a 'marooning party', as Forster called it. A beach with a stream and a supply of dry wood nearby was selected and a tent (oars propped against a tree and covered with a sail) was set up in a sheltered spot; a fire was lit and fish and birds broiled over it, to be served up on the boat's gangboard with the accompaniment of some grog. Sailors' tales were shared around the fire until tiredness took over and they retreated to their tent and slept on a bedding of twigs or ferns, huddled under their boat-cloaks until morning. A rosy picture, unless of course it happened to rain. It was probably on occasions like these that they met up with that ubiquitous hazard in Fiordland exploration – the sandfly. Innumerable and vicious, at times they virtually crippled Forster.[26]

But in spite of the vicissitudes of the weather, their accommodation, the sandflies, and the general wear and tear that the hardships of exploration far from home exact, their enthusiasm for their work is everywhere evident. Forster's journal is crammed with zoology and botany and virtually every entry tells of a new plant or animal discovered, another described, another proving satisfactory to the tastebuds.[27]

Of the 36 bird species observed here, some 30 were described and 18 still retain Dusky Sound as their type locality. George Forster illustrated 19 species.[28] This was a considerable improvement on the record of Cook's first voyage, where no New Zealand land birds were described, and only one sea bird was drawn. Included in this list of Forster descriptions are several birds now extinct or rare, for example the South Island sub-species of Thrush, Saddleback, and Kokako. Amongst the currently common or well known birds were the Paradise Shelduck, the Blue Duck, the Western Weka, and South Island Kaka. The fish were also well represented – upwards of 23 species, 12 of which were painted there by George Forster, and eight retaining Dusky Sound as their type

25. *Ibid.*, p.258. Lake Forster. George Forster (1777) briefly described the Giant Kokopu *(Galaxias argenteus)* from here (see Note 59 Chap. 4).

26. Wrongly classified by Forster in the genus *Tipula*. It is probably *Austrosimulium ungulatum*, no doubt the insect most commonly encountered, but not described until the 20th century.

27. Compared with Banks, 1962, Forster, 1982 has more detail on natural history. Together with the annotations of Hoare and his colleagues it is an invaluable zoological document.

28. 36 species of birds were observed at Dusky Sound, of which some 19 were drawn by George Forster. There are 18 species for which Dusky Sound is the type locality. The birds are:

* Variable Oystercatcher *(Haematopus unicolor)*
†*• South Island Robin *(Petroica australis australis)*
†*• Western Weka *(Gallirallus australis australis)*
†*• South Island Fantail *(Rhipidura fuliginosa fuliginosa)*
Pied Shag *(Phalacrocorax varius)*
*• Southern Blue Penguin *(Eudyptula minor minor)*
* • Blue Duck *(Hymenolaimus malacorhynchos)*
South Island Kokako *(Callaeas cinerea cinerea)*
†*• South Island Thrush *(Turnagra capensis capensis)*
• New Zealand Kingfisher *(Halcyon sancta vagans)*
Yellowhead *(Mohoua ochrocephala)*
*• New Zealand Pigeon

(Hemiphaga novaeseelandiae novaeseelandiae)
†*• Red-crowned Parakeet *(Cyanoramphus novaezelandiae novaezelandiae)*
*• New Zealand Dotterel *(Charadrius obscurus)*
New Zealand Shore Plover *(Thinornis novaeseelandiae)*
• New Zealand Falcon *(Falco novaeseelandiae)*
*• South Island Kaka *(Nestor meridionalis meridionalis)*
South Island Saddleback *(Philesturnus carunculatus carunculatus)*
*• Paradise Shelduck *(Tadorna variegata)*
Broad-Billed Prion *(Pachyptila vittata vittata)*
New Zealand Tui *(Prosthemadera novaeseelandiae novaeseelandiae)*
†* Bellbird *(Anthornis melanura melanura)*
*• New Zealand Scaup *(Aythya novaeseelandiae)*
*• Red-Billed Gull *(Larus novaehollandiae scopulinus)*
New Zealand Pipit *(Anthus*

novaeseelandiae novaeseelandiae)
Yellow-Breasted Tit *(Petroica macrocephala macrocephala)*
*• South Island Bush Wren *(Xenicus longipes longipes)*
*• Brown Creeper *(Finschia novaeseelandiae)*
†*• South Island Rifleman *(Acanthisitta chloris chloris)*
• Grey Duck *(Anas superciliosa superciliosa)*
Black Shag *(Phalacrocorax carbo novaehollandiae)*
Little Shag *(Phalacrocorax melanoleucos brevirostris)*
White Heron *(Egretta alba modesta)*
Brown Teal *(Anas aucklandica chlorotis)*
Wandering Albatross *(Diomedea exulans exulans)*
Australian Gannet *(Sula bassana serrator)*
* Dusky Sound, type locality.
• Drawn by Geo. Forster.
† Authored by Sparrman. Some of the above were drawn or described later from Queen Charlotte

Sound or elsewhere. Only three (the Variable Oystercatcher, Red-Billed Gull and Southern Blue Penguin) still have Forster's name as author. For further detail see Medway's notes *in* Forster, 1982.

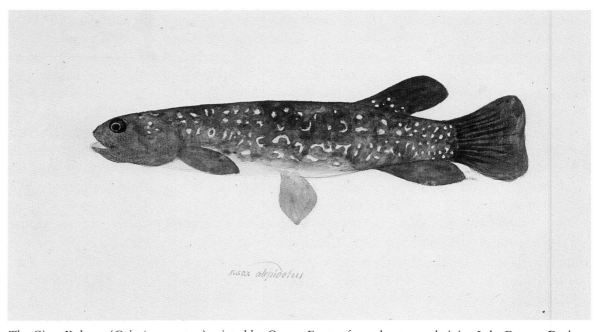

fisax alepidotus

The Giant Kokopu (*Galaxias argenteus*) painted by George Forster from the stream draining Lake Forster, Dusky Sound (vol. 2, pl. 235). The first New Zealand freshwater fish to be illustrated by a European. Published by permission of the Trustees of the British Museum. *(see p.25)*

locality.[29] Here too were some new and interesting additions including the Hagfish, caught unintentionally in a lobster-pot; the Giant Kokopu; a Ling caught while devouring another fish on a line; and of course the Jack Mackerel shot by Forster. A single mammal, the Fur Seal, was drawn and described, although it was not distinguished from a Northern Hemisphere species. It was a small specimen (32½ lbs — about 15 kg) but it does not appear to have been preserved. Added to these were many invertebrates: shells, including mussels, turbines, paua, and chitons; tunicates (sea-squirts) and corals.[30] The only insects mentioned specifically were the pestilential ones, the sandflies and the human lice.[31] Summer had passed and in the autumn this far south butterflies would be scarce, the noise of the cicadas had died out, and other insects were less obvious. Every expedition had to have its phantom and in this case it was a mysterious quadruped to match Spöring's long-tailed bird. Some seamen had seen a yellowish-coloured animal the size of a rabbit, while others observed a low-legged long animal with a pointed snout like a jackal. Had capture taken place this vision would probably have turned out to be the ship's cat on the prowl, or a ground-dwelling bird scuttling for cover.

30. The Fur Seal was *Arctocephalus forsteri*. The identification of most of the invertebrates can only be guessed at, as none was drawn or described.

31. *Pediculus humanus*, Forster, G. 1777, v.1, pp.215-216; Sparrman, 1953, p.42, for a medical man had some peculiar ideas on lice and the habit of chewing them.

29. Fish from Dusky Sound:
• Hagfish *(Eptatretus cirrhatus)*
*• Giant Kokopu *(Galaxius argenteus)*
*• Ling *(Genypterus blacodes)*
*• Saury *(Scomberesox saurus)*
 • Scorpion Fish *(Scorpaena cardinalis)*
• Jack Mackerel *(Trachurus declivis)*
• Tarakihi *(Nemadactylus macropterus)*
*• Common Trumpeter *(Latris lineata)*
• Moki *(Latridopsis ciliaris)*
*• Yellow-Eyed Mullet *(Aldrichetta forsteri)*
• Spotted Stargazer *(Genyagnus novaezelandiae)*
* Scarlet Wrasse *(Pseudolabrus miles)*
Blue Cod *(Parapercis colias)*
* Spectacled Triplefin *(Gilloblennius tripennis)*
*• Topknot *(Notoclinus fenestratus)*
Red Cod *(Pseudophycis bachus)*
Rockfish *(Acanthoclinus quadridactylus)*
Opalfish *(Hemerocoetes monopterygius)*
Witch *(Arnoglossus scapha)*
Barracouta *(Thyrsites atun)*
Red Gurnard *(Chelidonichthys kumu)*
Spotted Smoothhound *(Mustelus lenticulatus)*
? Spotted Spiny Dogfish *(Squalus acanthias)*
Others, such as the Clingfish *(Haplocylix littoreus)*, Silver Warehou *(Seriolella punctata)*, the Butterfly Perch *(Caesioperca lepidoptera)* and the Spotty *(Pseudolabrus celidotus)* may either have come from here or Queen Charlotte Sound. Of the 23 or so fish observed, a number were also seen at Queen Charlotte Sound, for example the Rockfish, Opalfish, Witch, Barracouta and Tarakihi were drawn from there. Eight (*) can be considered to have Dusky Sound as type locality, and at least twelve (•) were drawn here.

A young Fur Seal (*Arctocephalus forsteri*) painted by George Forster from Dusky Sound (vol. 1, pl. 2). This rather crude illustration demonstrates the erratic quality of some of George Forster's work. Published by permission of the Trustees of the British Museum. *(see p.26)*

The traffic of animals was by no means one-way: five geese were introduced at Goose Cove[32] and others set foot on land briefly, such as the sheep and goats which were put ashore only for the duration of their visit, and the spaniel, procured at the Cape of Good Hope, which was frightened by the sound of guns and had a 13-day sojourn in the bush before rejoining the ship. As noted, one of the cats on board had moments of freedom, playing havoc with small birds unfamiliar with predators.[33] Unseen and more permanent in their settlement were the rats which could easily make their way ashore from the ship. Sparrman, with a conservationist's sensitivity, was mortified by the thought that they should be the cause of introducing such animals.[34]

Towards the end of April there were signs from Forster that he had had enough and was anxious to move on, despite the natural beauty and teeming life that had enriched their collections. The variety of the resource was, for the time being, almost exhausted, while Queen Charlotte Sound and reunion with Furneaux and the *Adventure* lay ahead.

They made Queen Charlotte Sound on Tuesday 18 May, anchoring in Ship Cove, where they were greeted by a party from the *Adventure* and a present of some Red Cod. They were to spend only 19 days here, once again with variable and sometimes indifferent weather, although this did not seem to hamper their work. An almost immediate and significant addition to their collections was a bat shot on the bush margin, their first terrestrial mammal and their only mammal from Queen Charlotte Sound apart from Cook's observation of a sea-lion.[35] They added the Kahawai, Barracouta, the Variable Triplefin and possibly the Hagfish to the first-voyage records of fish from Queen Charlotte Sound, and repeated some Dusky Sound discoveries — George drew the Red Cod. Forster found the '. . . feathered tribe here is pretty numerous', observing some 25 species. Nine of these had their type localities in the Sound including the Morepork, New Zealand

32. Forster, 1982, p.265. Never heard of again.

33. Forster, G., 1777, v.1, pp.128; Cook, 1961, p.135.

34. Sparrman, 1953, p.42. The idea of the purity of the natural environment was not always well developed in those days. The Black Rat (*Rattus rattus*) was probably the one they left behind.

35. *Chalinobus tuberculatus*, the Long-tailed Bat, which was drawn. The 'sealion' was probably a Fur Seal (*Arctocephalus forsteri*). It would be unusual to see a sea lion this far north.

The Long-tailed Bat (*Chalinolobus tuberculatus*) painted by George Forster from Queen Charlotte Sound (vol. 1, pl. 1). Published by permission of the Trustees of the British Museum. *(see p.27)*

Falcon, and two species of shag, and seven were drawn by George Forster.[36] It was left to William Bayly, astronomer on the *Adventure*, to mention the 'few small harmless lizards'.[37] A 'cuttlefish' was the only invertebrate specifically recorded in Forster's journal, although they collected a number of shells. From the *Adventure* and Furneaux came a number of bird skins collected while the ship was in Tasmania.[38]

There was considerably more contact with the Maoris than at Dusky Sound and the Forsters were able to observe their dogs '. . . of different colours, some spotted, some quite black, and others perfectly white'.[39] Queen Charlotte Sound was seen as a more suitable place for the introduction of some of their livestock and on 2 June Cook, with Forster and Furneaux, released a male and a female goat on the eastern side of the Sound to add to the boar and two sows put ashore by Furneaux in Cannibal's Cove.[40] These might be considered the first introductions of mammals to the South Island — the unfortunate sheep which had recuperated at Dusky Sound died some days earlier on 23 May after being put ashore and eating a poisonous shrub.[41] Sparrman recalled that

37. Medway, 1983, MS. *The Natural history observations of William Bayly, Astronomer on Cook's second and third voyages.* Unpublished paper from The History of Science in New Zealand, Conference, February 1983.

38. Hoare, 1976, p.94. The 'cuttlefish' was probably a small version of the Long-finned Squid *(Notodarus sloani)* which would remind them of the Northern Hemisphere forms.

39. Forster, G., 1777, p.219.

40. Forster, 1982, p.293. Cook, 1961, p.169 mentions only one sow, which may be incorrect.

41. Possibly tutu, Beaglehole, *in* Cook 1961, p.167.

36. The fish:
*• Kahawai *(Arripis trutta)* • Red Cod *(Pseudophycis bachus)*, Barracouta *(Thyrsites atun)*
*Variable Triplefin *(Tripterygion varium)* and possibly the Hagfish *(Eptatretus cirrhatus)*.
The birds:
*• King Shag *(Leucocarbo caranculatus caranculatus)*
*• New Zealand Pipit *(Anthus novaeseelandiae novaeseelandiae*
*• New Zealand Falcon *(Falco novaeseelandiae)*

* Fluttering Shearwater *(Puffinus gavia)*
*• Spotted Shag *(Stictocarbo punctatus punctatus)*
*• Morepork *(Ninox novaeseelandiae novaeseelandiae)*
*• Tui *(Prosthemadera novaeseelandiae novaeseelandiae)*
• New Zealand Shore Plover *(Thinornis novaeseelandiae)*
* South Island Kokako *(Callaeas cinerea cinerea)*
* Yellowhead *(Mohoua ochrocephala)*
* Queen Charlotte Sound type locality.

• Drawn by Geo. Forster. Other birds observed were: Australasian Bittern *(Botaurus stellaris poiciloptilus)*, Black-Backed Gull *(Larus dominicanus)*, Black-Fronted Tern *(Sterna antarctica)* and others listed earlier: South Island Kaka, Australian Gannet, Red-Billed Gull, Red-crowned Parakeet, New Zealand Pigeon, Tui, South Island Saddleback, Bellbird, Yellow-Breasted Tit, South Island Fantail, Variable and South Island Pied

Oystercatcher. For arguments in favour of type locality designation see Medway, quoted by Hoare in Forster, 1982. See notes 44, 45, 47 for further records of this voyage from Queen Charlotte Sound. The Spotted Shag was described by Sparrman.

the abandoned huts which used to quarter the marines guarding Bayly's observatory were full of fleas, and that rats were abundant in the area.[42]

By 4 June they were making ready for sea, pausing to celebrate the King's birthday that evening, and on Monday 7 set sail in company of the *Adventure*, with Forster using some Pilot Whales for target practice as they left. They made a circular sweep through the Pacific that took in Tahiti, the Cook Islands and Tonga before returning once more to Queen Charlotte Sound, but not before Cook had put off more pigs, fowls, and seeds into Maori canoes off Black Head, which lies between Cape Kidnappers and Cape Turnagain. Further fowls were handed over at the entrance to Port Nicholson (Wellington Harbour). Although they expected the Maoris to make use of the progeny of the introductions, their major purpose was the provide future navigators with fresh food. The expeditions anchored again in Ship Cove on 3 November, to discover that the goats left on their previous visit had been killed and eaten (although this story was later contradicted) and the boar had become separated from the sows. However, encouraged by visible evidence of one of the sows, Cook released another breeding pair and some more cocks and hens. A few days later more pigs and poultry were put ashore at West Bay, but the loss of a male goat continued Cook's run of bad luck in his attempts to introduce sheep and goats to New Zealand. Viewed later, such difficulties seem remarkable.[43]

Forster found the season (early spring) insufficiently advanced for the satisfactory collection of plants and animals. Few fish were obtained — the seine was hauled once or twice but with no great success, although a net bought from the Maoris achieved some results. The Leatherjacket was the only fish species new to them (but recorded from the first voyage) and could easily have been amongst the catches traded to them by the Maoris, as was the Barracouta. A 'fine skate' was caught alongside the ship, possibly the same species of skate seen on the first voyage.[44]

With James Patten, the surgeon, Odiddy (from Raiatea in the Society Islands) went on a hunting expedition ashore, his first opportunity to test his aim. His previous experience with muskets had been restricted to discharging those of the others on their return from hunting and now he shot 'a fine green new Cockow', subsequently known as the Shining Cuckoo. The Northern Diving Petrel was their only other new addition.[45] A gruesome extra for the shipboard collections, bordering on natural history, was the head of a Maori boy who had been killed and eaten by a raiding party. It was brought on board and became the object of some controversial experimentation by Cook. Later Lieutenant Pickersgill traded a nail for the head, which was preserved in spirits for John Hunter's anatomical collection in London.[46]

When they began their departure from Queen Charlotte Sound on 25 November, the *Adventure* had still to make rendezvous but Cook, after conducting a search, decided to proceed without her and headed south to the highest latitudes of any of his voyages. Confronted with the peripheral ice of the Antarctic continent he turned north to close a vast circuit of the South-West Pacific that took them up to Easter Island, the Marquesas Islands, Tahiti, and across to the New Hebrides and New Caledonia before returning to Queen Charlotte Sound via Norfolk Island. When they anchored on Tuesday 18 October, 1774, they had been gone almost 11 months. The naturalists were quickly into the usual round of shooting and plant collecting and a pig spotted in the bush restored sufficient confidence in their programme of introductions for Cook to release another pair. George brushed up a few drawings and duplicated others, providing some specimens with 'a more picturesque twin'. Although many of the fish had already been encountered

42. Bayly was the astronomer on the *Adventure*. Sparrman mentions lice, but it is unlikely that human lice *(Pediculus humanus)* would be found free in an abandoned hut. The fleas may have been left there from the sailors from the first voyage (see Miller, 1971, p.132). The rats were probably Black Rats *(Rattus rattus)* — Sparrman's observation seems to confirm that they were left by the first voyage.

43. Cook, 1961, pp. 287, 296-297.

44.* The Leatherjacket *(Parika scaber)*, Barracouta *(Thyrsites atun)*, and possibly the Rough Skate *(Raja nasuta)* (* type locality).

45. *● Shining Cuckoo *(Chalcites lucidus lucidus),* *● Northern Diving Petrel *(Pelecanoides urinatrix urinatrix)* (* type locality, ● drawn by Geo. Forster.)

46. Cook wished to observe cannabilism at first hand; Cook, 1961, p.293; Sparrman, 1953, p.105.

The Shining Cuckoo (*Chalcites lucidus*), shot by Odiddy and painted by George Forster from Queen Charlotte Sound (vol. 1, pl. 57). Published by permission of the Trustees of the British Museum. *(see p.29)*

by Banks and Solander, there were still species not previously seen by Forster's party; a Snapper (presumably the one that weighed 11¼ lbs [5 kg] and was hauled over the side) was described, as were a Hapuku purchased from the Maoris, Butterfish, Rock Cod, Clingfish, and others. Some had been collected before but only now was there time for their illustration.[47] There was less to do with birds, apart from the usual hunting, and some botany was undertaken, with the weather once again indifferent. Expectations of new plants or animals were not high and clearly it was time to be going. They left Ship Cove for the last time on 10 November, in preparation for moving out of the Sound the next day on a voyage that would take them to Tierra Del Fuego, the Cape and then home.

On his departure from Dusky and Queen Charlotte Sounds, Forster had written his impressions of the fauna, flora and geography of each locality,[48] but they were modest

48. Forster, 1982, pp.275-278, 297-298. There is a manuscript *Observationes . . .* which contains G. Forster's notes from Dusky Sound 26 March-11 May 1773; MS189, Bibliothèque, Centrale, Muséum National d'Histoire Naturelle.

47. Fish observed on the third visit to Queen Charlotte Sound:
•* Rock Cod *(Lotella rhacinus)*
•* Hapuku *(Polyprion oxygeneios)*
• Rockfish *(Acanthoclinus quadridactylus)*,
•* Snapper *(Chrysophrys auratus)*,
•* Butterfish *(Odax pullus)*,
• Variable Triplefin *(Tripterygion varium)*,

•* Clingfish *(Trachelochismus pinnulatus)*,
• Kahawai *(Arripis trutta)*,
• Moki *(Latridopsis ciliaris)*, Conger Eel *(Conger verreauxi)*, Blue Cod *(Parapercis colias)*, also the Red Cod *(Pseudophycis bachus)*, and a Dogfish (not able to be identified further). The 13 fish observed represented their best haul yet from Queen Charlotte Sound —

in part contributed by the Maoris (e.g. Hapuku). Five of the above have Queen Charlotte Sound as their type locality (*), nine were drawn by George Forster (•). Five other species were observed from Dusky and Queen Charlotte Sounds but in the latter case are not referrable to a particular visit. These are: Tarakihi *(Nemadactylus macropterus)*,

Spotted Stargazer *(Genyagnus novaezelandiae)*, Opalfish *(Hemerocoetes monopterygius)*, Witch *(Arnoglossus scapha)*, Clingfish *(Haplocylix littoreus)*. The Red-crowned Parakeet, South Island Kokako, and New Zealand Pipit were drawn on the third visit.

statements, not the definitive summaries intended for later. At this stage of the proceedings we are left to furnish our own record of progress and achievements and what we would guess to be his considered view of the New Zealand fauna at the time of their leaving.

It is tempting to categorize this period in New Zealand as one that gave us the natural history of Dusky Sound and more or less founded ornithology in New Zealand, but there is more to it than that. To begin with, the time spent in Queen Charlotte Sound was of a similar order and the number of species recovered there made a significant contribution to the overall total. The ornithological results of this voyage had bettered those of the first by a considerable margin, but ichthyology too did better with more than half their records new additions to those made on the *Endeavour*.[49] Many specimens of fish and birds were preserved and accompanied them back to England.[50] But their efforts in terms of descriptions and drawings were clearly directed at the vertebrate animals and the treatment of the invertebrates was rather more haphazard. Apart from *Haliotis* (a species of Paua) mussels, and possibly the Snake-skin Chiton, it is not possible to recognize the shells from the unspecific terminology used in the journals, but it can be deduced that several new species were discovered. At least 11 specimens of the famous Star or Imperial Sun Shell came on board the *Resolution*,[51] including one from Cloudy Bay. Others were possibly picked up on the beach or caught in a trawl, but one notable specimen in the possession of one of the *Adventure's* officers was hauled up on the anchor cable from a depth of 60 fathoms, somewhere in Cook Strait.[52] Others, such as the Cats-Eye, Cook's Turban and the Tiger Shell, were easily picked up on the beach. But molluscan anatomy was of no interest to the naturalists and shells containing live animals were mostly cleaned.[53]

The insects were not to fare too well. Forster would later comment: 'No countries of the world produce fewer species of insects than those of the South-Sea'; in this category he undoubtedly put New Zealand.[54] While it has been acknowledged that the insect fauna of the country is not as large or colourful as some others, it is clear that Forster neglected this group in New Zealand. By comparison, Banks and Solander collected more insects than they did other animals, possibly because their preoccupation with plants automatically brought them into closer contact with the insect fauna. Forster gives the impression of devoting most of his time to collecting birds and other animals, leaving much of the botanising to his son and Sparrman. Also, the season at Dusky Sound was unsuitable for insects while the spring and early summer periods at Queen Charlotte Sound although better were a shade early, with weather conditions that were not ideal. But he didn't neglect insects entirely: a female cicada was collected at Queen Charlotte Sound and it is just possible that a few beetles and the Common Copper Butterfly were also taken here.[55]

Presumably the other invertebrates — the starfish, cuttlefish, tunicates and corals — were put into spirits and joined the rest in bottles and casks with the plant collections and the artifacts that formed the substantial New Zealand collection. This, together with their acquisitions from other countries, their notes and their drawings, were the stock with which Forster hoped to make his reputation. Linnaeus would not have expected more. One specimen did not make it back home — a jellyfish they caught in New Zealand waters and named *Medusa vesia* was sketched roughly in pencil by George and accompanied by a terse catalogue note, 'did eat it'.[56]

In terms of knowledge and specimens the Forsters took a great deal with them — it would be a long time before anyone took more — but in significant part, the voyage should also be remembered for its bird and mammal introductions to New Zealand:

49. In the order of 36 bird species noted at Dusky Bay, 23 described and most drawn by George Forster; some 27 noted and 12 described in Queen Charlotte Sound. A further two were named in *Descriptiones Animalium* but were subsequently described from different localities. The fish: 23 noted and 14 described or drawn from Dusky Bay; 23 noted and 18 described or drawn from Queen Charlotte Sound. 3 fish species are of uncertain locality. There is some overlap in species recovered from Dusky and Queen Charlotte Sounds. Begg & Begg, 1968, p.147, noted the value of the Forster contribution to the New Zealand ornithology, Wheeler, 1981, lists the fish.

50. Still extant are: the South Island species of Kokako, Kaka, and Thrush, Tui, also the New Zealand Falcon, Red-crowned Parakeet and Spotted Shag (the last three collected by Sparrman; Medway 1976, pp.120-131) and the Butterfish and Clingfish *(Haplocylix littoreus)* (Wheeler 1981, pp.794, 798).

51. Seymour to Poulteny, 28.11.1775, quoted in Dance, 1966, pp.108, 109.

52. Dance, 1966, p.110. The circumstances quoted are not beyond doubt; Wilkins, 1954, p.10 quotes a source favouring *Discovery*.

53. Respectively: *Turbo smaragdus, Cookia sulcata* and *Maurea tigris*. Apart from the well-documented Imperial Sun Shells, it is difficult to identify specific second voyage shells, although Forster was an enthusiastic collector. Dance, 1971, *continued overleaf*

p.376, lists most of those collected by the Cook voyages in New Zealand.

54. Forster, 1778, p.196.

55. The cicada, *Kikihia muta*, Dugdale and Fleming, 1969 p.930 – probably not variety *subalpina*, D. Lane, *pers. comm*; the Common Copper *(Lycaena salustius)*, was more likely to have been collected on the 3rd voyage, but see Note 10, Chap. 4.

56. Whitehead, 1978a, p.45.

57. Forster, 1778, p.185 *et seq.*

58. Parted from them before the second visit to Queen Charlotte Sound.

59. Forster, 1982, p.728.

60. Hoare, 1976, p.153.

61. Beaglehole *in* Cook, 1961, pp.cxii,cxiii.

62. Cook sent four casks of specimens and some shells to Solander at the British Museum, directing that the shells go to Lord Bristol (August John Hervey, a senior naval officer and natural history collector.) Edward Terrell (Tyrell) and John Marra (Able Seaman and Gunner's Mate respectively) offered specimens to Banks, Beaglehole, *in* Banks, 1962, pp.105, 106, 108; Whitehead, 1978, p.78; Medway, 1976, p.49. Forster gave his insects to Solander for equal distribution to the British Museum, Royal Society, Banks, Marmaduke Tunstall, and Sir Ashton Lever; Whitehead, 1969, p.163. Clerke gave Banks a New Zealand Shore Plover in spirits of wine; Fleming, 1982a, p.242.

63. The drawings (including NZ specimens) sold to Banks for 400 guineas; Hoare, 1976, p.164. Forster eventually

the geese, fowls, pigs and goats, and no doubt a share of rats. It marks the beginning of a (mostly) deliberate policy of animal introduction which was subsequently to have important biological consequences for the country. When Forster's assessment of the New Zealand fauna eventually came to light[57] it was made in terms of the South Pacific fauna as a whole: its teeming numbers of animals and its lack of variety. Although there are references to New Zealand animals, there is little said about them that is profound. He had described many new species, but given the descriptive state of the science, the limited coastal exploration, and the fact that some key New Zealand animals were not observed, the essence of New Zealand zoology would elude him, just as it had eluded Banks.

Sparrman with his collections left them at the Cape of Good Hope, where he was to work for another year before returning to Sweden, and it was here that they heard of the fate of the *Adventure*, which had reached the Cape 12 months before.[58] Immediately prior to their arrival there and with the burdens of their voyage almost behind him, Forster began to consider his future. Speculating morosely on the loss of friends and patrons while he was away, he wrote, 'I must begin life as if it were again . . .' and started to renew his contacts.[59]

The public acclaim after their return on 30 July, 1775, was directed more at the sailors than the scientists[60] although some interest overflowed to the naturalists and their specimens. This was not obvious from Banks, though, who was away yachting and left Solander to enthuse over the homecoming.[61] Banks did not have to be at the *Resolution*'s berth at Woolwich, at the Forster residence at St Pancras, or wherever else specimens were unloaded or unpacked. Sooner or later the collections would come to him: from Cook, Clerke, Furneaux, the men of the *Resolution* including Edward Terrell and the plausible opportunist John Marra, and eventually from Forster himself.[62] Despite his temporary fall from grace, Banks had friends on board and he was still pivotal to natural history interests in England.

While all round him the dispersal of the specimens collected on the voyage was already taking place, Forster must have remained confident that the scientific record, with his detailed descriptions and George's drawings, was intact. Surely these would all be published soon. At the time, the fate of the specimens must have seemed less important. Poised to publish by virtue of having prepared his manuscripts while at sea, Forster ran headlong into opposition from Cook over the former's diagnoses of the plant genera, *Characteres generum plantarum . . .* (1776) which would precede any other official account of the voyage. Although eventually allowed to go ahead, this was a warning of the difficulties that lay in front of him. He fell out first with Sandwich, First Lord of the Admiralty, and then, barely a year after their return, with Cook, causing the prospect of the joint publication of the voyage's science and official narrative to vanish almost completely. If this wasn't enough, a short time later his pressing financial circumstances forced him to sell George's drawings to Banks. And so it went on; the detail of the decline being described elsewhere, we can simply record that by late 1778 George Forster was abroad trying to arrange for the extraction of his unhappy parent from England.[63]

In spite of everything, some publications incorporating the voyage's natural history did emerge. George Forster's *A Voyage Round the World. . .* appeared in March 1777, making several zoological references to New Zealand. The following year his father's *Observations. . .* made similar references, but of a more general nature. The highly regarded *Historia Aptenodytae. . .*, read in 1780 and published in 1781, contained a description of the Southern Blue Penguin, collected at Dusky Bay; although not illustrated, this was a detailed and valid scientific description and the first of a New Zealand bird. A paper on the 'Tyger-Cat' from the Cape of Good Hope was followed in 1785 by one on some second voyage's albatrosses,[64] which included descriptions and illustrations of two species that breed on the sub-Antarctic islands and are sometimes seen in New Zealand waters. The *Enchiridion...*, published in 1788, was a collection of generic diagnoses rather than the descriptions of species. For zoologists, J. R. Forster's most valuable work was to be the *Descriptiones Animalium...* which contained the names and scientific descriptions of animals collected on the voyage. Its publication was delayed until 1844, long after his death, but in the interim the manuscript, which includes descriptions of 37 New Zealand birds, 32 fish and two mammals, was used opportunistically by many contemporary and later zoologists. This fate also befell George's valuable zoological drawings, few of which would ever be traced out by the engraver's burin.[65] Other works to remain unpublished were Anderson's *Characteres breves...*, which contained descriptions of birds written up after the voyage, and also there exists a series of paintings by an unknown hand, including several of New Zealand birds.[66]

Extract from Anderson's manuscript *Characteres breves Avium .....*, describing the South Island Kaka (*Nestor meridionalis meridionalis*). Published by permission of the Trustees of the British Museum. *(see above)*

The reasons for the failure of the voyage's natural history to be published as intended lie in the morass of personal failings — obstinacy, jealousy and pride being among them — of those involved. Whoever bore the major responsibility (and Forster himself was by no means guiltless) does not matter a great deal as far as zoology is concerned. What does matter is that after so much effort a scientifically desirable objective was not achieved; means had suddenly become more important than ends, and there was no individual or institution big enough to rescue the situation. Substantial powers and influence would have been needed to unite the conflicting forces which, complicated by Forster's personality and chronic insolvency, produced the almost inevitable result.

Meanwhile, slipping England and the Forster toils, Cook was undertaking yet another expedition to the South Seas with *Resolution*, this time in company with the *Discovery*. With the past voyage and the problems of natural historians fresh in their minds, Cook and the Navy did not seem particularly well disposed to the idea of a voyage naturalist. Instead one of the surgeons on the *Resolution*, William Anderson, was to represent natural history interests, with his '. . . reagents, a blow-pipe, a microscope and spirits of wine'.[67]

took up an appointment at the University of Halle. For preceding detail, see Hoare, 1976, p.151 *et seq.*

64. *Mémoire sur les Albatros*, Forster, 1785; Grey-headed Mollymawk *(Diomedea chrysostoma)* and Light-mantled Sooty Albatross *(Phoebetria palpebrata)*. Forster's 1777 bare and inadequate description of the Broad-billed Prion *(Pachyptila vittata)* should not be the basis for this name (D. Medway, *pers. comm.*).

65. Exceptional were an engraving of the Tui in Cook, 1777, and copies by Latham (1781-1785). The names of three birds (Variable Oystercatcher, Red-billed Gull and Fluttering Shearwater) from *Descriptiones animalium* are still current as are the Rock Cod and Long-tailed Bat.

66. Anderson describes 23 species from NZ (e.g. the Variable Oystercatcher, New Zealand Pigeon, Red-crowned Parakeet, *etc.*): in *Characteres breves Avium (in itinere nostro circum orbe visa) adhuc incognitarum anni 1772, 1773, 1774 et 1775.* MS. Zoology Library, British Museum (Natural History). The unknown artist is discussed by Lysaght, 1959, p.260-262. It is tempting to speculate that the manuscript and the drawings represent a shadow natural history effort — perhaps encouraged by Banks and Cook.

67. With some natural history experience from the previous voyage, he has yet to receive the recognition he deserves (D. Medway, *pers. comm.*). Beaglehole *in* Cook, 1967, p.1xxiv.

68. Beaglehole, *ibid.*, p.1478.

69. Beaglehole, 1974.

70. *Sphenodon punctatus* — NZ's largest living reptile, mostly restricted to offshore islands, was not encountered by Europeans until many years later. Skinner, 1964, reviewed the crocodile and lizard in Maori myth and culture. It is not easy to relate myth to specific reptiles, and as there is some evidence of their having eaten Tuataras, their general fear of reptiles cannot have been absolute. The Giant Lizard may indeed relate to crocodile mythology. There was also a Giant Gecko *(Hoplodactylus* sp.) believed extant at the time of Cook (A. Bauer, *pers. comm.*).

71. Anderson *in* Cook, 1967, p.796-819.

72. The 'scorpion-flies' were cicadas. The 'Dragon-flies' *Uropetala* spp. or Damselflies. The Maoris 'eat without scruple the vermin in their heads . . .' (Anderson, 1967, p.812, referred to the common practice of crushing lice). Fleas; Anderson *in* Cook 1967, p.800.

73. The White-fronted Tern *(Sterna striata)* a new record and Ellis painted the type of the species.

74. Cook, 1967, p.66; Samwell *in* Cook, 1967, p.995.

There was also David Nelson on board the *Discovery*, a Kew gardener enlisted by Banks as a botanical collector, although he appears on the muster lists as a servant to William Bayly,[68] the astronomer, who also made some natural history observations. William Ellis, the surgeon's second mate on the *Discovery*, was responsible for some of the zoological drawings, and so too was John Webber, the official artist to the expedition, whose job was to paint landscapes. So on the face of it, a fair compromise was made between the convenience of the Navy and the interests of science and the collectors at home. Cook had written to Banks indicating his continued desire to collect for him: in spite of everything, Banks was still the focal point of natural history in England.[69]

New Zealand, or more particularly, Queen Charlotte Sound, was to be a port of refreshment en route to the Pacific and the North-West Passage. So when they anchored there on 12 January, 1777, it was familiar territory, not without its interests and attractions, but lacking the focus of attention that it had on the previous voyages. Cook's journal for the period of their stay is largely devoid of comment on the animals encountered, although he recounts an interesting tale told him by a Maori about giant lizards and snakes that burrowed in the ground. Although reminiscent of the mythology of the ngarara or taniwha, there is perhaps something here of the Tuatara that is recognizable. The Maoris of Queen Charlotte Sound may well have been familiar with this, the most fearsome-looking of the New Zealand reptiles, as they are known to inhabit the offshore islands of the Marlborough Sounds.[70]

But it is to Anderson's journal[71] that we must turn for an account of the natural history of New Zealand as observed during this voyage. He describes some 22 birds and notes the frequency of their sighting: thus the Shining and Long-Tailed Cuckoos were 'scarce' but the South Island Thrush (now extinct) was found to be 'frequent'. More than 18 species of fish were noted together with shellfish of '. . . ten or twelve sorts such as periwinkles, wilks, limpets, and some very beautiful sea Ears and another sort which sticks to the weeds, . . .'. There were also oysters, cockles and two species of mussels. The crayfish are, once again, found delicious and starfish and sea eggs are mentioned. In spite of its being high summer, they still found insects to be rare: '. . . two sorts of Dragonflies, some Butterflies, small grasshoppers, several sorts of Spiders, some small black Ants and vast numbers of Scorpion flies which . . . fill the woods with their chirping'. Sandflies were, once again, numerous. Two or three sorts of small harmless lizard sufficed for the reptiles. Why these inoffensive creatures had to wait decades to be described, or even collected, is anybody's guess — being neither highly coloured nor edible placed them in some dull category beyond the interest of early natural historians. A few rats, a dog, fleas and headlice complete Anderson's list.[72]

Ellis painted five New Zealand land birds — the Red-crowned Parakeet, South Island Kaka, Tui, New Zealand Falcon and Yellow-breasted Tit. The White-fronted Tern and Wandering Albatross were completed at sea, off New Zealand.[73] Webber painted only the South Island Kokako. No fish or other animals appear to have been illustrated.

The practice of introducing animals continued. Pairs of goats and pigs were put ashore here, and also two pairs of rabbits; other animals made a brief appearance — horses, cattle, sheep, goats, turkeys, geese, ducks and a peacock.[74] They left Queen Charlotte Sound on 25 February, 1777, passing through Cook Strait to the Pacific. On the way out, somewhere off the south-east coast of the North Island, a tern flew aboard, which they captured and Ellis painted. It was a farewell gesture; the last animal collected in New Zealand waters for some years, the last new species for decades.

The remainder of the voyage was blighted by the tragic deaths of Anderson and then

The Red-crowned Parakeet (*Cyanoramphus novaezelandiae novaezelandiae*) painted by William Ellis from Queen Charlotte Sound. (pl. 12). A good example of this artist's meticulous style. Published by permission of the Trustees of the British Museum. *(see p.34)*

Cook. Anderson succumbed to consumption on 3 August, 1778, and if there had been any hope of some account of the natural history of this voyage being written and published, it died with him. There were testimonials to his character and skill. Cook noted that he had 'acquired much knowlidge' and Clerke, commander of the *Discovery*, acknowledged his 'attention to the Science of Natural History'.[75] There was no doubting his ability, but the downgrading of the natural history effort had commenced long before his death. The killing of Cook at Kealakekua Bay in Hawaii on 14 February, 1779, virtually closed a chapter in Pacific exploration in which New Zealand and its flora, fauna and ethnology had a prominent place.

The return on 4 October, 1780, was marked by indifference or disappointment at the animal collections. Private collectors and dealers were moving in, although Banks, even at a distance, was able to attract specimens collected on this voyage — Anderson was only too pleased to collect for him and Cook himself had been prepared to use his '. . . best endeavours to add to your Collection of Plants & Animals . . .'.[76] But in spite of this, the harvest from the third voyage was neither large nor well curated. There was

75. Cook, 1967, p.406.

76. Anderson to Banks, and Cook to Banks, quoted by Beaglehole *in* Cook, 1967, pp.1519, 1521.

even a suggestion that collecting, beyond their duty to Banks, was not encouraged: 'Geography has been the grand Object of our voyage, there we shine — that we shall not appear to equal Advantage with respect to natural History there are many Reasons that prevented us ...'.[77] There were complaints about the number and quality of specimens, and indignation at the manner of disposal of some of them. John White, writing to the collector Anna Blackburne, was disappointed at the 'one baskett of shells & not a Single bird ... not an Insect, or Animal could I find Except one Starved Monkey'.[78] Anderson was apparently the only one who had a collection of birds and these, now in an imperfect state, were reserved for Banks. The fish that were brought back were badly treated, being thrown into a common cask.[79] Bayly had a few animals of which he had given Ashton Lever first choice for his recently established museum, but it was thought beneath a man of Bayly's stature and income to advertise his collection for sale in a newspaper.[80]

Somewhere in the middle of this potpourri were the New Zealand specimens — some shells, possibly one or two insects, and a few birds including the Tui and Kokako.[81] The documentary natural history record was limited to the journals of Anderson and, to a lesser extent, Bayly, and a small notebook of Anderson's entitled *Zoologia nova...*,

Extract from Anderson's manuscript *Zoologia nova* .... describing the White-fronted Tern, (*Sterna striata*). Published by permission of the Trustees of the British Museum. *(see above)*

describing new animals encountered on the voyage. There were of course the careful finished paintings of Ellis and Webber, but neither the manuscript nor the paintings were published and the zoological results of the third voyage slipped into virtual obscurity.[82]

Thus ended a remarkable series of explorations, each with its scientific component having a unique character. From a natural history and a New Zealand perspective we have the promise and excitement of the first voyage, the determined, awkward, but productive nature of the second, the anticlimax of the third. With regard to the broad sweep of science, the legacy of the voyages was vast in the terms of their times. Within this panorama the natural history might be seen as disappointing in respect of achievement

77. Samwell to Gregson, quoted by Beaglehole *in* Cook, 1967, p.1561.

78. John White to Miss Blackburne, *ibid.*, p.1560.

79. Whitehead, 1969, p.163.

80. Samwell to Gregson, quoted by Beaglehole *in* Cook, 1967, p.1561.

81. Medway, 1976, pp.126, 128. The Tui was possibly the one painted by Robert Laurie in 1784 (see Murray-Oliver, 1969).

82. Anderson, 1776-1777, described the Shining Cuckoo (*Chalcites lucidus lucidus*) and the White-fronted Tern (*Sterna striata*): in *Zoologia nova seu characteres & Historia Animalium hactenus incognitoram qui in itinere*

in what we might call the hard currency of zoology. It was disappointing in part because it could not bring lasting credit to those who deserved it — only the name of Forster stands to any great extent as an author in the catalogue of animal species. The naturalists, particularly Banks and Solander, enjoyed contemporary reverence in England and abroad, but there would be little to remind future generations of scientists of what they had done. That their contribution was made in different ways need not lessen our respect for any of them. Banks was the focal point for material from all three voyages and without that degree of centralization things would have been even more chaotic. Solander provided the scientific record for Banks and in the New Zealand context at least, Forster emerges as the voyages' most productive naturalist and a faithful adherent to the Linnaean ideal. Sparrman and Anderson also emerge with credit, and so do the natural history artists whose work grew in importance as the publications failed to proceed.

Lest the problems of personal disputes, unpublished findings and lost or destroyed specimens are seen as symptomatic of these voyages and their naturalists alone, it pays to recall briefly the fate of some contemporary expeditions. Commerson's paintings, manuscripts and specimens went in various directions after Bougainville's expedition and were never made into a collective publication,[83] and Linnaeus grieved at the fate of the rich haul made by Forsskål on the Danish expedition to Arabia. When the collections reached Denmark in 1776, 'the alcohol evaporated from the containers, the animals rotted, the dry fish were attacked by the damp, the stuffed birds were transformed into colonies of fleas and moths'[84] and the manuscripts suffered similar indignities. There are many parallels to the voyages just discussed. Clearly, an outcome of such expeditions that might be taken for granted 50 years later was not the rule in the late 18th century. But regrets and might-have-beens aside, we can see that the Cook voyage naturalists had achieved something of real scientific value, if not for themselves, then for others who would later use what remained of their collections, manuscripts and illustrations.

nostro videbantur 1776 in Linguis Latinus & Anglicis traditus, MS. Zoology Library, British Museum (Natural History). Medway, 1976, published some of these drawings.

83. Brosse, 1983, p.32.

84. Hansen, 1964, p.278.

Chapter Four

'THOSE DESTINED TO DESCRIBE THEM MAY DIE . . .'

Linnaeus

I T WAS AS IF THE long-dead specimens from Cook's voyages had suddenly burst back into life, moving as they did from hand to collector's cabinet, from cabinet to auction, from auction to museum, between cities and countries, giving pleasure as curiosities or as the objects of scientific scrutiny. The relatively durable items such as the insects or more particularly the shells survived the handling intact, unless of course the shells had been polished or painted in an attempt to improve on nature. The dried specimens, or those preserved in spirits, were more vulnerable: faulty seals allowed the spirits to evaporate and the contents to dry out; imperfect preservation and attacks by mould, damp and insects rendered many of the items and their labels useless. Those from the first voyage, collected by Banks and Solander, were the more tightly held, but eventually they joined those from the succeeding voyages and the private accumulations of the officers and seamen, to be scattered to the four winds. In many, probably the majority of cases, the trail has been lost, but here and there there are signs: auction catalogues, species descriptions, paintings, a letter, sometimes even the specimen itself to give us a clue as to the path they took before reaching their final destination or disappearing forever.

In many respects the insects are the best group to start with as they were promptly attended to and present a less labyrinthine picture than do the other animal groups. This was because the ultimate fate of many of the insect specimens lay in the hands of a man who in the autumn of 1767 rode down to London from Edinburgh on horseback, following a collecting tour of the highlands of Scotland.

Johann Christian Fabricius, the son of a Danish physician, was yet another pupil of Linnaeus.[1] A modest and likeable man who regarded his former master with great respect and affection, Fabricius had entered virtually every city in Europe to look at the collections of other naturalists. Now he was to meet Solander and in due course Banks, Hunter, Drury,[2] Fothergill and other members of the English natural history establishment. He grew very fond of London – his fellow naturalists made him welcome and gave him duplicates of their specimens which he was able to send to Copenhagen. When the *Endeavour* left with two of his new colleagues on board, Fabricius went back to Europe for a period, timing his return to London with the departure of Forster and the others on the *Resolution* in 1772.[3]

During that summer and the three that followed Fabricius was able to work on the specimens collected on the first expedition by Banks and Solander; in 1775 he published the results in his *Systema Entomologiae. . .* including the names and descriptions of several New Zealand species collected on the first voyage. Here, for the first time, without fuss, delay, or great expense, were proper zoological descriptions: the first occasion on which animals from the *Endeavour* voyage were validly described and the first animals named from New Zealand.[4] There were no plates, sumptuous or otherwise, just a few lines of Latin text including a brief but adequate diagnosis, habitat, and a note of the collection

1. A short biography in Zimsen, 1964, Tuxen, 1967.

2. A silversmith with a large insect collection, including the Common Copper *(Lycaena salustius)* from New Zealand. (See Note 10, this chapter.)

3. Tuxen, 1967, p.3, has details of Fabricius's itinerary.

4. Some NZ molluscs were published in *Der Naturforscher* in 1774 and 1775, but were not given valid scientific names.

truncatus. **12. Sc. fcutellatus, thorace retufo cornu brevi truncato,
capite mutico**
Habitat in nova Zelandia. *Muf. Bancks.*
**Statura et magnitudo S. naficornis, fupra niger fub-
tus piceus. Caput truncatum muticum. Thorax**
retu-

Part of the Latin text of the first New Zealand animal to be scientifically described, The Large Sand Scarab (*Pericoptus truncatus*), by J. C. Fabricius in Systema Entomologiae . . . 1775

in which the specimen was held. Thus on page six appears *Scarabaeus truncatus* from New Zealand, from the collection of Banks — the Large Sand Scarab, holding a unique place in the country's faunal history as the first of its animals to be validly named in a published work. Now known as *Pericoptus truncatus* it is a common beetle found on New Zealand beaches; it might also have been, rather appropriately, one of the first encountered as they left their boats and stepped ashore. A log kicked aside or lifted above high water mark might have revealed a brown beetle of good size, interesting but not remarkable, and in no obvious way destined for future historical elevation.

In a similar unambiguous way, some 37 other New Zealand species were described, including some of the country's better known insects: the Giraffe Weevil, the Two-toothed Longhorn Beetle, the Lemon Tree Borer, the Manuka Beetle, the Red and Yellow Admiral butterflies, and three species of cicada.[5]

Following the appearance of *Systema Entomologiae. . .* Fabricius returned again to Europe and a professorship at the University of Kiel, taking his collections with him. Early in 1776 he wrote to Banks asking if Forster and Sparrman had brought back any new insects, hoping that he might have the opportunity to see them. Forster had of course given his insects to Solander, who distributed them to Banks, the British Museum and other collectors.[6] It was not until the summer of 1780 that the industrious Fabricius had once

6. Dawson, 1958, p.316. Lever was among the collectors favoured, Whitehead, 1969, p.163. Drury correspondence suggests that Forster had little in the way of insects (D. Medway, *pers. comm.*).

5. New Zealand insects collected on Cook's Voyages:

HEMIPTERA (Bugs) *Clapping Cicada *(Amphipsalta cingulata),* *Red-tailed Cicada *(Rhodopsalta cruentata),* *Variable Cicada *(Kikihia muta); Oncacontius vittatus;* Fabrician Lygeid Bug *(Rhypodes clavicornis).* LEPIDOPTERA (Butterflies & Moths) *Red Admiral *(Bassaris gonerilla),* *Yellow Admiral *(Bassaris itea),* Common Copper *(Lycaena salustius).* DIPTERA (flies) *Beach Stiletto Fly *(Anabarynchus bilineatus),* *Hover flies *(Helophilus trilineatus; *Pilinascia cingulatus),* New Zealand Blue Blow-Fly

(Calliphora quadrimaculata) HYMENOPTERA (Ants, bees, wasps) *Pterocormus lotatorius *Degithina sollicitorius *Levansa decoratorius *Priocnemis fugax;* *Golden Hunting Wasp *(Priocnemis nitida).* COLEOPTERA (Beetles) *Tiger Beetle *(Neocicindela tuberculata),* *Devil's Coachhorse *(Creophilus oculatus),* *Large Sand Scarab *(Pericoptus truncatus),* *Tanguru Chafer *(Stethaspis suturalis),* *Kekerewai Chafer *(Pyronota festiva),* *Plain Manuka Chafer *(Pyronota laeta); Dasytes minutus; Phymatophoea violacea; Phycosecis limbata;* *Pristoderus scaber;* *Orange-spotted Ladybird *(Coccinella leonina);* *Thelyphassa lineata;* *Selenopalpus cyaneus;* Bronze

Beetle *(Eucolaspis brunnea);* *Hybolasius cristus;* *Xylotoles lynceus;* *Grey Longhorn *(Xylotoles griseus);* *Lemon Tree Borer *(Oemona hirta);* *Two-toothed Longhorn *(Ambeodontus tristis);* *Variegated Longhorn *(Coptomma variegatum),* *Striped Longhorn *(Navomorpha lineatum),* Striped Longhorn *(Navomorpha sulcatum),* *Flower Longhorn *(Zorion minutum)* *Two-spined Weevil *(Nyxetes bidens);* *Tysius bicornis; Cecyropa modesta;* *Catoptes interruptus;* *Rhadinosomus acuminatus; Agathinus tridens;* *Stephanorrynchus attelaboides;* *Giraffe Weevil *(Lasiorhynchus barbicornis);* *Saprinus detritus.* (* collected on the first voyage)

In addition *Sitophilus oryzae,* the Rice Weevil, was nominally from New Zealand, but was actually a pest, probably introduced with the ship's stores — possibly the specimen was never landed but described on board. *Alphitobius laevigatus,* the Cosmopolitan Meal Worm is likewise introduced. *Tribolium castaneum *(Tenebrio ferrugineus* — see Radford, 1980, p.183) was also possibly recorded from New Zealand. List prepared with assistance from G. Kuschel (MS. and *pers. comm.*) and G. Gibbs.

more the opportunity of visiting London and seeing the specimens, but few were from New Zealand. Again, a publication was forthcoming: the *Species Insectorum* ... (1781), which included eight New Zealand species some of which might have been collected by the Forsters, but at least one having been left over from the first voyage.[7] Fabricius's later published works continued to make references to New Zealand species[8] including the Common Copper butterfly which came into the hands of Dru Drury — a silversmith and a noted entomological collector and dealer who allowed Fabricius to examine his collection — the description of which was published in 1793.

On one of his early visits to London when Fabricius was the guest of the wealthy Chelsea painter William Jones, he gave himself the opportunity of considering the illustration of his insects, an aspect neglected in his published work. Jones had amassed a considerable fortune by the time he was 35, leaving him free to follow his interests in natural history and painting.[9] The two worked closely together, Jones using the text of Fabricius's work to support his superb paintings of insects, which Fabricius later employed to assist his descriptions. It is evident that some of the paintings, complete with brief supporting text, were completed prior to Fabricius's publication; New Zealand species painted by Jones were the Red and Yellow Admiral and Common Copper butterflies.[10] The *Jones Icones*, as the bound collection of paintings was called, was never published — if there was a plan to do this as a supplement to the Fabrician works, then there is no evidence of it. However the contents of the *Icones* were extensively used by later authors, compilers of natural history works such as Edward Donovan, who had copied both the Red and Yellow Admiral in his *An epitome of the Natural History of the Insects of New Holland, New Zealand...* published in 1805. His copies had some artistic merit but were zoologically bad — he used strong body colours and relied on notes rather than a direct copy for parts of his illustration, leading to inaccuracies in the final work.[11]

Fabricius also described crustaceans — the term 'insects' was fairly broadly used in those days — and a few New Zealand species were included. A large sea louse, commonly seen in the mouth or gill chamber of freshly-caught fish, was collected on the first voyage and described in *Systema Entomologiae* in 1775.[12] The Hairy Spider Crab or Camouflage Crab, probably collected on one of the later voyages, he described in 1798,[13] years after

---

7. *Catoptes interruptus*; Kuschel, 1970, p.204.

8. Only four new species from these later works. The works were *Mantissa insectorum...*, 1787; *Entomologica systematica...* 4v. 1792-1798; *Systema eleutheratorum...*,1801; *Systema Rhyngotorum...*, 1803; *Systema Antilatorum...*, 1805.

9. Poulton in Waterhouse, 1938, p.15.

11. Westwood, 1872, p.107. Donovan, 1805, also illustrated the Bronze Beetle *(Eucolaspis brunnea)*.

12. Now known as *Codonophilus imbricatus*.

13. This is now *Notomithrax ursus*, first described by Herbst, 1788, p.217, Pl.14, fig.86, then by Fabricius, 1798, p.498.

---

10. The set containing the New Zealand species was bound and dated 1785 (Hope Entomological Library, Oxford University). For some time it has been assumed that *Lycaena salustius* (the Common Copper) was based on specimens of *Lycaena* that are part of the Banks collection currently in the British Museum. But the Jones illustration (which is accompanied by the Fabrician description and the name *salustius*) appears to closely resemble a male of a different species currently referred to as *Lycaena rauparaha*. Thus it seems likely that the original *salustius* was based on this, rather than the putative

Banksian *Lycaena*. (Fabricius's written description is — unfortunately — broad enough to cover both forms.) The specimen originally described by Fabricius and painted by Jones was from Drury's collection (not Banks's) and possibly did not reach the British Museum. Instead there is evidence that Lycaenids from the Drury collection were bought in a sale (in 1805) by Alexander Macleay and taken to Australia, and it is just possible they are still there in the Macleay Museum (P.R. Ackery, *pers. comm.*). A few Drury specimens reached the British Museum through the Milne collection

but there do not appear to be any types (Zimsen, 1964, p.16.). Thus it would appear that the species known as *Lycaena rauparaha* should be *salustius* and the species represented by the specimens from the Banks collection should be renamed. It appears *both* these *Lycaena* species could have been collected at times and localities as follows: the first voyage (Thames — Bay of Islands, late November — early December; Queen Charlotte Sound, January); Second voyage (Queen Charlotte Sound, November — less likely) or Furneaux's visit to Poverty Bay in 1773 (less likely); and the third voyage, (Queen Charlotte

Sound, February). *Lycaena salustius* was not described until 1793 (13 years after the return of the last voyage) suggesting that it could have come from any of the above. The Drury specimen may have been seen by Fabricius and painted by Jones at the time of the former's visit to London in 1780 or 1782. (I am grateful to Dr G. W. Gibbs for identification of specimens and discussions leading to these conclusions.)

The Red Admiral
(*Bassaris gonerilla*);
The Yellow Admiral
(*Bassaris itea*);
The Common Copper
(*Lycaena salustius*)
(left).
The original paintings
by William Jones in
the 'Jones Icones';
Papiliones Nymphales
..... 1785, Plates 35,
40, 59 respectively.
Published by
permission of the
Hope Entomological
Library, Oxford
University. *(see p.41)*

Fabricius N°361   Gonerilla   Sᵣ Josᵖʰ Banks

Alis dentatis nigris albo maculatis, fascia communi rufa, posticis ocellis quatuor   habitat in nova Zelandiã

Fabricius N°362   Itea   Sᵣ Josᵖʰ Banks.

Alis dentatis nigris, anticis fascia punctisque flavis, posticis disco rufo Ocellis quatuor. —

habitat in nova Zelandiã

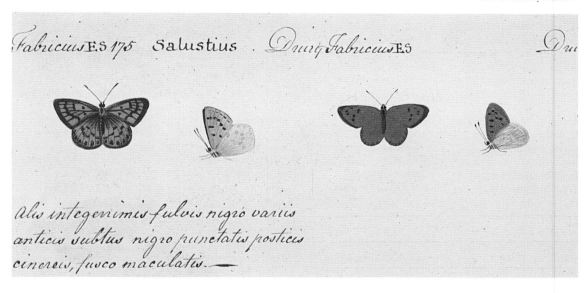

Fabricius ES 175   Salustius .   Drury Fabricius ES   Dru

Alis integerrimis fulvis nigro variis anticis subtus nigro punctatis posticis cinereis, fusco maculatis. —

The Red Admiral (*Bassaris gonerilla*) a hand-coloured engraving from Edward Donovan's *An epitome of the Natural History of Insects of New Holland, New Zealand* . . . . 1805 (unnumbered plate). The lower butterflies are not from New Zealand. Published by permission of the Alexander Turnbull Library. *(see p.41)*

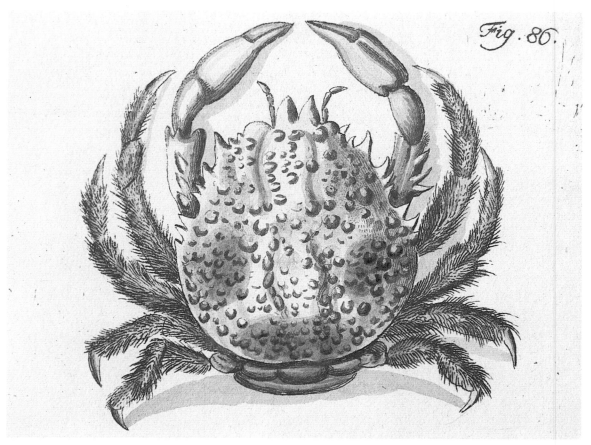

The Hairy Spider or Camouflage Crab (*Notomithrax ursus*) a hand-coloured engraving from Johann Herbst's *Versuch einer Naturgeschichte der Krabben ....* 1788 (pl. 14). Published by permission of the Australian Museum. *(see p.41 & below)*

it was collected, but a pastor in Berlin had already beaten him to it. Johann Herbst, whose ecclesiastical career included both a chaplaincy in the Prussian Army and an Archdeaconry in Berlin, had obtained a specimen of the crab — possibly the same one seen by Fabricius — and described and illustrated it in 1788; the first New Zealand crab to be so treated. Herbst's specimen is now in the zoological museum of the Humboldt University.[14] The paucity of specimens of the larger Crustacea reflects a general lack of interest in some invertebrate groups. One might ask what happened to the Rock Lobster so frequently mentioned and so obviously enjoyed by those on the voyages, since it remained undescribed for many years.

In his later years Fabricius remained in Europe, still travelling widely. In a letter to Banks written in 1805, he records how his faith in things English had been shaken by the younger English entomologists turning against him, even though he wished to return.[15] In 1807, the English highhandedly bombarded Copenhagen and captured the Danish fleet, a final act of perfidy by a people he was fond of, and a saddened Fabricius died on 3 March the following year. So ended one of the healthier chapters in the history of the voyage's zoology; Linnaeus, Fabricius's old mentor, would have nodded approvingly at the achievements of his former pupil, and few would do as well.

Fabricius was not the only Scandinavian to examine the voyage insects in those early years. In 1785 the Swedish entomologist Nils Swederus came to London, where he had access to Banks's collections. He wrote to Banks to say that the insects had been treated with great carelessness; their legs, antennae and heads were missing, and 'all are covered with dust and are ill-arranged'.[16] Despite the recent presence of Fabricius, it seemed that

14. Fabricius took some specimens with him to Kiel — easily accessible to Herbst. Most NZ specimens remained in the Banks collection which went to the Linnean Society and then to the British Museum.

15. Fabricius to Banks, 8.2.1805 in Dawson, 1958, p.317.

16. Swederus to Banks, 31.3.1785 in Dawson, 1958, pp.800-801.

even this group of animals had suffered from poor curation. Swederus then offered to put the collection in order; in so doing he must have examined some New Zealand specimens for in 1787 he was responsible for a publication that appeared in Sweden and described two New Zealand flies, one being the New Zealand Blue Blowfly or Bluebottle.[17] Swederus eventually returned to Stockholm and wrote to Banks that he had heard favourable comment regarding his rearrangement of the collections.

A short while later, in 1789, armed with the recommendations of his friends and sponsors, a French entomologist named Guillaume Antoine Olivier came to London to examine the Banks collection. One of the dissidents who founded the Linnean Society of Paris, he began a French association with New Zealand entomology that was to last well into the 19th century. He was working on his *Entomologie, ou histoire naturelle des insectes...*, a multi-volume treatise being prepared under the patronage of the wealthy Jean-Baptiste Gigot d'Orcy, himself an entomologist. Olivier's previous work, on the natural history of the Paris region, was curtailed by the French Revolution; the *Encyclopédie méthodique...* to which he contributed a section on entomology suffered from the loss of a large number of subscribers to the same cause. Both this work and his *Entomologie...* contained references to New Zealand insects, and he was able to illustrate the latter with Banks's insects seen at first hand. In 1797, Pierre Latreille, a conservative priest who was a friend of Olivier and Fabricius, provided plates of New Zealand insects for the *Encyclopédie*, following a spell of imprisonment by the Revolutionary authorities. By now it was evident that through the efforts of the French entomologists and others, elements of the New Zealand fauna were becoming well entrenched in standard European reference works in spite of the civil upheaval.

The Frenchmen were however preceded in their illustrated compilation of the Cook voyage entomology by Johann Herbst in Berlin, whose illustrated monographs on beetles (*Natursystem der Käfer*) were published between 1783 and 1795. Although he may have visited London, it is more likely that Herbst saw duplicates given to Fabricius by Banks and taken by the former to Kiel, which is relatively more accessible from Berlin. His illustrations of New Zealand insects are the earliest published. Two other German authorities, Friedrich Weber (1795) and Carl Schönherr (1806-1817) included references to New Zealand insects in their compilations.

The insect specimens (including several from New Zealand) remained in Banks's collection and eventually passed to the British Museum via the Linnean Society. Others, belonging to Fabricius, went to the University of Kiel and then to Copenhagen. Those in the Leverian Museum were auctioned in 1806. Some of these might even have reached Australia in the Macleay collection.[18]

The voyage insects were thus compactly and efficiently dealt with in comparison with other groups of animals, and largely due to the efforts of one man. This was something of a record for publication of expedition results and certainly enviable for those times. Fabricius was aided in his task by the fact that the insects were relatively easily preserved and were not in quite the same category as the molluscs, largely represented by shells, which were tremendously popular as collectable items: the fact that relatively few insects were bought and sold meant that the collections stayed largely in one place — at least until they were described.

The same cannot be said for the shells, but their durability and popular appeal helped ensure that their movements were frequently recorded. There were many collections of shells on board the *Endeavour* when she finally tied up in the Thames: those of Banks, Solander and Parkinson, as well as the small caches held by the officers and crew. While

17. New Zealand Blue Blow-Fly, *Calliphora quadrimaculata*; Swederus to Banks, 26.4.1788, in Dawson, 1958, pp.800-801.

18. Whitehead, 1969, pp.58,59,75,76. There is possibly a specimen in the William Hunter (brother of John) collection at Glasgow University. (*Dermestes navalis* = ? *Tenebrio ferrugineus*) Radford, 1980, p.157.

the larger and more 'official' collections were taken up river by sailing barge, others were no doubt part of the busy scene on the wharf and nearby where transactions in voyage 'curiosities' were already taking place. In among the wellwishers and the curious, we have reason to suspect, were George Humphrey, a prominent dealer and collector who had a particular interest in shells, and possibily also the Forster brothers, Jacob and Ingham, who likewise were dealers and collectors of natural history objects and coincidentally relatives by marriage of Humphrey.[19] The dealers could dispose of their stock wherever they wished, and for shells there was a market abroad as well as in England.

The territory for the salesmen abroad was substantial. On the Baltic coast of East Prussia was the free city of Danzig, the home of the Society of the Friends of the Natural Sciences (Der Gesellschaft der Naturforschender Freunde) and its curator Friedrich Zorn von Plobsheim. Inland to the south and west were Halle, Jena, and Schwarzburg-Rudolstadt, where within a short radius were a flourishing natural science journal edited by J.E.I. Walch and a famous shell collection owned by one of the local aristocracy. Moving north to Denmark and Copenhagen, still within the sphere of German influence and where Johann Chemnitz and Lorenz Spengler collected and described shells, and we have sketched out a triangle about the size of England. To these names and places can be added Strasburg and Berlin, Hermann, Martini, and threading his way through this tapestry via Danzig and Halle, the figure of J. R. Forster.[20]

Then there were the go-betweens; Woide (a life-long friend of J. R. Forster) and Fabricius, providing the direct link between German Europe eager for news of the discoveries, and Banks, Solander and the dealers in South Seas shells. In these hands and in such places the New Zealand shells came to light and for the first time became part of the permanent record. It is strange indeed that shells from the rocks and beaches of Queen Charlotte Sound and the North Island's East Coast of this little known country should be so dispersed — into the hands of pastors, theologians, Royal cabinetmakers and Thuringian princes — and that these people should be the first to try and explain them to the learned world.

Because he brings us the story in a scientific journal of which he was editor, we will begin with Johann Ernst Immanuel Walch of Jena, a professor at the university there. Trained in Semitic languages, natural science and mathematics, he held several chairs principally in the Faculty of Theology, where his father and brother were similarly engaged. In 1774 Walch took over the editorship of *Der Naturforscher*, a journal that published a range of scientific articles including natural history from authors domiciled largely in German-speaking Europe.[21] Conveniently near Jena was the principality of Schwarzburg-Rudolstadt, where a member of the ruling family, Friedrich Carls zu Schwarzburg-Rudolstadt, possessed a remarkably fine cabinet of shells which he allowed Walch to examine. It appears that a dealer by the name of Forster (most likely one of the Forsters of the wharfside trading with the naturalists and crew of the *Endeavour*) had travelled abroad to sell his wares, having satisfied the English market and probably aware of the potential interest and the wealth of shell collectors in Europe. Forster was apparently successful in the case of Friedrich Carls, because Walch described and illustrated some new shells from his cabinet in *Der Naturforscher* late in 1774. In the absence of Linnaean species names and precise locality information, the identities of these shells remain uncertain, but one of the hand-coloured engravings strongly resembles the New Zealand species *Cantharidus opalus*, the Opal Top shell. This is possibly the first published illustration of an animal from New Zealand, and the first paper to describe animals of any sort collected on the Cook voyages. Walch noted the eagerness with which European

19. Walch, 1774, pp.33,35 and (Zorn) 1775, pp.152-153 and other works of this period suggest that Humphrey and one of the Forsters were early on the scene, most likely Ingham, who had more to do with conchology than Jacob (see Whitehead, 1973, pp.361-363). Walch, 1774, p.35, refers to a Forster son living in London. He also suggests that Forster's shells were obtained in exchange from Banks and Solander.

20. Although surprisingly the contemporary German authors do not often mention him.

21. *Die Historische Commission*, 1896, v.40, p.652.

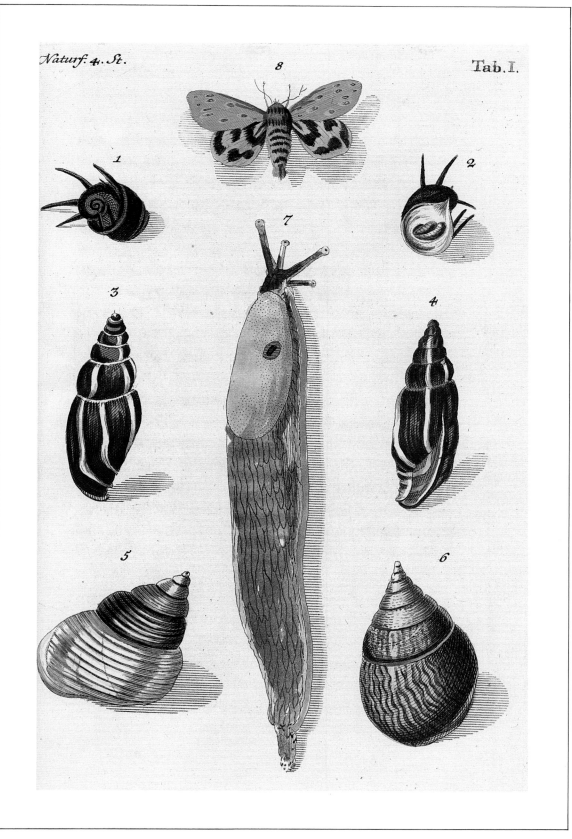

The Opal Top Shell (*Cantharidus opalus*) (figs. 5, 6), a hand-coloured engraving from J. E. I. Walch's *Beschreibung einiger neuentdeckten Conchylien* in *Der Naturforscher*, 1774 (pl. 1). Probably the first New Zealand animal to feature in a published work. The specimen on the left (fig. 5) appears to have been muricated or polished. The remaining specimens are not from New Zealand. Published by permission of the Trustees of the British Museum. *(see p.46)*

naturalists awaited the findings of Banks and Solander: they were to wait for some time.[22]

In the same year another link with the shells from the *Endeavour* was being made, this time in Danzig. It was done in the name of the Society of the Friends of the Natural Sciences, an exclusive group with a limit of 20 members whose interests were in the physical and life sciences.[23] With the object of acquiring specimens, the curator of the Society's natural history collections and other rarities, Friedrich August Zorn von Plobsheim, contacted Karl Gottfried Woide, then pastor to the Dutch and German communities in London. Woide had been a classmate of J. R. Forster at the Joachimsthal-Gymnasium in Berlin[24] and after leaving there to study theology arrived in London, where he mixed well with the community of English and European scholars. He was an ideal person to act as an agent for scholars abroad. Later he helped nurse Forster and his family through some of the crises they endured while in England. But for the moment he was acting on Zorn's behalf in approaching Banks and Solander for shells. In the face of previous generosity, the pair proved reluctant — no doubt thinking of their future plans for publication — leaving Woide to negotiate instead with Humphrey the dealer. As a result some 167 shells reached the Danzig society in 1774, including many from New Zealand, and Zorn von Plobsheim described and illustrated part of this collection in *Der Naturforscher* in 1775. Four from New Zealand were identified by Humphrey's obscure names (for example, the 'lesser Beauty') and one called 'The Smooth Emerald' has been identified with the help of an accompanying plate as the Cat's-eye, *Turbo smaragdus*. The same species was part of another contemporary collection, that of Dr J. F. Bolten, a Hamburg physician.[25] There is no doubt at all that we are now dealing with New Zealand shells, a long way from their point of discovery and indeed, a long way from their discoverers.

Interest in the South Sea shells did not stop with Walch and Zorn von Plobsheim. In Copenhagen, Lorenz Spengler, who came originally from Schaffhausen, was cabinetmaker by appointment to the King of Denmark and later Keeper of the Royal Art Collections. His position of influence and income were enhanced by his knowledge of the manufacture of false teeth, thus enabling him to support his interest in zoology and collections of shells and insects.[26] He was also a correspondent and customer of Humphrey, who wrote to Spengler on 15 October, 1775, apologizing for the delay in sending him shells and explaining that he needed to be on hand to take care of transactions on the return of the *Resolution*.[27] It may have been shells from this source that he described and illustrated in *Der Naturforscher* in 1776. Among them the Brown Top Shell, *Maurea punctulata*, and a possible Silver Paua, *Haliotis australis*, are New Zealand specimens, as is Cook's Turban, *Cookia sulcata*, described and illustrated for the first time in the same volume by Walch. In 1782 Spengler described the 'greater wavedlip bell or Nonpareil of New Zealand', and the 'lesser wavedlip bell', two familiar New Zealand shells now

23. Banks, Solander and Forster were later admitted as honorary members (Whitehead, 1978b, p.66).

24. Hoare, *in* Forster, 1982, p.4.

25. (Zorn) 1775, pp.151-168. The New Zealand shells were described more fully by Zorn, 1778. *Turbo smaragdus* identified by Dance, 1972, p.357 (as *Lunella smaragda*). The shells in the Plate are shown in polished and unpolished form. Other recognizable species are the Periwinkle *(Littorina (Austrolittorina) cincta)* and also the Red Top Shell *(Cantharidus purpureus)*. Bolten's work was published by Röding, 1798, in which some New Zealand shells received their valid scientific names, e.g. *Austrofusus glans* (p.56).

26. Cernohorsky, 1974, p.144.

27. Spengler, 1776, p.147.

22. Walch, 1774, pp.33-56 & Pl.I. Dance identified the shell as *Cantharidus purpurea*, but it is more like *C. opalus*, the Opal Top Shell (B. Marshall, *pers. comm.*). There are some inconsistencies in its history: the dealer (Forster) who sold the shells to Schwarzburg-Rudolstadt claimed that they were obtained on exchange with Banks and Solander and that

they came from the Falkland Islands. As the *Endeavour* did not visit the Falklands, Forster either misrepresented the offering, or was mistaken. Dance's identification of other shells on the Plate (from South America and the Philippines area) does not necessarily connect them with Cook as these localities were visited by earlier voyages. But if

the identification of *Cantharidus opalus* is correct, then at least part of Forster's merchandise contained *Endeavour* shells. It is more likely he bought them from a crew member than received them on exchange from Banks and Solander — simply using these names to dress up the sale. (Zorn) 1775, p.157, Pl.II, Figs. C1, C2, figures a shell from

New Zealand (via Humphrey and Rudolstadt) which he says he is spared from describing because of Walch's earlier paper — in their minds the *Cantharidus* from the 'Falklands' and New Zealand are the same.

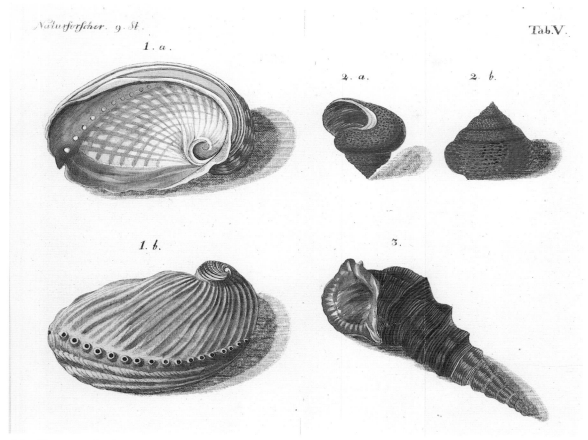

The Silver Paua (*Haliotis australis*) a hand-coloured engraving from Lorenz Spengler's *Abhandlung von den Conchylien der Südsee* . . . . . . in *Der Naturforscher*, 1776 (pl. 5). Published by permission of the Trustees of the British Museum. *(see p.48)*

commonly known as the equally perplexing Large and Small Ostrichfoot.[28]

This practice of using vernacular names was continued in 1778 by Zorn von Plobsheim in a major publication of the Danzig Society in which he catalogues the shells sent by Humphrey — some 59 different shells from New Zealand bear titles such as 'the Lesser or Painted mouth-zebra' and 'the Wrinkled Ear from New Zealand'.[29] Here we learn that Spengler, Zorn von Plobsheim and Chemnitz were engaged in a debate over the identity of the 'Brown deep mussel from New Zealand'. This matter was to be settled later by Johann Hermann from Strasburg, a widely published naturalist who in 1782 described and illustrated *Mytilus impactus*, found 'in Oceano Australi', thus giving us the first valid scientific description of a New Zealand shell.[30]

Johann Chemnitz was another German-born collector of long standing and pastor to the German garrison parish in Copenhagen. In his search for exotic shells from the South Seas he had secured the help of the English Consul-General and on a visit to Sweden he was able to hear more about molluscs collected on the second voyage from Sparrman, Forster's able assistant. In London Fabricius acted as an agent for Chemnitz, purchasing 10 guineas worth of shells from Humphrey, thus enabling Chemnitz to join the growing list of authors who published on New Zealand shells in *Der Naturforscher*. This was in 1783, and if before there were hints, by now there was actual concern expressed by the European collectors at the lack of publications from Banks and Solander.[31]

The regular appearance of *Der Naturforscher* was not the sole source of information for the Europeans. In France, Jacques de Favanne and his son Jacques Guillaume published

28. Walch, 1776, pp.203-204; Spengler, 1782, pp.24-31.

29. Zorn, 1778, pp.247-288.

30. Hermann, 1782, p.152, Pl.III.

31. Chemnitz, 1783, pp.177-178.

Naturf. 17.St.   *Tab. II.*

Capieux. fi. 1782.

The Large and Small Ostrich Foot (*Struthiolaria papulosa* and *Pelicaria vermis*) here referred to as the 'greater and lesser wavedlip bell', hand-coloured engravings from Lorenz Spengler's *Beschreibung zwoer seltenen neuen Gattungen Südländicher Conchylien* in *Der Naturforscher*, 1782 (pl. 2). Published by permission of the Trustees of the British Museum. *(see p.48)*

32. Dance, 1971, p.371. A fine bound set of originals by Jacques Guillaume de Favanne is in the Zoology Library British Museum — some New Zealand shells e.g. Cook's Turban Shell *(Cookia sulcata)* and the Imperial Sun Shell *(Astraea heliotropium)* are included.

33. e.g. Martini & Chemnitz, 1781, v.5, p.272, the 'notched snail from New Zealand' — it was a non-Linnaean work. Martini's widow persuaded Chemnitz to carry on the series. Volume 5 had a new artist, but like his predecessors, his work was below expectation (Cernohorsky, 1974, pp.144-145).

34. Whitehead, 1978b, p.64. The large Portland collection was studied by Solander and after the Duchess's death in 1785 was catalogued by Lightfoot, 1786, and sold at auction. See Dance, 1966, p.105. Calonne included several New Zealand shells in his collection.

the third edition of Dezallier d'Argenville's *La Conchyliologie* in 1780 with a number of references to New Zealand shells,[32] and Friedrich Martini in Berlin — who had been a contemporary of Walch at Jena — published a series of volumes entitled *Neues systematisches Conchylien-Cabinet*, which was carried on after his death in 1778 by Chemnitz. Volume five and subsequent volumes refer to New Zealand species.[33]

The French appear less frequently in the story of the voyage shells and perhaps were in a less favourable position to publish owing to the effects of the Revolution and the events surrounding it. However some of those wealthy enough to indulge an interest in conchology were obliged to flee the country, and for those who sought refuge in England there was an opportunity for even better access to voyage specimens. Charles Alexandre de Calonne had a fine shell collection but also the misfortune to be responsible for French financial affairs prior to the Revolution; his policies offended everybody and he was eventually forced to leave. He had bought many shells from the Duchess of Portland's sale, the New Zealand shells in his collection (for example an Imperial Sun Shell) probably coming from that source.[34]

Thus the picture of European natural history in the area of conchology at least is one of vigour, particularly at the individual level — an enthusiasm that was not grafted on to larger nationalistic or commercial considerations but was sustained by a network of kindred interests. Questions are raised by the involvement of the protagonists in this story. How was so much cooperation achieved considering the distance that separated them? Why were they able to publish their work, when those in England who had the specimens at hand found it so difficult? Was there something in their personalities or backgrounds that brought them together? Finally, was their achievement as significant as the publications suggest?

An impressive feature of the European conchologists is their apparent ease of communication, given the distance from each other and from a new and important source of shells. Language did not appear to be a barrier, there was a willingness to travel, and the texts and scholarly journals such as *Der Naturforscher* as well as societies like the one in Danzig were important vehicles. In England, naturalists waited on the appearance of grand works, richly illustrated with engravings, long in the preparation (too long as it turned out) and expensive in production. A short article in *Der Naturforscher* or the journal of the Danzig Society might have been less spectacular, the quality of paper poorer, the engravings not so rich, but it would appear in short order at reasonable cost, and would also be widely circulated. The naturalists and collectors in England, with one or two exceptions, were complacent — the materials were at hand and they were numerous enough to talk to each other. Let all Europe come to London and the dealers travel abroad if they wish.

It is also initially puzzling that the set should be so full of characters with theological backgrounds or who were active clergymen, at least among the Germans. In England and particularly in France, natural history was not an exclusive province of the clergy and its followers were fairly represented by those who had trained in medicine — a useful background, particularly if advances into fields such as anatomy were to be made. Part of the answer at least can be found in the state of the medical faculties of the German universities of the day which were held in very low regard and were even openly ridiculed. As a result they attracted small numbers of students while places in the faculties of theology and law were sought by the students with prospects.[35] A theological education was in any event no handicap to the study of natural history — plants and animals had long been the raw material for religious philosophy and, whatever its faults, the German system catered for polymaths like Walch, Forster and Woide, who could turn their minds to virtually any subject. One possible consequence of this preponderance of theologians and pastors might have been a general reluctance, out of training or belief, to accept the Linnaean system of classification. (Forster was a notable exception.) This of course would have had a considerable effect on the ultimate value of their work.

This brings us to an evaluation of this period — how important was its contribution to the description of the South Sea shells, including those from New Zealand, and to the science of conchology? Sadly, it was less than it might have been. The persistent use of vernacular names limited the zoological significance of the publications and only Hermann's paper, amongst the early works, proposed a valid name for a New Zealand shell. But these papers are of considerable historical significance for in spite of their difficulties in obtaining specimens, they give us an early insight into the fate of the shells collected on Cook's voyages and the participants in their redistribution, and they were a vanguard of serious European interest in the natural history of this distant part of the world.

35. McClelland, 1980, pp.29-31.

It is time now to take up the story in England where the berthing of the *Resolution* and *Adventure* was eagerly awaited by Humphrey, Dr George Fordyce (a noted Scottish collector) and others who wished to purchase curiosities, and by the crew members, no less eager, who wished to sell them. By this time they were competing with resales of first voyage material. Earlier in the year someone posing as a collector, Sir John Dalston, had sold up 'before leaving for America'. Several lots of New Zealand shells were included in the sale such as 'Lot 71 the iris, a beautiful kind of ear shell from New Zealand.'[36]

Mindful of his £150 investment in second voyage shells, Humphrey immediately set about trying to turn a profit.[37] Some shells, as we have seen, probably went to Europe, while others were bought by English collectors such as Henry Seymer and the wealthy Reverend Cracherode, who was unwilling to attend auctions and would rather pay the premium that Humphrey would extract from his private clients. A pair of common tiger shells from New Zealand at four guineas seems exorbitant, even for those days.[38] In 1779 Humphrey held a major sale that included a large amount of Cook material. But of the fate of Forster's own collection of shells not much is known. It is possible that soon after their return he sent some shells to Linnaeus; three years later in September 1778 he was prepared to offer his shell collection to Banks, but was turned down.[39]

With the return of the *Resolution* and *Discovery* from the third voyage there was substantial competition in the bidding on the quayside from Thomas Martyn, who in advance had commissioned those on board to collect for him and who bought more than two-thirds of the shells on offer for 400 guineas.[40] A natural history author, dealer, and collector, he was to make a substantial contribution to the voyage conchology in a rather unusual way.[41]

Martyn was something of an enigma in British natural history circles. His narrow view of the appropriate way to describe animals for the learned and interested world — in works composed of illustrations, with no text beyond a preface and introduction — and his unorthodox means of achieving this end, were to put him in a category of his own. His 'Magnum Opus', and no more appropriate words can describe it, was to be a work on shells starting with the recent discoveries from the voyages of Byron, Wallis and Cook, and later portraying every shell that was currently known. Like certain other publishing ventures in natural history it was to succumb to the weight of its own ambition, but not before its early parts appeared, equal to the finest iconographies of shells yet produced. Of undoubted artistic merit, and a useful historic record of some of the shells brought back from the voyages, it would be rescued from future zoological obscurity by its adherence, in part, to the Linnaean system of classification.

The means of going about this enormous venture, with their elements of philanthropy and practical considerations as to cost, are described in detail in the introduction to *The Universal Conchologist*. Martyn says '. . . the author conceived the project of endeavouring to accomplish his design by means as simple as they were new: he had to find for the execution of his purpose such hands as, possessing abilities adequate to the end, could not, from their situations in life, be more profitably employed in other occupations. The labour of boys he knew is always cheaper than that of men: and he concluded, that where nature had plentifully sown the seeds of genius for any particular pursuit, very little art would be required to cherish and rear the plant to maturity.'[42] Thus, spurning the services of professional artists, he began to establish an academy of young men 'born of good but humble parents', trained initially by himself, but eventually able to pass on their experience to new recruits, until there were nine young artists in service. Supervising the training and execution of their work himself, he foresaw a uniformity

36. Anon., 1775. There is doubt that Sir John Dalston existed — the name may have been a front for a seller who wished to remain anonymous (D. Medway, *pers. comm.*). Several lots contained New Zealand shells, mostly 'Ear Shells' or paua *(Haliotis* spp.), apart from a 'Fine Trochus'.

37. Dance, 1971, p.366.

38. Wilkins, 1955, p.77.

39. Whitehead, 1969, p.186 (Note 7); Forster to Banks, 26.9.1778, in Dawson, 1958, p.339. Tunstall sent shells to Linnaeus, but few if any from New Zealand.

40. Dance, 1971, pp.366, 370.

41. He may have dealt a little in shells but did not seem to be a dealer on the same scale as Humphrey (Dall, 1905, p.415).

42. Martyn, 1784-1787, p.28. See Dall, 1908, p.187, regarding the dates of publication.

The White Rock Shell (*Thais orbita*) a hand-coloured aquatint from Thomas Martyn's *The Universal Conchologist*
. . . . 1784-87 (pl. 54). Published by permission of the Alexander Turnbull Library. *(see p.54)*

of output, not achievable by other means.[43]

Martyn was a perfectionist, but he was not a workhouse beadle — while his apprentices lived a simple spartan life, there is nothing to suggest that they were badly treated. Seventy copies of the two-volume work were completed in 3½ years — involving some 6000 duplicate paintings — when Martyn decided to scrap the work and start again. His dissatisfaction extended even to the copper plates which formed the basis for the later colouring work, and henceforth these would be prepared within his academy rather than at the shops of outside engravers.[44]

The plates were prepared by the process known as aquatint, one that was no doubt carefully chosen as the results give tonal effects with all the delicacy of a wash or soft pencil drawing, ideal for shells. Powdered resin was put on the plate which was then heated and etched, the resin providing points of acid resistance. Finishing touches could be made by further etching or engraving. In this case the work was particularly delicate and was then hand-coloured by the apprentices to the point where the aquatint ground was scarcely visible, and the finished product looked as if it were wholly done by hand. Each plate offered two views of the shell, and the first two volumes comprised 81 plates in total. Martyn drew from his own collections, as well as others including Humphrey's and the Duchess of Portland's. Several of the collectors are individually thanked in the preface to the work. Amongst the four or so who loaned Martyn New Zealand shells was a Miss Fordyce, daughter of Dr George Fordyce who had stood with Martyn and Humphrey in a bid for *Resolution* shells at the end of the third voyage; it was she who lent the Imperial Sun Shell.

Volumes I and II containing the first 81 plates were ready in 1784, and Volumes III and IV were finished early in 1787. Of the shells depicted, 32 were said to be from New Zealand, 23 of these being currently accepted species; only nine have kept Martyn's original names, owing to liberties he took with the Linnaean classification. The remainder were

43. Some of the work was signed by the artist (e.g. W. Lewin, 1784-87) in the unbound plates (uncat.) Knowsley Hall library.

44. Dall, 1905, p.417.

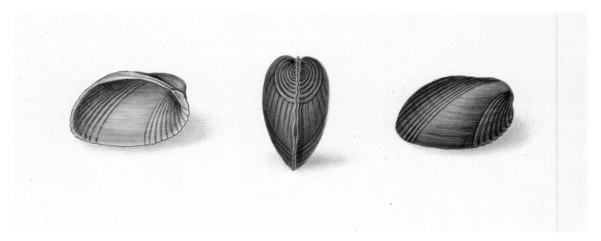

The Nesting Mussel (*Modiolarca impacta*) a hand-coloured aquatint from Thomas Martyn's *The Universal Conchologist* . . . . 1784-1787 (pl. 77). It was, earlier, the first New Zealand shell to be scientifically described and named. Published by permission of the Alexander Turnbull Library. *(see p.49 & below)*

45. The nine species are indicated in Note 46.

corrrectly named later by other authors, who used Martyn's illustrations (and sometimes his names) as a basis. Among the nine that still carry his name as author are: the Large and Small Ostrich Foot, the Star or Circular Saw Shell (Imperial Sun Shell), the Opal and Brown Top Shells, the Southern Cats Eye, two whelks and a limpet.[45]

Without a doubt *The Universal Conchologist* was one of the finest settings that New Zealand shells would ever enjoy, and the first major work to give scientific names to a substantial selection of shells from that country. Along with the European publications and the English sale catalogues it gives us a fair indication of the range of shells collected on Cook's voyages.[46] (The period 1779 to 1806 was noted for its sales of major collections as well as some smaller ones, such as that of the *Discovery* officer in 1781. It, like the others, offered New Zealand material such as the paua and an eclectic Lot 98 containing

46. New Zealand shells recorded from Cook's voyages. The following appeared in Martyn (1784-87). Those marked (†still bear his name as author): *†Lined Whelk (Buccinulum linea linea);* Speckled Whelk *(Cominella adspersa);* †Spotted Whelk *(Cominella maculosa); Lepsithais lacunosus;* White Rock Shell *(Thais orbita);* Dark Rock Shell *(Haustrum haustorium);* Knobbed Whelk *(Austrofusus glans);* *†Opal Top Shell (Cantharidus opalus);* Red Top Shell *(Cantharidus purpureus);* †Brown Top Shell *(Maurea punctulata);* Tiger Shell *(Maurea tigris);* †Southern Cat's Eye *(Turbo (Modelia) granosus);* Cat's Eye *(Turbo smaragdus);*

†Imperial Sun Shell also (coarsely) the Circular Saw Shell *(Astraea heliotropium);* Cook's Turban *(Cookia sulcata);* Arabic Volute *(Alcithoe arabica);* †Small Ostrich Foot *(Struthiolaria (Pelicaria) vermis);* *†Large Ostrich Foot (Struthiolaria papulosa);* Paua *(Haliotis iris);* Mud Snail *(Amphibola crenata);* *Nesting Mussel (Modiolarca impacta);* *Green Mussel (Perna canaliculus);* †Limpet *(Cellana denticulata).* Other species recognized as voyage shells are: Virgin Paua *(Haliotis virginea);* Radiate Limpet *(Cellana radians);* Cunningham's Top Shell *(Maurea selecta); Cellana stellifera; Maurea pellucida;* Green Top Shell *(Trochus (Thorista) viridus);*

Common Top Shell *(Melagraphia aethiops);* Black Nerita *(Nerita (Melanerita) atramentosa melanotragus);* *Hairy Trumpet (Monoplex parthenopeus);* *Ribbed Mussel (Aulacomya ater maoriana);* *Common Cockle (Chione stutchburyi);* Pipi *(Paphies australis); *Notirus reflexus;* *Periwinkle (Littorina (Australittorina) cincta),* (Zorn von Plobsheim, 1775, p.167, Pl.II) *Silver Paua (Haliotis australis),* (Spengler, 1776, p.150-152, Pl.v.) *Duck-bill Limpet or Shield Shell *(Scutus breviculus)* (Painted by Parkinson) *A brachiopod, probably from the first voyage *(Terebratella sanguinea),* (see Wilkins, 1955, p.85) and almost certainly from Queen

Charlotte Sound. With the exception of *Maurea tigris,* and the last three species, the above were listed by Dance, 1971, p.376. (*) indicates specimens collected on the first voyage. Apart from *Monoplex parthenopeus* all were originally described from New Zealand. The Portland Catalogue (Lightfoot, 1786) lists only six species from New Zealand, including the Southern and Common Cat's Eyes, White Rock Shell, Cook's Turban, Imperial Sun Shell, and the brachiopod *Terebratella sanguinea.*

'a small trumpet, a muricated Trochus, N.Z. and 2 Surinam toads' which went for 2s 6d.[47]

Some of Martyn's New Zealand shells would be renamed in other works, for example by Bruguière in the *Encyclopédie Méthodique*,[48] but several would feature in one of the most notorious compilations of the century – that of Johann Friedrich Gmelin who, making free use of many non-Linnaean works of the day, revised Linnaeus's *Systema Naturae*, beginning in 1788. By introducing new and acceptable binomials he was able to obtain authorship of numerous species, and many New Zealand molluscs and fish collected on Cook's voyages bear his name. The pauas are a notable example. Other compilers would follow: George Perry's *Conchology...* published in 1811, was a highly controversial work with illustrations varying between 'excellent' and 'atrocious'. Even his names were ridiculed at first, although they were eventually accepted. Among them two New Zealand trumpet shells, Spengler's Trumpet and the Australasian Trumpet, have scientific names that survive today.[49]

In spite of the substantial documentation of the shells that came back from the South Pacific, only a relatively small proportion of the specimens collected are known to exist today. Eight New Zealand shells form part of the Banks collection in the British Museum, and at least two specimens of that famous New Zealand rarity, the Imperial Sun Shell, are still extant. One of these was the so-called 'pink variety' of Imperial Sun, which was drawn up on the anchor cable of either the *Adventure* or *Discovery* and bought by Sir Ashton Lever from Cook or Cook's widow.[50] The shell was sold in 1806 to J. J. A. Fillinham for 23 guineas and from him it went to the Duke of Bourbon, who was then living in England. It was sold again on the latter's return to France and later reached the British Museum, where it remains today.[51] During its time in the Leverian Museum it was painted by Alexandre Chevalier de Barde, a noted French natural history painter and another fugitive from the Revolution[52] and was illustrated yet again by Donovan in *The Naturalist's Repository*.

Another specimen, taken from Cloudy Bay, was purchased by the collector Henry Seymer from a sale of second voyage specimens; it was recently rediscovered in the Museum of Zoology, University of Cambridge, complete with its operculum and a note of Seymer's showing that he had paid £2 17s for it[53] – not a bad price, considering that of the pink version above. Specimens of the Nesting Mussel collected from New Zealand on the first voyage are still in existence and a member of the brachiopods (the so-called lamp shells) collected on the second voyage, and at some time in the possession of the Reverend Cracherode, is still in the British Museum.[54] There are probably many other and less distinguished New Zealand shells scattered over England and Europe, unrecognized as being from this period of discovery. A great many became lost for all time as the 19th century wore on and they were auctioned off to anonymous buyers and the market became saturated with specimens from later voyages.

The fish were to suffer a fate somehow disproportionate to the importance given them in the journals and paintings of the collectors and artists, where they were among the most frequently mentioned and illustrated of the animals. Solander worked on the first voyage collection, the New Zealand part of which was described in his manuscript *Pisces Australiae*. Unlike insects or shells, there was little dealer interest and few foreign naturalists showed up to take a look at them or beg specimens for their own collections – at least not immediately. The fish were added to by the Forster and (scanty) third voyage collections and at some point specimens reached Lever's and Bullock's Museums, but it is not clear how or when.[55]

From 1780 to 1782 Banks was visited by the Frenchman Pierre Broussonet, whose

47. Anon., 1781, quoted in Dance, 1971, p.368. Muricated means treated in hydrochloric acid to reveal the nacreous layers beneath.

48. e.g. *Cominella adspersa*, Bruguière, 1789, p.265.

49. Dance, 1966, p.120. Spengler's Trumpet (*Cabestana spengleri*); Australasian Trumpet – *Ranella australasia*.

50. Wilkins, 1955, p.97 and Wilkins, 1954, pp.7-12. Dance, 1966, p.110 mentions a source stating that it was the *Adventure's* anchor cable whereas Wilkins' (1954) account favours the *Discovery*. If the former, it is possible that Cook himself passed the shell to Lever.

51. Dance, *ibid*. p.110.

52. The originals now in the Louvre; Whitehead, 1978, p.58.

53. Dance, 1966, p.109.

54. Wilkins, 1955, p.86, 98.

55. Whitehead, 1969, p.180.

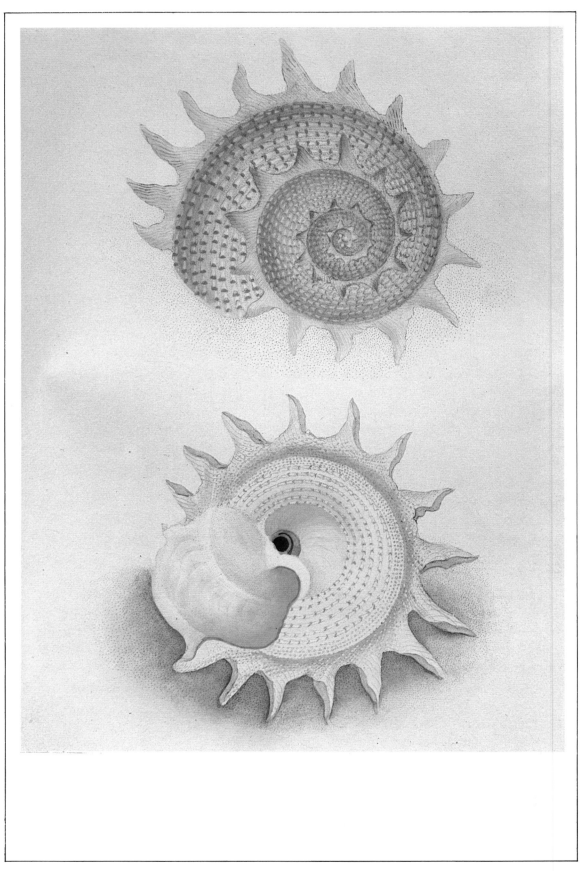

The pink variety of the Imperial Sun Shell (*Astraea heliotropium*) a hand-coloured lithograph from Edward Donovan's *The Naturalist's Repository* . . . . 1834 pl. 11. Published by permission of the Australian Museum. *(see p.55)*

interests were akin to Banks's own in that he was a botanist who was also drawn to agriculture, including sheep breeding, and other branches of science. He divided his time between Banks's home and the British Museum, working on fishes under Solander's supervision. It was not long before his first publication appeared in which he described 'Isabella' — a vernacular name for the Carpet Shark, taken from New Zealand waters and painted earlier by Parkinson.[56] In 1789 the unfortunate Broussonet was forced to leave Paris in the wake of the Revolution, eventually becoming a physician to the United States Embassy in Morocco, before being allowed to return to France.[57] But his 'Isabella' was meanwhile given a valid scientific name by Bonnaterre in the *Encylopédie Méthodique* in 1788, and so became the first properly named New Zealand fish.[58]

The Carpet Shark was not however the first New Zealand fish to feature in a publication. That honour went to the Giant Kokopu described by George Forster in his *A Voyage Round the World...* (1777) thus: '... a small species of fish (*Esox*), without scales, resembling a little trout; its colour was brown, and mottled with yellowish spots in the shape of some ancient Asiatic characters'.[59] This brief account was enough for the German compiler Gmelin (1789) to take up and add the specific name *argenteus* to the one of *Esox* mentioned above. It carries that specific name to this day.

George Shaw, a natural history writer and future keeper of zoology in the British Museum, published some of the Cook voyage fishes in 1792 in the *Museum Leverianum*, but no New Zealand fishes appear. Again, it was activity taking place in Europe that was to lead to more extensive publication of the New Zealand and other fish species. This time it was in Berlin, where at the age of 56 a wealthy physician named Marcus Bloch[60] began to take a belated interest in fishes at about the time of the return of Cook's third voyage. His highly regarded work on German fish was followed by a 'less trustworthy' account of foreign species, and eventually by a compilation that was unfinished at his death in 1799.[61] This last work, edited by J. G. Schneider and published in 1801 as *M. E. Blochii... Systema Ichthylogiae iconibus cx illustratum...*, describes and validly names all but a few New Zealand fish collected by the Forsters, including most of the well known species. Bloch and Schneider saw neither specimens nor drawings,[62] and their reliance on the material of others — including the manuscript *Descriptiones Animalium* — meant that descriptions and locality information were sometimes incorrect, although in the case of the New Zealand species there was limited scope for error. Although the work was illustrated with a few engravings, there was none of the New Zealand fish.

Curation at the British Museum during the early years of the 19th century was at its lowest ebb under its zoological keeper, George Shaw.[63] Many fish and other species no doubt vanished during this period, placing an even greater reliance on the drawings and paintings of the voyage artists. One to make use of these was Georges Cuvier, who was to become a dominant figure in French zoology during the early 19th century.[64] Cuvier and his assistant, Achille Valenciennes, did not actually see the voyage drawings; instead, they were to be plentifully supplied with tracings and sketched copies by a remarkable traveller, writer and artist called Sarah Bowdich. Together with her husband Thomas Bowdich, she had spent three-and-a-half years in Paris where they became close friends of Cuvier and his colleagues, and Thomas Bowdich wrote some travellers' guides to Cuvier's arrangement of animals. The Bowdich copies of fish drawn by George Forster and the other artists are in pencil on now-faded scraps of paper, and include a number of New Zealand species. They are bound in with the original Cuvier manuscript from which emerged the enormous *Histoire naturelle des Poissons* (1828-49) in which some further fish from the Cook voyages were properly described and named — including the Yellow-

56. Broussonet, 1780, p.648.

57. Anon *in* Encyclopaedia Brittanica, 1910, v.4, p.656.

58. *Cephaloscyllium isabella*, Bonnaterre, 1788, p.6.

59. Forster, 1777, p.159. McDowall, 1978, p.176 notes the incongruity of the name *Galaxias argenteus* – possibly resulting from Gmelin's confusion with a neighbouring illustration in Forster's Folio, *Esox argenteus* from Tahiti. The earliest published record was from the lake although it was recorded (and probably illustrated) from the stream lower down (Forster, 1982, p.252).

60. A founder member of the Gesellschaft der Naturforschender Freunde zu Berlin in 1773 – Spengler and Chemnitz belonged and must have been acquainted with Bloch (Karrer, 1978, pp.129, 143-144).

61. Boulenger, 1910, v.14, p.245.

62. Wheeler, 1981, p.782.

63. Whitehead, 1969, p.188.

64. Wheeler, 1981, p.784.

65. *Aldrichetta forsteri* (often called 'Herring') and *Rexea solandri*, the specific names commemorating Cuvier's illustrious predecessors.

66. Richardson, 1843, pp.12-30.

Eyed Mullet and the Gemfish from New Zealand.[65]

Some time later John Richardson, Inspector of Naval Hospitals at Haslar Hospital (Portsmouth), named further species using Solander's manuscripts and Parkinson's drawings as their basis:[66] the Scorpion Fish and Sand Flounder are examples from New Zealand. It seems remarkable that these fish, among the first seen by Cook's naturalists in that country, should take more than 70 years to receive this treatment. The belated publication in 1844 of Forster's *Descriptiones Animalium* formalized the naming of many of the remaining species, leaving a tortuous trail to be followed by those interested in fish systematics from this period.

Retracing our steps to a point earlier in the 19th century, we find the first published anatomical work on a New Zealand animal which, in all likelihood, concerns the same hagfish that was captured in a lobster-pot in Dusky Sound on the second voyage. The paper, by Sir Everard Home was published in the *Philosophical Transactions of the Royal Society of London* in 1815, lengthily entitled: 'On the structure of the organs of respiration in animals which appear to hold an intermediate place between those of the Class Pisces and Class Vermes, and in two genera of the last-mentioned class.' In 1792 some animals from Banks's collection were given to John Hunter's museum, and it was some time after this that William Clift, the Museum's curator, described 'Myxine Glutinosa' in a manuscript that began: 'the great Lamprey from the South Seas differs from the English Lamprey. . .'.[67] This manuscript may have been compiled from Hunter's notes and almost certainly refers to the New Zealand Hagfish, placing it within Hunter's collection and the Royal College of Surgeons.

67. Royal College of Surgeons Library, M.S. 275-h-3 (5). John Hunter's Museum became incorporated in the Royal College of Surgeons in 1800; Whitehead, 1969 p.165.

68. Bettany, 1908, v.9, p.1121-1122.

Sir Everard Home was an ex-pupil and close associate of John Hunter and later took over his lecturing duties and became a trustee of the collections. After Hunter's death Home became embroiled in an unpleasant scandal when it became evident that he had taken unpublished records and manuscripts from the museum and burnt them, presumably to destroy evidence of his plagiarism of Hunter's work.[68] It is possible that John Hunter took part in the hagfish dissection and in the observations that followed as he, in the first instance, would have been the recipient of the material. However to do justice to Sir Everard, the presumed arrival date of the specimen at the museum was not long before Hunter's death in 1793, leaving the latter with only a short time to make use of it. The involvement of the Home family in New Zealand zoology was to be continued in the 19th century by Sir Everard's son.

Of the current existence of the fish specimens from the voyages, not much can be said. A very few are known from the Naturhistorisches Museum in Vienna and the Cuming Museum in London; others are known from Paris, in the Muséum National d'Histoire Naturelle (via Broussonet) and in the British Museum (Natural History), London. In the last institution are the Butterfish and one of the 'Clingfish' collected by J. R. Forster in New Zealand.[69]

69. Whitehead, 1969, p.180; Wheeler, 1981, pp.781-802.

Of all the vertebrates, the birds were to fare the best after the voyages were over. Those that were collected had to survive the appetites and culinary experiments of the naturalists and crew, but if they avoided that pitfall there was a fair chance of their reaching England in a reasonable state, especially those preserved in alcohol. Here, most of them ended up in the hands of Banks, including the first voyage specimens, the collections of the Forsters and Furneaux from the second voyage, and those of Anderson and Clerke from the third. A few birds went directly to other collectors — to Ashton Lever (who had a New Zealand Falcon and a Tui)[70] and possibly to another museum owner, Daniel Boulter from Great Yarmouth. There may well have been others. Of the fate of

70. Lysaght, 1959, p.287.

The Tui (*Prosthemadera novaeseelandiae*) here called 'The New Zealand Creeper', a hand-coloured engraving from Peter Brown's *New Illustrations of Zoology* . . . . 1776 (pl. 9). The butterfly is not from New Zealand. Published by permission of the Trustees of the British Museum. *(see p.60)*

71. Medway, 1976, p.44 *et seq.*

72. Brown, 1776, p.18, Pl.IX.

73. Murray-Oliver, 1969, p.xxii.

74. Compared with engravings where the ink lies in the grooves cut by the burin.

75. One of Buffon's pupils.

76. Stresemann, 1975, p.90.

77. *Ibid.*, p.88.

78. *Eudyptula minor minor.*

79. Medway, 1976, p.52

80. Stresemann, 1975, p.79.

Parkinson's bird collection, nothing is known.[71]

Banks gave at least two first voyage birds — the Tui and the South Island Kokako — to Marmaduke Tunstall and in 1776 it was this Tui, '. . . a stuft specimen in tolerable preservation . . .', that was illustrated in *New Illustrations of Zoology. . .* by Peter Brown, a Dane who worked for Thomas Pennant.[72] It was the first published illustration of a New Zealand bird, but unaccompanied by a written description or a scientific name, it is of no great zoological importance.

The Tui was to be New Zealand's most illustrated animal in those early years. An engraved copy of George Forster's paintings of a Tui accompanied Cook's *A Voyage Toward the South Pole. . .* published in 1777. The engraving, showing the Tui in an uncharacteristic pose, was too faithful to George Forster's original, which was of an obviously dead bird. A more lifelike attempt was made by Robert Laurie, who devised a means of printing mezzotints in colour.[73] Intended as part of an atlas of illustrations and maps to accompany various voyage reports, it has particular interest as very few illustrations of New Zealand animals were made by the mezzotint process. This was achieved with a copper plate thoroughly burred by a 'rocking-tool' and the ground then worked to modify the capacity of the burrs to hold ink and thus create lighter tonings.[74] One of the more bizarre settings for a Tui was in a work of the notable French ornithologist François Le Vaillant,[75] entitled *Histoire naturelle des Oiseaux d'Afrique* (1799-1802). Its presence there was owed to Gigot d'Orcy, the patron of the entomologist Olivier who has been mentioned earlier in connection with the illustration of New Zealand insects. It is possible that Olivier, after visiting Banks, took the specimen back to France with him and passed it on to his associate who then had it painted by Audebert.[76] How it came to be regarded as African in origin remains a mystery, but it might well have been a result of Le Vaillant's eagerness to pad out his material from other collections.[77]

George Forster's *A Voyage Round the World. . .* (1777) contained brief descriptions of a few New Zealand birds — for example an unidentified petrel from Dusky Sound — but none was validly named in this work. In 1781 J. R. Forster published his study on the penguins, *Historia Aptenodytae. . .*, the first zoological work of any consequence to describe species from the voyages, and in it was a Latin description of the Southern Blue Penguin[78] collected from Dusky Sound, the first New Zealand bird to be described.

In the same year, the first of a series of works by John Latham began to appear. Latham was a prosperous physician who practised in a small town outside London. An ex-pupil of John Hunter the anatomist, and a correspondent of Thomas Pennant, he was well acquainted with the natural history establishment and his interest in collecting birds and in Linnaean systematics inevitably led him to Banks's library and collections. Banks kindly allowed him to copy some of George Forster's bird paintings, including the Shining Cuckoo and the South Island Bush Wren from New Zealand; he possibly even gave him some specimens, as Latham was known to have a kaka, kokako and Tui. Further assisted by examination of J. R. Forster's manuscript notes, a few bird drawings[79] and visits to Lever's museum, Latham went about preparing his *General Synopsis of Birds* (1781-85). As the first major work to include birds from Cook's voyages it should have had considerable significance. But despite his laborious efforts, which included preparing his own engravings, the book was not wholly successful. It suffered from dealing with the material at second hand, with the result that details of descriptions and localities are often incorrect.[80] And in spite of his interest in Linnaean classification, his conservatism got the better of him at the last and he applied English names to all his species. The other errors he could be forgiven, because they were part and parcel of many similar works of the period, but his avoidance of Latin binomials for the species names laid the work open to the predations of Gmelin (1788-93)[81] who, by correctly renaming Latham's species, entitled himself to their authorship. Latham was to recant in 1790 in his *Index Ornithologicus* but by then it was too late and most of the damage was done. This explains why so many New Zealand birds discovered in those years have Gmelin's name as author: the Shining Cuckoo, the Tui, Morepork, South Island Kaka — the list is long.

Forster's paper *Memoire sur les Albatros* (1785) described two species seen in New Zealand waters, although they do not breed there, leaving the next work of consequence to New Zealand ornithology to another participant of the second voyage. After leaving the Forsters and the *Resolution* at Capetown, Anders Sparrman continued his work in Africa before eventually returning to Sweden, together with a collection of South Pacific birds that Forster had allowed him to keep. These were deposited in a private museum owned by Johann Gustav von Carlson and later described and illustrated by Sparrman in his *Museum Carlsonianum* (1786-1789). Today nine species of New Zealand birds are recognized from this work[82] which was the most substantial on New Zealand birds to that date. It was also one that escaped Gmelin's revision because of its adherence to Linnaean principles.

It is not clear what prompted Sparrman to publish, unless it was impatience or a realization that unless he intervened, the voyage ornithology would continue to be squandered in works like Latham's. From what we know of his character, it is unlikely that Sparrman would have wilfully pre-empted Forster unless he thought there was very little hope that the latter would have his work published in full.

Gmelin's revision of *Systema Naturae* (1788-1793) took care of many of the birds not covered by the previous work of Forster and Sparrman by renaming some 27 species — the majority in fact of those collected by Cook's naturalists. Included were the Tui, the Paradise and Blue Ducks, the White-fronted Tern, the Kakas, the Shining Cuckoo and the Morepork; many in fact that had figured largely in the early accounts and illustrations of the voyages. Gmelin appears here as a opportunist, snatching credit from those who had spent long hours in the field or at the writing desk. But to be fair, he was doing a job that would sooner or later have to be done, as the prospect of the original authors revising their work was, apart from a few cases like Latham's, slight.

It was by now 1790, and most of the New Zealand birds collected between 1769 and 1777 were already described and properly named — not a bad effort considering that there would be 19th century expeditions with worse records. George Shaw's *Museum Leverianum* (1792-1796) illustrated the South Island Kokako and the Kaka but added little to previous work. The publication of Forster's *Descriptiones Animalium* in 1844 formalized the naming of four more birds including, oddly enough, the Variable Oystercatcher which was so familiar from their journals; the Common Red-billed Gull, and a bird that breeds in New Zealand but was discovered south-west of Cape Horn — the Mottled Petrel.[83]

Unfortunately many of the specimens described have disappeared. More so than fish, birds were collectable, particularly once stuffed and mounted, and seemed to have been favoured by those with large collections or museums. But they lacked the durability of shells and many probably quickly succumbed to vermin and other forms of decay. When the Leverian Museum was sold in 1806, Leopold von Fichtel bought four New Zealand birds for the Imperial Collections in Vienna[84] which were deposited in the Naturhistorisches Museum where the Spotted Shag, New Zealand Falcon, the South

81. In a revision of Linnaeus's *Systema Naturae*, 13th ed.

82. The Bellbird *(Anthornis melanura melanura)*, the South Island species of Thrush *(Turnagra capensis capensis)*, Robin *(Petroica australis australis)*, Fantail *(Rhipidura fuliginosa fuliginosa)* and Rifleman *(Acanthisitta chloris chloris)*, the Western Weka *(Gallirallus australis australis)*, Spotted Shag *(Stictocarbo punctatus punctatus)*, Long-tailed Cuckoo *(Eudynamis taitensis)* and Red-crowned Parakeet *(Cyanoramphus novaezelandiae novaezelandiae)*.

83. *Pterodroma inexpectata*; Fleming, 1982a, p.89.

84. Whitehead, 1978b, p.68.

Tanagra capensis.

The South Island Thrush (*Turnagra capensis capensis*), now presumed extinct; a hand-coloured engraving of a specimen from Dusky Sound in Anders Sparrman's *Museum Carlsonianum . . . .* 1786-1789 (pl. 45). Published by permission of the Trustees of the British Museum. *(see p.61)*

Island Kaka and Red-crowned Parakeet are still to be found. At the same sale a Tui and a South Island Kokako were purchased for Lord Stanley's (later the Earl of Derby) collection, from where they passed to the Merseyside County Museum in Liverpool.[85] With the price of shags at the sale varying between eight shillings and £2/13s and kakas being knocked down for as little as 10/6d it can be seen that birds were not the most highly prized of collectables. Two of the birds described by Sparrman — the Red-crowned Parakeet and the South Island Thrush — went by different routes to the Naturhistorika Riksmuseet in Stockholm where they have remained. These and the species mentioned above are the only surviving type specimens of New Zealand birds that date from the Cook voyages.[86]

Although two New Zealand mammals were drawn by George Forster it would be years before the illustrations and descriptions were published. The bat[87] was described in *Descriptiones Animalium* in 1844 and kept Forster's name as author. Its first published illustration was a hand-coloured lithograph by G. H. Ford, accompanying a paper by R. F. Tomes in the *Proceedings of the Zoological Society of London* of 1857. The Fur Seal was described by René Primivere Lesson, the naturalist who accompanied the first French scientific expedition to New Zealand. Happily, he remembered the Forsters' association with the animal and named it *Arctocephalus forsteri*. Because of its size the seal was unlikely to have been kept, and the whereabouts of the bat is unknown.

We have spoken of the inability of Cook's naturalists, or those who retold the story of their discoveries, to reveal some of the real essence of New Zealand zoology. There was an accounting and a systematizing, there was description, but there was very little analysis or interpretation. They had not discovered the zoological elements that would identify New Zealand as a country cast adrift before advanced and predatory forms of animal life could leap aboard; a faunal cul-de-sac, an ark for a fortunate few species whose relatives were doomed to extinction elsewhere. The naturalists had not, except in rather crude terms, seen the fauna as that of a large oceanic island. Only Forster was able to begin to piece together his botanical and zoological observations to note biogeographical relationships between islands and neighbouring land masses, pointing to similarities between the avifaunas of New Zealand and Norfolk Island.[88] It would have been altogether surprising had they done more, for several decades had to pass before all the New Zealand ingredients were adequately known and described, and it was not until the second half of the 19th century that such matters as biogeography received any serious attention.[89] So we can be grateful to them for the discovery, recording or description of around 165 animal species from New Zealand — of birds, fish, molluscs, and insects, the odd crustacean and a couple of mammals. Along with the theories of zoogeography, the lesser zoological lights would have to wait.

We can now note — perhaps with some surprise — as we leave them, that the effort that brought about this result was an international one with the participants crossing national and linguistic boundaries in order to share their findings and make their discoveries known. In the end, the contributions of the French, German and Scandinavian students of Linnaeus mattered almost as much as those of the English naturalists and explorers. The French can be remembered for their involvement in New Zealand entomology and its incorporation into their tradition of encyclopedic works; the Germans for their efforts in conchology, ichthyology, and of course their gift of the Forsters. And what could be more solid and respectable than the work of Fabricius, Sparrman, and Solander — the students of Linnaeus? We can only marvel that they all carried on in spite of the wars and revolutions that troubled Europe throughout the period — an

85. Medway, 1976, pp.120-131.

86. Medway, *Ibid.*, p.125, 131.

87. Now *Chalinolobus tuberculatus*.

88. Forster, 1778, pp.174, 198, 199.

89. P. L. Sclater wrote a pioneering paper on zoogeography in 1858, followed by another in 1860 by A. R. Wallace. The latter's *Geographical Distribution of Animals* appeared in 1876.

SCOTOPHILUS TUBERCULATUS Forster sp.

The Long-tailed Bat (*Chalinolobus tuberculatus*), the first published illustration of the specimen originally described by J. R. Forster; a hand-coloured lithograph from R. F. Tomes *On Two Species of Bats* in *Proceedings of the Zoological Society of London*, 1857 (G. H. Ford, the illustrator). *(see p.63)*

indication that the spirit of enquiry can persist in the most difficult times. It would be a number of years before the New Zealand fauna was brought to English and European attention in so forceful and compelling a way as it was in the late 18th century, and when it was it would be with a somewhat different perspective. The new emphasis would be on the elements of the country's strange relict fauna, with the animals of the past mattering as much as those of the present. The beginnings of that phase were not too far off.

EXPLORERS, SEALERS AND KIWIS

APART FROM WILLIAM BLIGH'S DISCOVERY of the Bounty Islands in September, 1778 — broadly within the region of New Zealand — the area was virtually ignored for a decade until Vancouver in the *Discovery* and Broughton in the *Chatham* entered Dusky Sound on 2 November, 1791. They were on their way to North America, but the ravages of dysentery and a need for planks and spars for the vessels caused Vancouver to seek out what was, for him, a familiar port.[1] The journals of Lt Archibald Menzies of the *Discovery*, who acted as botanist, and Edward Bell, a clerk on board the *Chatham*, provided a sketchy zoological record of their short stay.[2]

It seems that 10 years after the Forsters, Dusky Sound was still much the same. There were fish that took every hook, and congregations of birds — more finding their way into pies than jars of spirits. Tuis, 18 of them, were the 'most delicious & savoury food we had yet used' and had they been greatly esteemed more parrots would have been taken.[3] The 'cole fish' were found equal in flavour and firmness to the 'codd', and as well there were mussels, limpets and crayfish, and the inevitable sandflies. The explorers saw no reptiles or mammals, nor any sign of the geese released on the previous voyage. Menzies, a friend and correspondent of Banks, was equally enthusiastic about plants but appeared to have even less interest in animals, apart from his enjoyment of hunting.[4]

Three weeks after their arrival they left Dusky Sound, the two vessels being separated by a storm the following day. Independently they sighted the Snares, populated by what seemed to be Sooty Shearwaters,[5] and the *Chatham* later discovered what came to be known as the Chatham Islands. There Bell had noted that the 'Ear shells' littered the surrounds of Maori dwellings and Broughton recorded several shore birds and made the first sighting of the Chatham Islands Pigeon.[6]

They returned to the Thames in October 1795, having left little mark on New Zealand zoology, but from among his souvenirs of the voyage, Joseph Mear of the *Discovery* produced a 'Paroquet from New Zealand: the Poapoee Birds from the same part', which he presented to Sir Everard Home, who was associated with John Hunter's museum.[7] There were, no doubt, other small collections of a similar kind brought back for sale or presentation.

Even more ephemeral contacts with New Zealand followed prior to the century's end. Some time before Vancouver's visit, Alessandro Malaspina, a Sicilian officer in the Spanish Navy, had written to Joseph Banks asking him what he thought would be a suitable subject for study on a forthcoming voyage to the South Pacific by three Spanish vessels, the *Descuvierta*, *Atrevida* and *Sutil*.[8] Whatever Banks's advice, zoology did not apparently gain much of their attention and when they arrived off New Zealand their contact was limited to a landing on Bauza Island in Doubtful Sound in February 1793.[9] They saw a few birds and some small limpets, but nothing else other than sandflies, and on the way out some seals. Bad weather forced them to abandon their efforts to enter Breaksea Sound, the northern connection with Dusky Sound where, had they pressed through, they would have found a sealing party left there by Captain Raven and the *Britannia*. It was the first and last Spanish scientific expedition of any size to visit that part of the world.

1. He had been with Cook on the second and third voyages.

2. Bell, 1791, QMS, Alexander Turnbull Library. Menzies, *in* McNab, 1907, Appendix C, pp.301-317.

3. Menzies, *in* McNab, 1907, Appendix C, p.313.

4. See Fell, *et al.*, 1953, Section V, for a summary of animals observed.

5. *Puffinus griseus, ibid.*, Section V.

6. The shells were Paua *(Haliotis* sp.). The pigeon was named much later *Hemiphaga novaeseelandiae chathamensis*; Broughton in Vancouver, 1984, v.I., p.387.

7. M.S. 275-h-3 (ii) c.1795, Royal College of Surgeons Library. A *Cyanoramphus* and a Tui. Joseph Mear (also spelled Mears) was Surgeon's second mate on the *Discovery* (Vancouver, 1984, v.I p.1649).

8. Dawson, 1958, p.569.

9. McNab, 1907, pp.50, 51; Hall-Jones, 1984, pp.17, 24.

The first published illustration of a kiwi (probably the South Island Brown Kiwi, *Apteryx australis australis*, although the Stewart Island Kiwi, *A. australis lawryi*, cannot be excluded); a hand-coloured engraving by R. Nodder for George Shaw and Elizabeth Nodder's *The Naturalist's Miscellany* . . . . . . , 1813 (v. 24, pl. 1057). The elongated stance of the bird results from it being drawn from a dried skin. Published by permission of the Trustees of the British Museum. *(see p.69)*

Not long after Malaspina, d'Entrecastaux and Huon de Kermadec, respectively of the *Recherche* and *Espérance*, sighted Three Kings Islands on 11 March, 1793, and traded briefly with the Maoris off the Northland coast, departing shortly thereafter. There were further visits to Dusky Sound by sealing parties and the *Mermaid*, one of the first whaling vessels to visit New Zealand, spent three months at Great Barrier Island.

While the passage of vessels and people between Port Jackson (Sydney) and New Zealand increased, there was during the last years of the 18th century and the early ones of the 19th, a general absence of major expeditions in New Zealand waters, resulting in a dearth of new specimens and a consequent reworking of old material by the compilers. One notable exception was to appear in the middle of this hiatus, a bird that was to become symbolic of the New Zealand fauna and eventually of New Zealand itself, and to puzzle zoologists for two decades both as to its relationships and its whereabouts. It first manifested itself as the Brown Kiwi, *Apteryx australis*, the wingless bird of the South.

To begin the story of its discovery it is necessary to go back to Dusky Sound in 1792 where the inevitable commercial exploitation of New Zealand's wildlife was taking place. The Sound had become the base for the first sealing gang; trade in sealskins had started between New Zealand and China and this, with the concession of the East India Company, was allowed to extend to England.[10] In the first few years of the 19th century the Bass Strait sealers, based at Port Jackson, turned their attention to New Zealand and once again Dusky Sound was the chosen spot. But the focus was starting to shift to the islands to the south, and by 1809-10 much of the activity was around Foveaux Strait and Stewart Island.[11]

It must have been a wretched and dangerous life for those who succumbed to the temptation of 'liberal encouragement' and joined up with the vessels of the Sydney companies. Apart from the seals and some spectacular scenery, the southern coast of New Zealand could offer little but often atrocious weather, indifferent and uncertain food supplies, and the chance of combat with the Maoris. In spite of such conditions and the frequent loss of ships, parties were put ashore for long periods before returning to the comforts of Sydney. On the face of it, the contributions of these gangs to natural history were negative ones, but now and then from such lonely outposts would come one or two spectacular zoological offerings that would eventually reach London and an excited scientific world.

It was against this background that a privateer named the *Providence* arrived at Port Jackson on 2 July, 1811. On board were 174 convicts, three officers and 36 men of the 73rd Regiment, and some general merchandise.[12] Her commander and part-owner was Andrew Barclay, a nuggety Scotsman who had spent the greater part of an adventurous life at sea.[13] With his cargo discharged he spent three months in port before being cleared to leave on 18 October. He sailed on the 20th, bound for China with a load of ballast and instructions to pick up tea for London.[14]

While Barclay was in Port Jackson, a bird skin from New Zealand came into his hands. The plumage, of a reddish-brown colour, soft-looking but slightly coarse to the touch, was unusual; the bird was fairly large, apparently wingless, and had a long, slender beak. Nothing like it had been seen before. The circumstances that brought the skin to Barclay can only be speculated on — perhaps a chance meeting with a sealer in a tavern, one who had spent some uncomfortable months on the southern coast of New Zealand and saw profit in hawking a rare skin around the port. Barclay would have undoubtedly spent much of his time in taverns during his three months' stay. Perhaps some money changed hands with the thought that a collector back in England would find the specimen valuable?

How the sealer came by the bird is also anybody's guess. Possibly it was tracked down by a dog from one of the ships, or it was an item of trade with the Maoris. We will

10. McNab, 1907, p.47.

11. *Ibid.*, p.101.

12. Bladen, 1901, vii, pp.593, 609.

13. Stancombe, 1966, v.1, pp.56, 57.

14. Bladen, 1901, vii, p.645.

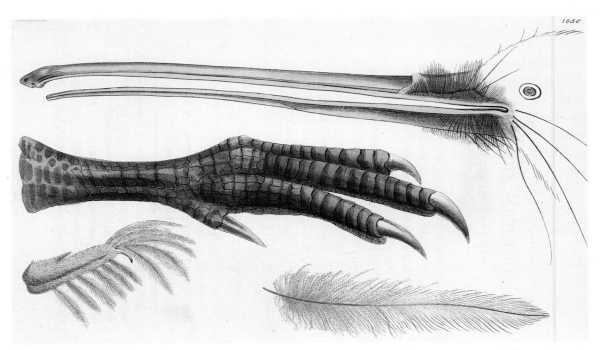

Beak, foot and feather of the kiwi referred to above, from *The Naturalists Miscellany . . . . . .*, 1813 (vol. 24, pl. 1058). Published by permission of the Trustees of the British Museum. *(see p.69)*

probably never know. Even the place of its capture is clouded with doubt. Some years later, 'Dusky Bay' was nominated, but the evidence is circumstantial and the southern coast of the South Island or Stewart Island are also possibilities. At that time, sealing activity had moved south to Foveaux Strait and Stewart Island, and two vessels known to have relieved sealing groups in the area returned to Sydney only a few weeks before the arrival of the *Providence* and Barclay. They were the *Boyd* (26 March, 1811) and the *Sydney Cove* (13 April, 1811). The kiwi skin might easily have been on board one of them, and although the possibility of their having calling at Dusky Sound cannot be excluded, there is no mention of their having done so.[15]

There is also no indication that Barclay himself stopped at Dusky Sound, or any other part of the south coast, on his way to China. Regrettably, nor is there trace of the ship's log, and Barclay's autobiography dictated to a friend when he was 77 and privately published 15 years after his death, tells us little.[16] Lloyds' list of 15 May, 1812, his arrival date back in England, states simply: *'Providence*, Master Barclay. January, arrived China from New South Wales.'[17]

After his arrival in London, Barclay no doubt gave thought to disposing of the skin, but fortunately through the mediation of a mutual friend, a Mr W. Evans, it passed into the hands of the Assistant Keeper of the British Museum, George Shaw. It was later alleged in Shaw's description of the bird that it was 'presented' to him, so perhaps Barclay did not make anything out of the transaction after all. The bird certainly became Shaw's personal property rather than an acquisition of the Museum — it is nowhere mentioned in the Museum's *Book of Presents* for that period.

Given Shaw's future reputation in taxonomy and the curation of specimens, one might now have feared for the fate of this exotic animal. But such apprehension would have been misplaced. The bird was examined by John Latham, an ornithologist already prominent in the description of New Zealand birds, but his experience let him down and he classified it with the penguins.[18] Shaw, on the other hand, correctly assigned the specimen to the same group as the Emu and other flightless birds, and commissioned

15. *Ibid.*, p.551; McNab, 1907, p.114. The more rufous appearance of this bird suggests a Stewart Island locality.

16. Barclay, 1854. His recall of sailings of the *Providence* does not tally with the shipping and other records and is unreliable. He settled in Tasmania and may have chosen to forget his role in transportation.

17. Lloyds List, 15.5.1812, National Maritime Museum, Greenwich.

18. Newton, v.15, 1911, p.842.

68

R. P. Nodder to prepare two plates to accompany a description in the 1813 edition of the *Naturalist's Miscellany*.[19] The plates were not uniformly successful, the first of them showing the kiwi standing oddly upright, almost penguin-like in appearance, with a background uncharacteristic of its environment. The other plate, showing detail of the beak, feet and feathers, was more acceptable. Shaw named the bird *Apteryx australis*, and followed the description with a note that Barclay had come by the bird on the south coast of New Zealand — no mention of Dusky Sound.

The Brown Kiwi was one of the last animals to be described by Shaw, as he died that same year; at the sale of his effects the skin was purchased by Lord Stanley (later the

The entry in the Earl of Derby's manuscript catalogue referring to the kiwi bought at auction from George Shaw's estate. Published by permission of the Merseyside County Museum, Liverpool. *(see below)*

13th Earl of Derby).[20] Apart from the occasional loan, it remained in the collections of the Earls of Derby for many years until it passed with the rest of the collection to the Merseyside County Museum in Liverpool, where it remains today. None of the labels attached to the specimen, including the original Derby label, offers any clue to the bird's locality, nor does the entry in the Derby manuscript catalogue.[21] But the specimen, given its age and travels, is in remarkably good condition.

It was but a single specimen, to their minds possibly a freak, but corroboration came shortly after its description was published, although not by an ornithologist, and from an entirely different locality. John Liddiard Nicholas, a traveller visiting northern New Zealand between 1814 and 1815 in the company of the missionary Samuel Marsden, observed: '. . . from the feathers which line the garments of some of the chiefs, it would appear, that there is here a species of the cassowary, but we did not see any in our excursions. The feathers are precisely the same as those of the emu in New Holland (Australia), except being somewhat smaller.'[22] Shaw's work had by then been published but evidently Nicholas had not seen it — even after his return to London and the publication of his narrative. In this light, his observation which was based only on the feathers was quite acute. Undoubtedly they were those of the North Island Brown Kiwi, described later as a sub-species separate from Shaw's. Four years later, another visitor

19. Shaw and Nodder, 1813, v.24, Pls.1057,1058 and text. Frederick Nodder died around 1800, and vols. 14 on were published by Shaw and Elizabeth Nodder (Frederick's widow) with engravings by their son, R. P. Nodder. The title page of volume 24 bears the names of Shaw and Elizabeth Nodder only so they presumably become the authors of *Apteryx australis* with the plates drawn, engraved and published by R. P. Nodder. (See Cust, v.24, 1909, p.531. for Nodder family history.)

20. Yarrell, 1833a, p.34. Fully described in Yarrell, 1833b, pp.71-76.

21. No. 180 in MS. catalogue of the collections of the Earl of Derby, 1830-1840, Merseyside County Museums. There are 6 labels attached to the bird and a 7th loose in the enclosing bag. The oldest label appears to be Derby's, bearing only a coat of arms and 'No 180'. No labels give detail on locality, other than 'Middle Island' (i.e. South Island).

22. Nicholas, 1817, v.2, p.255.

was to make a similar observation. Major Richard Cruise, during a 10-month stay in New Zealand, observed an 'Emu-feathered mat' on which he commented: 'the Emu is found in New Zealand, though we were never fortunate enough to meet with one. The natives go out after dusk, with lights, which attract their attention, and they kill them with dogs. Their feathers are black, smaller and more delicate than the emu of New Holland and a mat ornamented with them is the most costly dress a chief can wear.'[23] This short description of the kiwi hunt was to be repeated almost verbatim by many subsequent explorers, including the French.

We next hear of the kiwi when the corvette *Coquille* commanded by Duperrey was anchored in the Bay of Islands; the ship was visited by the missionary Thomas Kendall in April, 1824. Kendall convinced them of the existence of a bird which he compared with the flightless birds of Australia, and told the naturalist Lesson and the others that the birds were called 'Kivikivi' and were hunted at night by the Maoris with the aid of dogs and torches.[24] Even more convincing to the French, however, were the assurances from the Maoris of the existence of a wingless bird and their presentation of some 'remains' — possibly a piece of skin with some feathers attached — but not much more.[25] These fragments were reminiscent of an Emu and were enough for Lesson to describe a new species, *Dromiceius novaezelandiae*, before he realised that the 'Kivikivi' and Shaw's *Apteryx* were one and the same species.[26] Many years later, in 1838-39, Lesson published an account of his voyage in which he illustrated his references to the Kiwi, not with his own plates, but with copies of Shaw's — a fair indication that the 'remains' given to him by the Maoris were insufficient properly to describe the bird.[27] It is just possible that he might have seen the original specimen, then in Lord Derby's collection, for he had visited England and was a correspondent with some notable British ornithologists.[28]

However even before Lesson's observations from the first French voyage had seen the light of day, members of a second voyage under d'Urville were gathering further evidence at Tolaga Bay on the North Island's east coast on 5 February, 1827. Here, during some friendly exchanges with the Maoris, they learned how kiwis were hunted, their size (that of a small turkey), and the fact that they were flightless. They saw also a cloak trimmed with kiwi feathers. By now, doubts about the kiwi's existence must have begun to recede. Scepticism in England retreated further when, in 1833, the Earl of Derby gave his specimen one of its rare outings and allowed the stuffing to be removed so that the ornithologist Yarrell, of the Zoological Society of London, could examine its skin and appendages more closely. A redescription followed and a much more credible illustration, the first since Shaw and Nodder's, was prepared by John Gould, the noted English ornithologist and illustrator.[29] Later that same year an appeal was made for more specimens, a request that was granted virtually by return mail.

Alexander Macleay, the Colonial Secretary of New South Wales, an entomologist and head of a prominent zoological family, was the ideal person to field the Society's plea. His enquiries to the Reverend William Yate of the Mission at the Bay of Islands met with the following response on 10 March, 1834: 'About six weeks ago I had one of these birds in my possession — the second I have seen in the Land. — I kept it nearly a fortnight, and in my absence it died. — One of my Boy's took off the Skin: The Legs rotted off. I have very great pleasure in sending you the Skin as it is . . .'.[30] Yate concluded with some observations on the kiwi's feeding habits. Macleay sent the skin by the *Redman* to London where it was received by Edward Bennett, Secretary of the Zoological Society, in 1835.[31] Although at that stage unrecognized, it was the first North Island Brown Kiwi to be seen by ornithologists.

23. Cruise, 1823, p.316.

24. Lesson and Garnot 1828, v.1(ii), p.418.

25. No record of this in the Ornithological Accessions Register, Muséum National d'Histoire Naturelle (Mammifères et Oiseaux) or in Lesson's list of species taken from New Zealand, MS. 354, Bibliothèque, Muséum National d'Histoire Naturelle.

26. Lesson, 1828, pp.209-211.

27. Lesson, 1838, p.348.

28. e.g. Lesson to Swainson, 28.9.1828; 2.10.1828; 10.5.1830; Swainson correspondence, Linnean Society of London. He sent Swainson specimens, including some collected by his brother in New Zealand, and spoke of visiting London.

29. Yarrell, 1833a, pp.24, 80; 1833b, v1, p.71, Pl.10

30. Macleay to Edward Bennett, 25.10.1834, MS. 663 Alexander Turnbull Library.

31. Bennett, 1835, p.61.

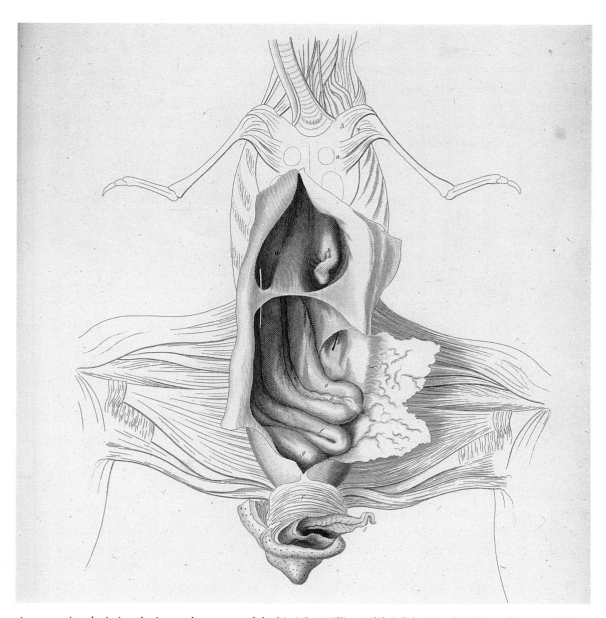

An engraving depicting the internal anatomy of the kiwi, by William Clift (of the Royal College of Surgeons) from Richard Owen's *On the Anatomy of the Southern Apteryx* in the *Transactions of the Zoological Society of London*, 1840, (pl. 49). *(see p.71)*

Following this, kiwis and information about them began to appear thick and fast. In 1837, reports of living birds came from Thomas K. Short of Tasmania along with the promise of specimens,[32] and Lord Derby purchased another kiwi, from John Smith in July of that year.[33] In March 1838, Derby gave the Society the body and viscera of a bird which was passed to Sir Richard Owen, the eminent anatomist, along with others from a Dr Logan and the Sydney zoologist, George Bennett. Owen used these specimens as the foundation of a series of fine anatomical studies that would make the kiwi one of the century's best known and most studied animals.[34] In April, 1838, the Sydney-based botanist and explorer Allan Cunningham sailed to New Zealand on the French corvette *L'Héroïne*.[35] While there he acquired a live kiwi for 28 shillings, which he preserved in the acetic acid carried by botanists for plant preservation. He sent the bird to England and on his return wrote to the Zoological Society of London telling them of his find.[36] It was Cunningham's last expedition — he returned from New Zealand

32. Short, 1837, p.24

33. M.S. Catalogue of the collections of the Earl of Derby, 1830-1840, No. 180. Merseyside County Museums. Short, 1837, p.24.

34. Owen, 1840, v.2, pp.257-302.

35. The vessel was engaged largely in protecting the French

continued overleaf

whale fishery, but
collected specimens from
the Bay of Islands and
Akaroa. Dunmore, 1969,
p.339.

36. Cunningham, 1839,
pp.63-64. The body was
said to have been
preserved in acid (Bagnall
and Petersen, 1948, p.74).
In this case it was acetic
acid normally used as a
preservative by botanists.

The North Island Brown Kiwi (*Apteryx australis mantelli*) collected at the Bay of Islands by Dumont d'Urville and J.-B. Hombron on the second voyage of the *Astrolabe* to New Zealand; the original watercolour painting by Werner signed by him in 1843 (later engraved for d'Urville's *Voyage au Pôle Sud . . . Atlas . . . Zoologie*, 1853). Published by permission of the Bibliotheque Centrale, Muséum National d'Histoire Naturelle (MS 585). *(see p.73)*

The Little Spotted Kiwi (*Apteryx oweni*) a hand-coloured lithograph by John Gould and H. C. Richter, from Gould's *Birds of Australia*, 1848 (vol. 6, pl. 3). From the early-mid nineteenth century the English-based lithographers were to excel in the "chalk-style" (i.e. use of a greasy crayon on stone instead of ink) and the use of lithography in natural history work would predominate. Published by permission of the Alexander Turnbull Library. *(see p.75)*

with an illness from which he never recovered, and he died in 1839.[37]

Eventually enough specimens were received in England for some to be redistributed to museums in Europe and the United States.[38] But the bird was still novel enough to be singled out in the instructions for d'Urville's last voyage to the South Seas. While at the Bay of Islands, d'Urville asked a Maori to find him a kiwi and on 1 May, 1840, a pair were presented to him — for a small charge.[39] In the Atlas of the voyage zoology published in 1853 there was an engraving of one of the birds and another of a skeleton. But by then the bird illustrator, John Gould, had already published a lithograph of the North Island Kiwi in *The Birds of Australia and the adjacent islands* in March 1838. He based his drawing on two specimens supplied by the newly formed New Zealand Association, but he also had access to the bird sent by Yate and Macleay.[40]

Separation of the North, South and Stewart Island subspecies of kiwi had not yet taken place when a further and very distinctive specimen arrived on the zoological scene — one that was found predominantly in the South Island. Its appearance was owed to the Australian collector and natural history dealer, Frederick Strange.

Strange was an important although rather obscure figure in the early history of antipodean natural science. Many Australian and New Zealand specimens must have passed through his hands before reaching a major collection in England, and his skill as a preparator ensured that his specimens were much desired.[41] When the ornithologist John Gould visited Australia in 1839, he became acquainted with Strange and a productive

37. Perry, in *Australian Dictionary of Biography*, 1966, v.1, pp.265-267. Allan Cunningham was originally sent to Australia by Banks. He later succeeded his brother Richard as Colonial Botanist of N.S.W.

38. Newton, v.15 1911, p.842.

39. Dumont d'Urville, 1846, v.9, p.184.

40. Gould 1837-1838, *(Apteryx australis)*. The same plate used again in the *Birds of Australia*, 1840-1869 (issued 1841). This was *Apteryx australis mantelli*. Mrs Gould first put it on stone in 3.11.1837 (Sauer, 1982, p.34) but the issue of the first book was withdrawn by Gould — it is now extremely rare.

41. Whittell, 1947, p.97.

The separation of the North Island Brown Kiwi (*Apteryx australis mantelli*) from the South Island Brown Kiwi (*Apteryx australis australis*); a lithograph by the renowned illustrator Joseph Wolf from A. Bartlett's *On the genus Apteryx* in *Proceedings of the Zoological Society of London*, 1852 (1850) (pl. 30). *(see p.75)*

liaison flowered eight years later, after the loss of Gilbert, Gould's official collector.[42]
Specimens were sent off to Gould who in turn would sell them to other collectors such
as the Earl of Derby. By 1843 Strange had moved to Sydney where he dealt in natural
history specimens, including many from New Zealand. His location put him in regular
contact with sealers who brought him information on rare New Zealand birds such as
the Kakapo — the ground parrot — and the kiwis. Thus it is probably in circumstances
rather similar to those described for Barclay and the original *Apteryx australis* that Strange
received a kiwi of a new kind and sent it off to John Gould, along with some other
New Zealand material.

This bird, the rare Little Spotted Kiwi was smaller than the others seen earlier, with
a greyish-white banding of the feathers. Gould named it *Apteryx oweni* in honour of
Richard Owen who had distinguished himself on the subject of kiwi anatomy some years
earlier, and plates of the new bird appeared in the *Transactions of the Zoological Society of
London* in 1848.[43]

By the late 1840s there had been a good supply of specimens from the North Island
of New Zealand, possibly because it was more densely settled and more frequently visited
than the South Island. In fact, apart from the recently described *Apteryx oweni*, virtually
all the specimens seen since Shaw's original one had come from the north, so at that
stage it was not particularly obvious that the birds from the two islands were in any
way different — particularly as the Shaw specimen was tucked away in the Earl of Derby's
collection. It was the Superintendent of the Zoological Society's gardens, A.D. Bartlett,
who eventually noticed the difference between the North and South Island birds, and
this was the result of being loaned one of the latter by Dr Gideon Mantell.

Mantell was the enthusiastic paleontologist who was to play an important role as a
middleman between his son Walter, in New Zealand, and the English zoological
establishment. It seems likely that this particular specimen was one of two that were
sold to Walter Mantell as a job lot by a sealer who had been working in Dusky Sound,
and which included the first known specimen of the *Notornis*, together with other skins.
Bartlett noticed the difference between this kiwi and the others that had been recently
arriving in England and was about to give it a new name when caution intervened and
he thought to borrow Shaw's specimen from the Earl of Derby's collection at Knowsley
Hall. He found this bird and the one supplied by the Mantells to be similar. He therefore
correctly surmised that all the North Island specimens, previously regarded as *Apteryx
australis*, must belong to another species of kiwi, which he proposed to call *Apteryx mantelli*
in honour of Walter Mantell's efforts in the cause of natural history in New Zealand.[44]
It may have been Bartlett's paper which lent substance to the belief that Shaw's kiwi
had come from 'Dusky Bay': in concluding, he remarked that J.E. Gray, the British
Museum's Keeper of Zoology, had said that Shaw's bird had come from this locality.
If there was ever any evidence for this conclusion of Gray's, it is now lost; the remaining,
albeit circumstantial evidence, suggests that it came from further south — possibly Stewart
Island.

In 1851, 40 years after the first kiwi was discovered, a live female specimen of *Apteryx
mantelli* was presented to the Zoological Society by Edward J. Eyre, who was then
Lieutenant-Governor of New Zealand. Gideon Mantell was an early visitor: 'saw the
live *Apteryx* just arrived from New Zealand: the poor bird was asleep coiled up in a ball
with its beak between its legs: on being disturbed it stood upright and kicked violently
uttering a hissing grunt, but did not attempt to strike with its beak.' They fed it by
burying pieces of mutton in a heap of soil, into which it plunged its beak. This particular

42. *Ibid.*, p.105.

43. *Ibid.*, pp.106, 108. In a
letter to the Reverend R.
Taylor (24.2.1850) Strange
says that his *Apteryx oweni*
came from the South
Island, but neither in his
catalogue nor later did he
add any further detail
regarding locality. It was
sent on the *Hamlet* in late
1847. See also Gould,
1847, p.93, and Gould,
1848a, p.379, Pl.57,
Gould, 1848, v.6 Pl.3.

44. Bartlett, 1852 (1850),
p.274. The accepted date
of publication of *Apteryx
australis mantelli* has been
1850. However the
Proceedings often appeared
later than the nominal
date of publication and in
the list compiled by
Waterhouse *(Proceedings,*
1893, p.436 *et seq.)* The
correct date of
publication of the above
description would appear
to be 1852.

J.G.Keulemans del.

T. Walter lith.

APTERYX HAASTII, POTTS. ♀

The Great Spotted Kiwi (*Apteryx haasti*) a hand-coloured lithograph drawn by J. G. Keulemans for G. D. Rowley's *Ornithological Miscellany*, 1876-78 (Vol. 1, pl. 3). Published by permission of the Alexander Turnbull Library. *(see p.77)*

bird lived for many years in the Society's gardens, laying its first egg in 1859 — in vain, it would seem, as it met its first male in 1865.[45] But it was most useful in providing new information on kiwi behaviour.

45. Bartlett, 1868, p.329; Mantell, 1940, p.278.

The first hints of a further species came from the sealers reporting to Frederick Strange in Sydney, who in turn passed the news on to John Gould. The sealers called the bird the 'Fireman' — why, is not very clear — and Strange's informants based their conviction of a new species on observation of eggs of two different sizes.[46] In the early 1860s Julius Haast, the provincial geologist of the province of Canterbury, found further evidence in the form of footprints from a large kiwi.[47] But actual discovery was still some way off, and it was not until the summer of 1871 that the Christchurch naturalist, Thomas Henry Potts, received a consignment of skins from a collector in Westland, including one of a new kiwi. A second specimen was obtained, thought to be from the ranges above Okarito, and Potts described his findings in the *Transactions and Proceedings of the New Zealand Institute* in 1872. Potts named his kiwi *Apteryx haasti*, the Great Spotted Kiwi, and noted that his collector had . . . 'literally slain his thousands of Apterygidae and through whose exertions colonial and foreign museums have been supplied with examples . . . .'[48]

46. Strange, 1847, p.51.

47. Haast, 1862, P.104.

48. Potts, 1872, p.205.

In 1891 Sir Walter Lawry Buller distinguished a large form of the kiwi from Stewart Island, which was renamed by the Honourable Walter Rothschild as *Apteryx lawryi*, taking Buller's name in honour of its discoverer.[49] Along with the North and South Island Brown Kiwis, the Stewart Island form was later relegated to a subspecies — one that was very difficult to differentiate from the neighbouring South Island subspecies — and therein lies the difficulty of determining the origin of the bird that began the kiwi story at the start of the century.

49. Rothschild, 1893, v.10, pp.61, 62.

By now it was clear that the Europeans, pastoralists and collectors alike, were accelerating a decline in kiwi numbers that would eventually reduce the birds to the status of rarities. Nocturnal and secretive, even the commonest of the species would seldom be seen outside of a museum or zoo. Almost from the time of the earliest discoveries, we have heard how the kiwis were hunted by the Maoris, with dogs and torches by night. There are several written accounts, but none so graphically told and at such length as that of Walter Buller which featured in the second edition of his book *A History of the Birds of New Zealand*:[50]

50. Buller, 1888, p.317.

'After having refreshed ourselves in the morning, we started on our first real Kiwi-hunt. We took a course down the side of the gully and were soon in a perfect labyrinth of supplejack (*Rhipogonum scandens*). These vines hung from the trees, ran along the ground, twisted around each other and crossed and recrossed, forming the most complete Chinese puzzle one could imagine, and so interlacing the underwood together that it was a matter of extreme difficulty to get through it even at a slow pace. Then when the little dogs took up the scent and disappeared down the gully it became necessary to follow quickly in the direction their bells indicated, so as to be "in at the death;" and then the hunt became as exciting as it was difficult — the kareao catching the feet and tripping one up or striking painfully across the shins — and so up and down, now swinging by a vine, now pushing on all fours through the tangle; forcing one's way through clumps of kiekie and dense beds of *Lomaria* down into the bottom of the ravine; then, as the scent led upwards, following the tinkling bells (the dogs being out of sight) up the tangled slope again, the course sometimes forming a complete circuit of the "field", and representing the erratic wanderings of the Kiwi upon the feeding ground the night before. Heated, out of breath, scratched in the face and hands, and with our shins aching from repeated contact with the kareao-vines, every now and then we halted to ascertain by the sound of the bells the position of the dogs, and then, full of excitement, resumed our novel chase again. At length, just beside a rough track on the hill-side, our dogs ran their quarry to earth, and began to tear with their paws at the opening to the "rua-kiwi". Calling the dogs off and closing in upon the spot, we drew from the cavity a fine male Kiwi, and then two vigorous young birds, all unharmed, but evidently much scared, and striking boldly with their claws. Our captives were soon secured in a Maori ket and we sat down to rest for a short time before taking up the scent again. I put my arm far down into the cavity, and found that, although the rounded entrance was just large enough to admit the bird, the chamber opened out inside, extending diagonally to a depth of about two feet, and wide enough at the bottom for the accommodation of two full-grown birds. I drew out the nest materials, consisting of shreds of kiekie-leaves and other dry litter, mixed with Kiwi feathers.

'We had not to hunt long before we came upon another bird, a fine adult female, and presumably the mate of the one we had just caught. She had taken refuge in a cavity under a rata-root, and one of the dogs, having unfortunately slipped his muzzle, killed the bird by breaking her neck. Other captures followed, and the aggregate result of the first day's hunt was ten Kiwis of all ages and one splendid egg.'

Chapter Six
THE FRENCH INTERVENTION

AT THE BEGINNING OF THE 1820s virtually no major voyage of exploration had visited New Zealand for three decades. This does not mean to say that New Zealand was neglected — on the contrary it was not — but Christianity and commerce were the preoccupations of those who came, and while the Church ministered to the indigenous population at one end of the country, the Sydney companies plundered the seals at the other. Traffic to and from Sydney was almost brisk, and along with the flax, seal skins, whale oil and Maori curios that went back to Port Jackson there were no doubt many bird skins and shells, perhaps even the odd butterfly. These were traded or sold to form a small but steady trickle of specimens that reached the English, European and American markets. Many an unsourced specimen in a European or American museum probably arrived this way, with only the occasional rarity, such as the kiwi, attracting much attention.

An example of this sort of collecting is seen in the records of the Sydney sloop *Snapper*, sent to collect flax on New Zealand's southern coast in November 1822. The sailors killed a large number of birds including 'linnets, pois, whattle birds and saddle-backs', and on their return to Sydney the following year their cargo (apart from flax) was said to consist of '. . . potatoes, skins of birds, articles of dress, and objects of curiosity belonging to the natives. The collection of birds was fairly large, the naturalists thinking that it contained many new kinds'.[1] But while the cabinets of a few collectors in museums may have been enriched by the efforts of sealers and flax gatherers such as these, there had been a lengthy hiatus in serious collecting and scientific scrutiny, the reasons for which are not hard to find among events taking place in Europe.

On the other side of the world the French had been preoccupied with the Revolution, the collapse of their Navy, and the Napoleonic Wars. The British position was no better. Ten years after the War of Independence, Britain went to war again, and for more than 20 years her economic development was warped by conflicts with the French.[2] The aftermath was almost worse: an economy artificially boosted by wartime conditions was suddenly exposed in all its deficiencies, and a deep and lengthy depression followed. Although the advance of science slowed in Britain it did not stop in mid stride; in the realm of natural history some branches, such as geology, even made progress.[3]

Zoology was not so fortunate. Its position was made worse by a succession of lacklustre Keepers of Zoology at the British Museum, who had failed to provide that much-needed focal point for collection and study: a substitute for Banks's collection. The unsatisfactory reigns of George Shaw (who in spite of his inadequacies played a small but important part in New Zealand zoology) and J. G. Children were interrupted only too briefly by that of the brilliant Leach.[4] Thus, even if the British had been disposed to sending further voyages of exploration to the South Pacific, it is doubtful that the resultant collections would have fared any better than they did in the previous century.

There were grounds for optimism however as the 1820s wore on. Social developments such as the advent of working-class movements, railways and new universities helped widen the base of education and science. More specifically, there was the emergence of new learned societies such as the Zoological Society of London (1826) and the

1. For this record we are indebted to Jules de Blosseville, midshipman on the *Coquille*, later to visit New Zealand. He had access to the diaries of Edwardson, captain of the *Snapper*, when he visited Sydney. McNab, 1907, pp.199, 206, 208.

2. Gregg, 1971, p.57.

3. Allen, 1976, pp.52-72.

4. Günther, 1980, pp.29, 37, 49, 55.

5. *Ibid.*, p.69.

6. Anon., 1910, v.12, p.391.

7. Richardson, 1846, p.72.

8. Owen, 1831, MS. 275-h-7, Royal College of Surgeons Library.

9. Limoge *in* Fox and Weitz, 1980, p.212.

10. See Note 8 above.

11. Limoge *in* Fox and Weitz, 1980. p.222.

12. *Ibid.*, p.222.

13. Swainson, 1840, p.73.

14. Dunmore, 1969, p.50.

15. *Ibid.*, p.109.

Entomological Society (1833)[5] and the employment, in 1824, of a promising assistant and future Keeper of Zoology, John Edward Gray. An habitué of Joseph Banks's Library and a pupil of Dr W. E. Leach, he would eventually rescue the collections and put the Museum on an equal footing with those elsewhere.[6]

The drought was eventually broken by Beechey's voyage in the *Blossom* in the northern Pacific, and later by the voyages of FitzRoy and Belcher. But it took some time to recover, and even as late as 1843 Sir John Richardson contrasted the meagre British collections with the 'magnificent accumulations' of the French, noting rather bitterly that many of their most prized specimens 'have been obtained on the coasts or in the very harbours of the English colonies'.[7]

In spite of the Revolution and the wars that followed, zoological science in France was initially much better off. Of all the establishments bearing the impress of royalty the Jardin du Roi suffered the least from the attentions of the Revolution. It was said that it survived by putting about the idea that its botanical endeavours were wholly concerned with medicinal plants and the chemical laboratory was involved in the manufacture of saltpetre for gunpowder.[8] Whatever the real reasons, the revolutionary authorities went along with the proposals for the official inauguration and development of the Muséum d'Histoire Naturelle,[9] although some of the museum's personnel were not quite so lucky and were compelled to flee Paris or to contrive some hasty subterfuge. Daubenton, originally an associate of Buffon and who had been experimenting on the growth of wool, escaped execution only by being represented to the Committee on Public Safety as a shepherd.[10] Ultimately some of those who fled Paris returned to continue their work, and in the four decades that followed the turn of the century the Museum entered upon its 'golden age'.[11]

Already favoured by Buffon and then buttressed by the celebrated figures of Lamark and Cuvier, the scientific traditions were ably carried on by the zoologists Geoffroy St Hilaire, Henri Ducrotay de Blainville and Constant Duméril; the Museum went from strength to strength, enjoying a status and financial support unequalled by any comparable institution in Europe.[12] In 1840 Swainson, the English naturalist shortly to embark for New Zealand — although at that stage in his thwarted career, not the most unbiased of observers — remarked that the Museum was '. . . the most celebrated in the world . . .'.[13] But by then the pendulum was beginning to swing the other way.

Fortuitously coincident with the early rise in the stature of the Museum was the economic resurgence of France and a need to reassert herself in the world — a circumstance that would once again allow the contemplation of voyages of exploration to the South Pacific.[14] Thus, not for the first occasion, the discovery and development of New Zealand natural history was about to be shaped by political, economic and scientific events abroad, this time with the rich irony that it was not the winner of the conflict between Britain and France who was poised to take the initiative, backed by an adequacy of economic and scientific resources, but the loser. The initiative was not quickly taken however with regard to New Zealand and more than two decades of the 19th century would pass before the first French expedition was solidly in view; some of the early impetus and enthusiasm was by then inevitably lost.[15] In the meantime, the country received a visit from voyagers from another and rather unexpected quarter.

On the face of it, it seems surprising that the Russians should suddenly take an interest in that distant part of the world, but the remnants of the influence of Empress Catherine II, the sometimes Napoleonic visions of Emperor Alexander, the development of settlements in Eastern Russia and deep interest and respect for the achievements of Cook and the British Navy have all been considered to have fired the need for this expedition.[16] The instructions issued to Captain Fabian Gottlieb von Bellingshausen, who was to lead it, stated that '. . . in zoology all relevant observations should be made and, where possible, specimens collected'.[17] Unfortunately the appointed naturalists failed to meet up with the ships in Copenhagen, so their efforts in this department were virtually crippled, much to Bellingshausen's disgust.

The *Vostok* under Bellingshausen and *Mirnyi* under Lazerev reached Cook Strait in late May, 1820. Their passage through the Strait was marked by observations of various sea birds (including a penguin) and some luminescent jellyfish, and two 'green parrots' which alighted on the *Vostok* and delighted the crew with their antics.[18] Anchored in Queen Charlotte Sound on 29 May they were almost immediately visited by Maoris who traded 250 pounds (113 kg) of fish. Siminov, the voyage astronomer, recorded flounder, mackerel and cod in the catch.[19] They appreciated the bird song that had entertained previous visitors to Ship Cove, saw some paua while ashore, and shooting parties took a few shags and some smaller birds, including a Tui. They left on 4 June — a short stay considering their unruffled exchanges with the Maoris — taking with them a purchase of two Maori dogs, thought to be part of the native fauna. Identifiable birds included the New Zealand Pigeon, a Parakeet and the New Zealand Falcon among ducks and various sea birds noted by Nikolai Galkin, the *Mirnyi*'s surgeon.[20]

They returned to Kronstadt on July 24, 1821, little over two years after their departure. It was a highly regarded voyage in most respects[21] and in spite of the absence of naturalists some drawings of birds and marine mammals were made by the expedition artist Paul Mikhailov, and some natural history collecting took place. There was little of zoological account from New Zealand, although from Macquarie Island in the south they collected the Macquarie Island Shag.[22]

Meanwhile back in France there was renewed interest in the Pacific as a means of furthering trade and national ambitions; although the voyages of Baudin and Freycinet were an indication that science was firmly in the minds of those who promoted and undertook exploration — even as an underpinning of commercial interests. As the century moved into the 1820s some of the impetus for Pacific exploration declined and for a time the remaining momentum was to swing even further in the direction of scientific research.[23] Expeditions even experimented briefly with carrying civilian scientists: Baudin's was on a grand scale with a corps of 22 civilians acting in various roles, and it seems that part of the disastrous muddle that was the hallmark of this voyage can be blamed on the size of this group and its personalities.[24] The lesson was not forgotten and on subsequent voyages the scientific duties were carried out by naval officers, some of whom had considerable skills in the natural and other sciences. Furthermore, the depth and breadth of talent among the officers on any one ship was often considerable. There are contrasts here with the earlier British expeditions.

So it was not surprising that when the participants of Louis-Isidore Duperrey's expedition to the South Pacific began to assemble, a number of them possessed an appropriate amalgam of naval and scientific experience. Duperrey himself was a competent hydrographer, and ultimately a member of the Académie des Sciences; Dumont d'Urville, who sailed as second-in-command, had trained in the natural sciences, was interested in botany, and would assist with entomology. The duties of Charles-Hector Jacquinot were in astronomy,[25] and Jules de Blosseville, a junior officer, would also be mentioned in connection with animal collections. Finally there were the two naturalists: René Primevère Lesson and Prosper Garnot, who acted as surgeons, although one of them,

16. Debenham, 1945, v.1, pp.xi,xii.

17. *Ibid.*, p.79.

18. *Ibid*, p.199. Barratt, 1979, p.55.

19. Siminov, *in* Barratt, 1979, p.51.

20. *Ibid.*, p.55 and Galkin *in* Barratt, 1979, p.70.

21. Debenham, 1945, v.1, p.xi.

22. Fleming, 1982a, p.129.

23. Dunmore, 1969, p.110.

24. Although in zoological terms it was ultimately quite successful. *Ibid.*, p.38.

25. He would later assist with entomology and with his younger brother Honoré would see through much of the published work of the third voyage.

Lesson, would earn a lasting reputation as an ornithologist.

These men sailed on the *Coquille* from Toulon harbour on 11 August, 1822. With the benefit of previous experience and sound preparation they progressed without major incident around Cape Horn, up the coast of South America, across the Pacific, around New Guinea and Australia to Sydney, where they arrived on 17 January, 1824. Here Garnot left the ship to return home, weakened by a debilitating dysentery contracted in South America. Further misfortune overcame him when he was shipwrecked en route to France, with the loss of part of his valuable natural history collection,[26] barely escaping with his life. But he survived to share with Lesson the task of publishing the natural history of the voyage. The loss of Garnot's services to the voyage was not as serious as it might have been, as collecting was shared with the other officers and Lesson retrained some of his assistants to help fill the gap.

They left Sydney in the company of a missionary, the Reverend George Clarke, with his wife and son, and two Maoris returning to New Zealand.[27] To the discomfort of everybody, including their passengers, the *Coquille* battled gales in the Tasman Sea for nearly two weeks before they sighted Doubtless Bay on 3 April, 1824, and anchored at the Bay of Islands. The presence of their passengers, and the sight of anchored whaling vessels, must have been a reminder, if they needed it, that this was no longer some little-known corner of the world. The aura of the potentially exotic southern continent had long since disappeared and the coastal reaches had become increasingly familiar to traders and missionaries. Only the interior retained some air of mystery.

There were friendly overtures from the Maoris from the very beginning, with some already showing the signs of corrupting European influences and becoming constant shipboard companions. Handsome offers were made for French guns and even their antiquated piece issued for the collection of natural history specimens was valued at 20 pigs, but the French refused to part with it.[28]

The time spent in New Zealand was not going to be long, so as often as the weather would let him, Lesson explored his immediate surroundings, observing the Maoris and collecting plants and animals. When confined to the ship he was able to prepare his specimens with the help of his medical orderly, whom he had trained in the art of preparing birds. Lacking a natural history artist, he was obliged to do his own sketches, particularly of delicate specimens that would perish during transport back to France.

Birds he shot when out in the company of surveying parties, or else he was supplied with specimens which fell to the guns of his fellow officers or Rolland, the master gunner, who collected for Lesson when the naturalist was detained elsewhere. Birds were still sought as additions to their diet and the New Zealand Quail and Pigeon were highly regarded. It was possibly an acknowledgement of the superior skills of the locals that they relied upon the Maoris for fish, which were brought on board or picked up as a result of a chance encounter, as was the case with the newly-caught Sea-horse given to them as they were walking back along the beach to the corvette.[29] Although the Maoris provided them with most of their fish, they collected the shells themselves, either on the sandy beach at Moturoa Island, or on excursions such as the one led by the junior midshipman, de Blosseville, when in search of pigs for the ship's provisions on 9 April. Accompanied by Gabert, the purser, and Lejeune, the expedition artist, they set out for Kerikeri by boat. On their way they were received by the missionaries King and Hall at Rangihoua before moving on to Kerikeri. There their former passenger George Clarke and another missionary, James Kemp, had established a mission station, complete with livestock — cows, goats and a horse — that had acclimatised well.[30] In exploring

26. Stresemann, 1975, p.137.

27. Dunmore, 1969, p.140.

28. Lesson, *in* Sharpe, 1971, p.58.

29. *Ibid.*, p.70.

30. de Blosseville, *in* Sharpe, 1971, p.114.

The Spotted Stargazer (*Genyagnus novaezelandiae*) collected by R. P. Lesson on the voyage of the *Coquille* to New Zealand; a hand-coloured engraving from Duperrey's *Voyage autour du monde . . . Atlas . . . Zoologie* 1826, *Poissons* (pl. 18). Published by permission of the Alexander Turnbull Library. *(see p.85)*

the falls from the nearby Kerikeri River,[31] de Blosseville found some small shells, while Gabert found a large land snail under the trees that bordered the river. This last was the Flax Snail[32] and was one of the first land snails to be described from New Zealand. Otherwise their collection of shells was oddly biased towards slipper shells and limpets, speaking more for their casual interest in this group, than the scarcity or the variety of molluscs. Indeed, Lesson was to recall later that a rich variety of shells awaited those who would come after them.[33]

In their ramblings about the bays and inlets they observed other invertebrate animals — crayfish, shrimp, crabs, starfish, sea eggs and others. Lesson claimed that he saw no insects, and it is possible that the bracken-covered slopes of the Bay of Islands were not particularly productive at that time of year, although one cannot help thinking that a little more effort might have produced something for the entomologists.[34] But in spite of these deficiencies, there was a useful accumulation of around 50 species overall (including duplicates from this locality, many of which were preserved for return to the Muséum d'Histoire Naturelle).

On 17 April, 1824, they left the Bay of Islands and headed into the Pacific where

31. Possibly the Waianiwaniwa or Rainbow Falls.

32. *Placostylus hongii.*

33. Lesson, *in* Sharpe, 1971, p.84.

34. Although the Tiger Beetle was later recorded from New Zealand with an illustration in the Zoological Atlas.

35. Southern Royal
Albatross *(Diomedea
epomophora epomophora).*
Fleming, 1982a, p.74.

36. Dunmore, 1969,
p.253.

Part of R. P. Lesson's list of animals he collected in New Zealand. Published by permission of the Bibliothèque Centrale, Muséum National d'Histoire Naturelle (MS 354). (see p.85)

The Flax Snail (Placostylus hongii) *collected by Gabert on the voyage of the* Coquille *to New Zealand; a hand-coloured engraving from Duperrey's* Voyage autour au monde . . . Atlas . . . Zoologie, *1826,* Mollusques *(pl. 7). Published by permission of the Alexander Turnbull Library. (see p.83)*

somewhere not far to the north of New Zealand they took a new species of albatross.[35] It was to be 24 March of the following year before they saw France again, but the smooth running of the voyage continued, unmarred by shipwreck or large-scale illness among the crew. At the end there was a handsome total of specimens: 264 birds and mammals, 63 reptiles, 288 fish and — to demonstrate their poor or non-existent collection of New Zealand insects was not entirely the result of apathy — 1200 insects.[36]

With their safe return and the scientific objectives of the voyage achieved, there was the expected official and public approval. After this one would normally look for the same delays that attended the scientific results of the earlier British voyages. But the quality, scope and promptness of publication, which was to be a hallmark of the French expeditions, was here evident. With the approval of King Charles X, publication commenced almost immediately and in 1826 the handsomely illustrated zoological atlas appeared, including hand-coloured engravings of eight New Zealand species[37] in which the artists used the preserved specimens or redrew the animals from Lesson's sketches. (Engravings of New Zealand animals were to be relatively uncommon in the 19th century as this process was about to be replaced by lithography as the preferred means of illustration.) In 1828 came the first part of the supporting text, written by Lesson and Garnot, while the remaining parts came later. Most of the scientific accounts — botanical, hydrographical, and zoological — were completed before the 1830s with only the final part of the zoological text delayed until 1838. Remarkably it was only Duperrey's narrative of the expedition which remained unfinished — his account ends at Chile, thus omitting New Zealand.

Returning to the zoology: more than half the birds brought back from New Zealand were new, but only a proportion of them were recognized as such by the authors. Some of these were North Island subspecies of birds already recorded in the South Island (the Fantail, Robin and Saddleback) and one or two, like the Fernbird, were recorded for the first time, but somehow overlooked by Lesson in publication. However the Whitehead and Southern Royal Albatross were new species attributable to Lesson and the White-headed Petrel was described for the first time by Garnot. The Kingfisher and Pied Tit were also new subspecies.[37, 38] Only a few fish were new. The Sea-horse and the Red Gurnard were first described in this work; the latter had been illustrated by Sydney Parkinson years earlier but somehow never given a scientific description and name until now. The Goatfish was ultimately described from a specimen collected by Lesson at the mouth of the Kerikeri River.[39] No reptiles were described in detail from New Zealand, but there is a tantalizing reference to a striped skink with a "blue" tail, a description which does not easily tally with anything currently known from New Zealand and it was possibly something seen elsewhere.

Contrary to Lesson's glowing and much later recollection there were no rarities in their shell collection and only 10 or so New Zealand species were described in the

37. Specimens collected or observed by Lesson and others at the Bay of Islands:
Mammals: Dog (this was the so-called 'Maori dog'); Rat (species not known).
Birds:
*†Southern Royal Albatross *(Diomedea epomophora epomophora)*, (at sea to North),
Red-billed Gull *(Larus novaehollandiae scopulinus)*,
*White-headed Petrel *(Pterodroma lessoni)*,
†Black Shag *(Phalacrocorax carbo novaehollandiae)*,
†Little Shag *(P. melanoleucos brevirostris)*,
Pied Shag *(P. varius)*,
Reef Heron *(Egretta sacra)*,
White-fronted Tern *(Sterna striata)*,
†North Island Kaka *(Nestor meridionalis septentrionalis)*,
Red-crowned Parakeet *(Cyanoramphus novaezelandiae novaezelandiae)*, Yellow-Crowned Parakeet *(C. auriceps auriceps)*,
Tui *(Prosthemadera novaseelandiae)*,
†New Zealand Quail *(Coturnix novaezealandiae)*,
•Bellbird *(Anthornis melanura melanura)*,
*†New Zealand Kingfisher *(Halcyon sancta vagans)*,
Long-tailed Cuckoo *(Eudynamis taitensis)*,
†North Island Fantail

(Rhipidura fuliginosa placabilis),
New Zealand Pipit *(Anthus novaeseelandiae novaeseelandiae)*,
•*†Pied Tit *(Petroica macrocephala toitoi)*,
•*†North Island Robin *(Petroica australis longipes)*,
*†Whitehead *(Mohoua albicilla)*,
•*†North Island Saddleback *(Philesturnus carunculatus rufusater)*,
New Zealand Pigeon *(Hemiphaga novaeseelandiae novaeseelandiae)*,
†North Island Fernbird *(Bowdleria punctata vealei)*,
†[North Island Brown Kiwi] feathers or part skin only *(Apteryx australis mantelli)*,
?†North Island Rifleman *(Acanthisitta chloris granti)*, (uncertain as referred to under a N. hemisphere name).
Pukeko *(Porphyrio porphyrio melanotus)*, (mentioned by name but confused with Kingfisher. *Zoologie* v.1(2) p.415).
† Undescribed at time of observation or collection; including several subspecies later overlooked.
* Described by Lesson or Garnot, Bay of Islands type locality.
• Illustrated in zoological atlas of the voyage.
Reptiles:
Only one (unidentified) skink 'striped with a blue

tail' possibly a *Leiolopisma moco* or *L. smithi*.
Fish:
Variable Triplefin *(Trypterygion varium)*,
Leatherjacket *(Parika scaber)*,
Snapper *(Chrysophrys auratus)*,
•Spotted Stargazer *(Genyagnus novaezelandiae)*,
Spotty *(Pseudolabrus celidotus)*,
Trevally *(Caranx georgianus)*,
†Goatfish *(Upeneichthys)* *lineatus* (or *porosus)*,
•*†Red Gurnard *(Chelidonichthys kumu)*,
*†Seahorse *(Hippocampus abdominalis)*,
Blue cod *(Parapercis colias)*,
New Zealand Mackerel *(Trachurus novaeseelandiae)*,
Also unidentified are a bully, what appears to be a young Kahawai, and a freshwater fish, possibly a *Galaxias*.
(•*† see under *Birds* above.)
Insects:
Only one recorded, Tiger Beetle *(Neocicindela tuberculata)*, from New Zealand and Port Jackson. Illustrated in Atlas, Pl.1 Fig.4, so presumably collected.
Molluscs: *Smooth Slipper Shell *(Maoricrypta (Zeacrypta) monoxyla)*,
*Circular Slipper Shell *(Sigapatella novaezelandiae)*,
Small Circular Slipper Shell *(Zegalerus tenuis)*,

Radiate Limpet *(Cellana radians)*,
Fragile Limpet *(Atalacmea fragilis)*,
•Pipi *(Paphies australis)*,
Cooks Turban *(Cookia sulcata)*,
Silver Paua *(Haliotis australis)*,
•*Flax Snail *(Placostylus hongii)*,
(•* see under *Birds* above)
Paphies, a new genus erected on the basis of *P. australis*, by Lesson. A brachiopod (*Zoologie* v.2(1) p.421) *Terebratella sanguinea* is also referred to. The Imperial Sun Shell is apparently referred to in error — see Quarmby, *in* Sharpe, 1971, p.84.
Other invertebrates:
Crustaceans, including crayfish, shrimps, portunid and grapsid crabs, a *Squilla*; Annelids (tube-worms), starfish and coelenterates — but none identified.

38. Lesson, MS., Collections conservées dans l'alcohol: Mollusques et Crustacés, Ornithologie; Collections rapportées du Brésil, de Tahiti, de Chile, de la Nouvelle Zélande, de la Nouvelle Hollande, de la Nouvelle Irelande. Muséum National d'Histoire Naturelle, Bibliothèque Centrale, MS. 354.

39. Cuvier and Valenciennes, 1829, v.3, p.455; Lesson, 1830, v.2, p.217.

The Bellbird (*Anthornis melanura melanura*) collected by R. P. Lesson on the voyage of the *Coquille* to New Zealand; a hand-coloured engraving from Duperrey's *Voyage autour du monde . . . Atlas . . . Zoologie* 1826, *Oiseaux* (pl. 21). The birds in this atlas look wooden compared with later work in which the birds appeared more lifelike and backgrounds were often sketched in. Published by permission of the Alexander Turnbull Library. *(see p.85)*

zoological record. Among them Lesson was able to recognize two new genera, represented by a Slipper Shell and the Common Pipi.[40] But the Flax Snail stands out as the most interesting of their discoveries among this group of animals. The only New Zealand insect collected and illustrated was a Tiger Beetle and it was possibly this specimen, or a duplicate of it, that was acquired by a private collector, le Compte Dejean, as it appears in a catalogue of his collection published in 1826.[41] In their attempt to make some analysis on New Zealand faunal relationships they found that in spite of the proximity of Australia and New Zealand, their faunas, with the exception of marine fish, were unrelated. What is more, Latreille, a prominent zoologist who worked on the voyage insects, regarded the geographical affinities of New Caledonia and New Zealand as American, but the zoological affinities as Asian.[42]

This was the first of a series of important French voyages to New Zealand and one is inclined to overlook the fact that on this expedition the time spent in the country was comparatively short, in an area not noted for the richness of its fauna. Under the circumstances their efforts were commendable. Although it was later remarked that they paid some attention to invertebrates,[43] it was obvious that these were largely secondary to Lesson's interest in ornithology. The next voyage however was to show a difference in the collecting programme, with the emphasis on the smaller and less spectacular animals.

It was little over a year later, on 25 April, 1826, that *La Coquille*, renamed *L'Astrolabe*, sailed again from Toulon — this time with Dumont d'Urville in command. They were to search for La Pérouse, missing since 1788 on a voyage to the Pacific, as well as undertaking scientific and hydrographic work. The focus of the voyage was to be on the islands of the Western Pacific; New Zealand was a later addition.[44] The naturalists, apart from d'Urville himself, were the Naval Surgeons Jean-Renée-Constant Quoy and Joseph Paul Gaimard, both seasoned from an earlier voyage under Freycinet in the Pacific. In addition, Quoy was an accomplished artist who would provide numerous sketches or watercolours of plants and animals. They were assisted by other officers with interests in scientific research: the Lieutenants Victor-Charles Lottin (who collected insects), Charles-Hector Jacquinot and Pierre-Adolphe Lesson, who collected for his brother, the naturalist on the previous voyage.[45] Once again, preparations were thorough; some idea of their extent can be gained from the four to six tonnes of apparatus and chemicals that were put aboard. There were 600 litres of spirits of wine, hundreds of wide-mouthed jars with stoppers, lead tags, numbered punches, cotton for stuffing, magnifying glasses, boxes made of tin and wood, devices for catching insects, pins, scalpels — the list goes on.[46] Gaimard, at d'Urville's direction, visited museums in England and Holland to study their natural history collections and to note any gaps in the French collections that might be filled during their voyage.

When New Zealand was sighted on 10 January, 1827, it was the north-west coast of the South Island from which they sailed north, turning into Cook Strait. At this point a drag-net cast from the ship yielded nothing. Five days later Quoy and Gaimard went ashore for the first time at Separation Point, which divides Golden and Tasman Bays, but, with the sailors, took away little save a few artifacts found in some huts on shore, earning d'Urville's anger for this injudicious souveniring.[47] On 16 January the *Astrolabe* found a sheltered bay, called Astrolabe Bight by the French, where they anchored and put ashore the following morning, having satisfied themselves that the Maoris had friendly intentions. In the company of Lesson and one of the sailors, d'Urville entered the bush that fringed the bay, where they were surrounded by birds; Tuis and Fantails among them, and seven or eight species were shot. A rat was seen, presumably the 'native' rat,

40. Quarmby, *in* Sharpe, 1971, p.84; *Paphies* and *Sigapatella* (see note 37).

41. Miller, 1956, p.92.

42. Latreille, *in* Lesson and Garnot, 1828, v.1, p.30.

43. Hombron and Jacquinot, *in* Dumont d'Urville, 1841, v.1, p.24.

44. Dunmore, 1969, p.180.

45. Lesson to Swainson, 28.9.1828, Swainson correspondence, Linnean Society.

46. Quoy and Gaimard, 1833, v.4, pp.379-81.

47. Wright, 1950, p.73.

but apart from a Tiger Beetle they saw no insects. Further exploration followed the next day, the apparent emptiness of the bush and forest a source of fascination to d'Urville even though he brought back several birds to the ship, including a specimen of the South Island Kaka. On 20 January, while the naturalists explored the adjacent Torrent Bay, d'Urville and an assistant shot more than 40 birds including a New Zealand Pigeon, two South Island Kokakos and several Tuis. Apart from an unidentified reddish insect, possibly a red spider wasp, only a few cicadas and crickets caught his attention but he did not bother to collect them.[48]

Although d'Urville had thought the place rather empty, the naturalists had been busy; when they left even he acknowledged that their collections were 'enriched every day by most interesting material'. Gaimard wrote: 'the chief birds that we have found during our stay in Astrolabe Bight are the following: The orange wattled crow, . . . the black oyster catcher, the pied oyster catcher and the small penguin, as well as several new species — the morepork, the south island thrush, the grey warbler, the tit, the rifleman, the fern bird, and the yellowhead. The molluscs, which were more numerous, were painted while still alive by M. Quoy . . .'.[49] In fact only two of the birds were new, the Grey Warbler and the South Island Fernbird, but they discovered a native land slug for the first time and found many new species of marine molluscs, such as the Brown Top Shell, Snake-skin Chiton, the Rock Borer and the Turret Shell, as well as marine slugs, starfish and sea eggs.[50] Collecting of invertebrate animals was beginning for the first time in earnest.

48. *Ibid*, p.82.

49. *Ibid.*, pp.83, 210. In the translation Wright used the common New Zealand bird names, rather than a literal translation of the French.

50. Specimens collected by Quoy and Gaimard and others on Dumont d'Urville's second voyage to New Zealand:

Mammals:
• Common Dolphin *(Delphinus delphis)*, They also took specimens of the New Zealand dog and rat (not native).
Birds:
South Island Thrush *(Turnagra capensis capensis)*,
South Island Saddleback *(Philesturnus carunculatus carunculatus)*,
South Island Rifleman *(Acanthisitta chloris chloris)*,
South Island Kaka *(Nestor meridionalis meridionalis)*,
• *South Island Fernbird *(Bowdleria punctata punctata)*,
South Island Kokako *(Callaeas cinerea cinerea)*,
Brown Creeper *(Finschia novaeseelandiae)*,
• *Grey Warbler *(Gerygone igata)*,
Yellowhead *(Mohoua ochrocephala)*,
• *New Zealand Quail *(Coturnix novaezealandiae novaezealandiae)*,
• *Wrybill *(Anarhynchus frontalis)*.

(* New species and • illustrated in atlas of zoology.) They also observed morepork and a rifleman.
A 'Lamprotornis' was attributed to New Zealand in error.
Fish:
Banded Kokopu *(Galaxias fasciatus)*,
Red Gurnard *(Chelidonichthys kumu)*,
Snapper *(Chrysophrys auratus)*,
Leatherjacket *(Parika scaber)*,
Kahawhai *(Arripis trutta)*,
Porcupine fish *(Allomycterus jaculiferus)*,
(none of these fish illustrated in the Atlas.)
Molluscs:
Some 80 species of molluscs were recorded, many new. The following is a selection of more common species, described for the first time by Quoy and Gaimard:
• Snakeskin Chiton *(Sypharochiton pelliserpentis)*,
• Slit Limpet *(Emarginula striatula)*,
• Brown Top Shell *(Trochus tiaratus)*,
• Turret Shell *(Maoricolpus roseus)*,
• Necklace Shell *(Tanea zelandica)*,
• Spiny Murex *(Poirieria zelandica)*,
• Octagonal Murex *(Muricopsis octagonus)*,
• *Oyster borer *(Lepsiella scobina)*,
• Siphon Whelk *(Penion dilatatus)*,
• Pink Turrid *(Phenatoma rosea)*
• Small Volute *(Leporemax fusus)*,
• Large Dog Cockle *(Glycymeris laticostata)*,
• Long Trough Shell *(Longimactra elongata)*,
• Scimitar Shell *(Zenatia acinaces)*,
• Rock Borer *(Myadora striata)*,
A• salp, a land slug *(Athoracophorus bitentaculatus)*, several marine slugs (including • *Onchidella patelloides* and • *O. nigrican)*,
Also an • Octopus was among the other molluscs recorded from New Zealand.
Crustaceans:
Among several crabs taken were: *Eurynolambus australis, Leptomithrax australis,* together with porcellanid and grapsid crabs and crayfish.

Insects:
• Chorus Cicada, *(Amphisalta zelandica)*,
Magpie Moth *(Nyctemera annulata)*,
Coleoptera:
Aphanasium australe; Hexatricha heteromorpha; Neocicindela tuberculata (in Accession list, Catalogue des Animaux articulées Crustacés, Arachnides, Insectes, v.1, 1829).
Insects discovered earlier (e.g. the Tanguru Chafer and Devil's Coachhorse) were collected on this voyage and illustrated in the Atlas.
Other Invertebrates:
• Soft corals Dead Man's Fingers, *(Alcyonium lauranticum)*,
• Fan coral *(Flabellum rubrum)*,
Cup coral *(Culicia rubeola)*, These corals were new species. Also they took sea anemones, salps, sea squirts (e.g. *Asterocarpa coeulea)*; sea eggs etc. and a Brachiopod, *Terebratella haurakiensis* (mis-identified as *T. sanguinea)*.
• illustrated in Atlas.

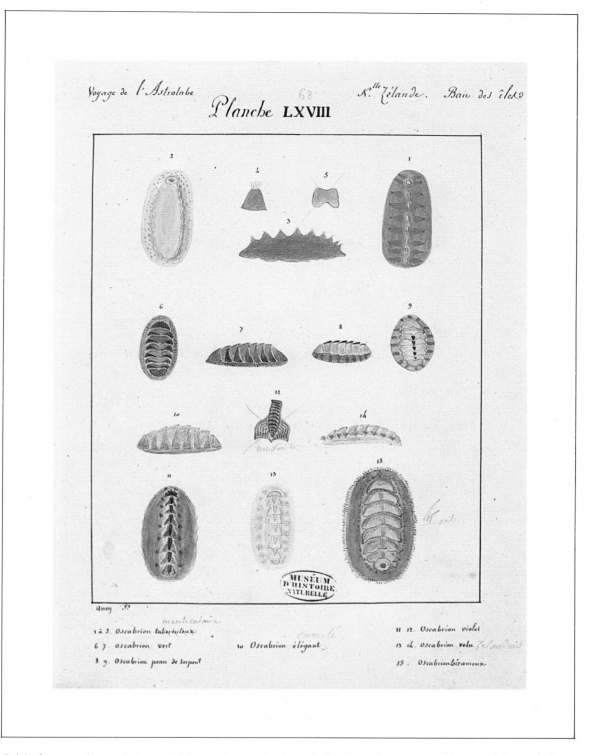

Original watercolour paintings of chitons taken at the Bay of Islands on the voyage of the *Astrolabe*, made by J. R. C. Quoy, expedition naturalist. Published by permission of the Bibliothèque Centrale, Muséum Nationale d'Histoire Naturelle (MS 105). *(see p.88)*

Leaving Astrolabe Bight they headed north-east towards some of the more eventful sailing of their voyage. Their hazardous navigation of the notorious French Pass, as it was later called, is well known and to reassure the nervous sailors d'Urville put the naturalists and the expedition artist ashore to foster the impression that all was normal. Some collecting must have taken place, as two birds, a New Zealand Thrush and a Kokako,

The Common Dolphin (*Delphinus delphis*) collected by J. R. C. Quoy and J. P. Gaimard on the first voyage of the *Astrolabe* to New Zealand; a hand-coloured engraving from Dumont d'Urville's *Voyage de la corvette Astrolabe . . . Atlas . . . Zoologie*, 1834, *Mammifères*, (pl. 28). Published by permission of the Alexander Turnbull Library. *(see below)*

are recorded from the Bay of Currents, immediately preceding French Pass. The subterfuge worked, and although all their skills of seamanship were called into play, they successfully navigated the passage and spent the last days of January passing through Cook Strait to head up the east coast of the North Island.

Following a brief and unproductive pass at Palliser Bay they sailed towards Castle Point, seeing petrels, albatrosses, terns and gannets south of Riversdale. Continuing north into Hawke Bay around the Mahia Peninsula, and across Poverty Bay to Gable End Foreland, they caught a dolphin that was considered unusual, and even later described as a new species, but which in fact was the Common Dolphin, *Delphinus delphis*.[51] It was taken on board for further study, and meat from the unfortunate animal was given to two Maoris who had joined the ship at Palliser Bay.[52] A shark was caught on the evening of the same day and suffered a similar fate. It was 4 February, 1827. The day following was spent at Tolaga Bay where during some friendly exchanges with the Maoris, d'Urville was able to find out more about the kiwi. He was probably already aware of the work of Shaw, Lesson, and other writers on the subject, and a continuing scarcity of complete specimens encouraged further enquiries into this mysterious bird's existence.[53] The naturalists went ashore and collected some plants and Gaimard recorded paua and a plover — perhaps the New Zealand Shore Plover or one of the dotterels.[54] Tolaga Bay was left behind on 6 February and d'Urville paused to reflect on what future New Zealand scholars would make of their finding that New Zealand was almost devoid of useful animals.[55]

By the 8th they were off East Cape where the sailmaker shot a gannet and two kingfishers and a boat was put off to pick up the dead birds. The kingfisher was probably the same species described by Lesson from the previous voyage, the gannet a representative of the large colonies that live on the North Island East Coast. Following a battle with a severe storm off East Coast they crossed the Bay of Plenty and passed the volcanic White Island wreathed in smoke and steam, to encounter an even more violent storm which almost brought an end to the voyage. But once again the crisis passed and they traversed the Hauraki Gulf to Whangarei where d'Urville went ashore, noticing the absence of beetles and butterflies and seeing only 'locusts, crickets, bugs, and flies, etc.'[56] The birds were shy, but the delicious Rock Oysters made up for these deficiencies.

Their next port of call in the Hauraki Gulf was the northern shore of Waitemata

51. Quoy and Gaimard, 1830, v.1, pp.149-151. *Atlas of Zoology*, Pl.28.

52. Wright, 1950, pp.115-116.

53. Smith, 1909, pp.130-139.

54. Wright, 1950, p.213. The New Zealand Shore Plover, *Thinornis novaeseelandiae*; Dotterels, *Chadrius* spp.

55. *Ibid.*, p.128.

56. *Ibid.*, p.149.

Harbour, opposite the site of the future Auckland. Then it was barely populated, but on the shore they shot some birds, including two new ones: the Wrybill and the New Zealand Quail – the latter would become extinct not many years later. These, added to the sea anemones, shells, and some fish,[57] made up their zoological collections from this area. The remainder of their journey up the north-east coast of the North Island was without major zoological significance and they reached the Bay of Islands on 12 March, 1827.

Here d'Urville visited Henry and William Williams at the Mission Station at Paihia after which the latter guided them around the Bay of Islands, demonstrating the local timbers, visiting Maori settlements and describing their culture. In the process, they collected some mud snails[58] in the estuary of a river. Generally however the Bay of Islands was to prove not particularly productive: the Red Gurnard was collected there and so were sea anemones and shells, including the Large Dog Cockle and the Small Volute. They did however discover a freshwater fish, the Banded Kokopu, which was drawn by Quoy and a copy later used by Cuvier for description.[59] It was only the second time that a freshwater fish had been recorded in detail by European scientists.

It was in the Bay of Islands that d'Urville discovered that his colleague and friend, the Sydney botanist Allan Cunningham, had left New Zealand only three months earlier. His visit, during which he explored White Island in the Bay of Plenty with the Williams brothers, was an early indication of scientific interest in this embryonic colony from a source much closer than Europe.[60]

The vague stories of large reptiles that had emanated from previous New Zealand informants took on more substance when William Williams told them of some 'fairly large lizards' seen on the offshore islands of the Bay of Plenty.[61] We can be sure that observations of Tuataras had already been made by Europeans, and it was possibly on this voyage that the French acquired a specimen of yet another large reptile which still remains a fascinating zoological riddle. In the museum at Marseilles is housed the only specimen of the world's largest gecko, 60 cm in length, and it is believed to have come from New Zealand.[62] If indeed this were so, how and when did it get to Marseilles? One of several possible answers relates to the early French expeditions, such as this one by the *Astrolabe*. She and the *Coquille* before her had docked at Marseilles to offload their natural history collections for Paris before finally moving on to the naval base at Toulon. Amongst the thousands of specimens that left the ship, one or two might have found their way into the hands of local people.

We can only speculate as to who might have given the French the gecko. The specimen is stuffed, suggesting that it might have been in that condition when presented to them in New Zealand; if it had been received alive it would more likely have gone into spirits.[63] William Williams was one of the few Europeans in New Zealand likely to have collected and prepared the animal as he had more interest in natural history than his colleagues at that time, and had a medical background that would have enabled him to prepare the specimen properly. On the other hand, it would have to have been smuggled onto the ship, since d'Urville had firm instructions regarding the disposal of the animal collections from Count Chabrol de Crousol, Secretary of the Navy. On no account, he said, were specimens to go anywhere other than to the King's collections; the excesses of the previous voyages where the officers and crew could acquire their own collections were to be avoided.[64] So perhaps it was someone from the *Coquille* who took the gecko back, although it is strange that none of the officers or naturalists mentioned having received a specimen.

57. Possibly those caught by the crew off Brown's Bay and the Northern Coromandel coast, including Kingfish, Snapper and Leatherjacket.

58. *Amphibola crenata.*

59. See Note 50 above. Cuvier's description of *Galaxius fasciatus* was preceded by Gray's.

60. Wright, 1950, p.180. Orchiston, 1969, p.139.

61. *Ibid.*, p.195.

62. Bauer, *pers. comm.*and *Dominion*, 10.9.1984. Buller, 1905, p.xxi, describes a large quasi-arboreal lizard – the Kawekaweau – that died out around 1870. It was about 60 cm long, beautifully marked with alternate bands of colour.

63. Although they did have the wherewithal for stuffing: Quoy and Gaimard 1830, v.4, p.380

64. Wright, 1950, p.231.

The South Island Kokako (*Callaeas cinerea cinerea*) collected by J. R. C. Quoy and J. P. Gaimard on the first voyage of the *Astrolabe* to New Zealand; a hand-coloured engraving from Dumont d'Urville's *Voyage de la corvette L'Astrolabe . . . Atlas . . . Zoologie*, 1834, *Oiseaux* (pl. 15). Published by permission of the Alexander Turnbull Library. *(see p.89)*

Returning to the remainder of their voyage: weary of the 'futility' of searching for botanical and entomological specimens, d'Urville stretched out one day on the bracken of a hilltop and gave himself over to reflecting upon the new France of the South and the role of the Bay of Islands as a busy port.[65] He was soon overtaken by reality however and returned to the ship to make preparations for their departure the following day. They left on 19 March, 1827, probably with some measure of relief. The weather on occasions had been atrocious and there had been moments of great anxiety, but from the point of view of natural history and hydrography their visit to New Zealand had been a great success. The naturalists Quoy and Gaimard had been hardworking; particularly Quoy, who had prepared many valuable paintings and sketches. In spite of all this work the two naturalists remain rather indistinct figures leaving us with little personal and informal comment on the New Zealand fauna, saving their remarks for the later published scientific works.[66] A Tui and a kaka that d'Urville had taken live from New Zealand died two weeks after their departure, much to his distress.

The expedition finally reached Marseille on 25 March, 1829, where the natural history specimens were despatched to Paris and the voyage finished at the point of departure, Toulon. The museum in Paris was overwhelmed with specimens and drawings: 1263 animals were illustrated; 500 species of insects had been collected by d'Urville and Victor Lottin, the third officer.[67] Some 520 birds were accessioned, and there were many botanical and geological specimens.[68] Recognition of their achievements seemed to be slow in coming and for a while it seemed that publication was going to languish along with the promotions for the officers, but the publication was finally approved 'on the most splendid scale'.[69]

The zoological and entomological volumes appeared between 1830 and 1835 and the atlas of illustrations in 1834, the latter with fine hand coloured engravings of the same high standard as the first voyage. Quoy and Gaimard described only a proportion of what was collected from New Zealand, but many of the hundred or so species dealt with were new and by far the greater majority were invertebrate animals, with shells predominating; others, such as corals and sea squirts, were described for the first time from that part of the world. In the sections devoted to entomology we find a few New Zealand insects — in spite of their initial despair at finding anything at all in this line from New Zealand — three of which were new species, including the colourful and familiar Magpie Moth, described by Boisduval.[70] In its broader approach this voyage was part of a definite turning point in zoological collecting and description, reflecting the developing interests of professional zoologists and the fact that vertebrate specimens from the South Seas were no longer novelties and that new ones were becoming increasingly harder to find.

In the period that followed the return of the *Astrolabe*, enthusiasm for major scientific voyages of discovery receded and the boundaries between expeditions based on science, trade and naval protection became blurred under the headings of national interest and imperialist expansion.[71] But in terms of major expeditions, France was still leading the field in the South Pacific when Laplace in *La Favorite* set sail for Indo-China. From there, following a period of exploration, he returned to France via New Zealand, stopping at the Bay of Islands on 30 September, 1831. If any natural history collecting was undertaken here there was no mention of it.

From that time, apart from French whalers, no major French expedition touched at New Zealand until the voyage of Du Petit-Thouars in the *Vénus* which left France at the end of 1836, and on 13 October, 1837, was anchored in the Bay of Islands. Their

65. *Ibid.*, pp.199-200.

66. Quoy gave his diary to d'Urville, who quoted it but there was little about natural history; *Ibid.*, p.54. Many of Quoy's natural history sketches remain in the Muséum, National d'Histoire Naturelle, e.g. M.S.105, 106, 107, 109. A record of these has been made by I. Ollivier, Blue Books III, IV, Alexander Turnbull Library.

67. Dunmore, 1969, p.224.

68. Catalogue des Mammiféres et des Oiseaux, Mai 1829, Muséum National d'Histoire Naturelle (Ornithologie), Paris.

69. Dunmore, 1969, p.225; copper plate engraving, an expensive method of illustrating books of this kind was still being used in France, whereas in Britain natural historians and their artists were turning to lithography to illustrate their works.

70. The Magpie Moth, *Nyctemera annulata* (see Note 50).

71. Dunmore, 1969, pp.228-230.

mission was largely concerned with the protection and support of the French whale fishery in the Pacific, along with the *Héroïne* under Cécille, also in New Zealand waters and bearer of Allan Cunningham, the Australian naturalist. The *Vénus* stayed in the Bay of Islands until 11 November, assisting a damaged whaler and allowing various excursions to be made in the surrounding area. Although there was no major concern with natural history, the senior surgeon Adolphe Néboux amassed a considerable collection of animals and plants, including a few birds.[72] But these by now familiar species (North Island Kokako, Whitehead, North Island Robin, Red-crowned Parakeet and others) can no longer have been of much interest to the overtaxed Muséum d'Histoire Naturelle, which by now had plenty of specimens from earlier voyages.[73]

In late 1836, d'Urville's proposal for another and final expedition to the South Pacific was accepted with modifications, the major one being that they should take an interest in Antarctic exploration. At the same time the British Government, with the support of the Royal Society, was organizing the *Erebus* and *Terror* expedition under James Clark Ross, while Congress was authorizing the launching of the United States Exploring Expedition under Charles Wilkes. The former had Antarctic exploration as its principal objective, the latter was to include it in a more general plan of Pacific exploration. Under the circumstances, a French presence seemed desirable.[74] But their interest in the Antarctic was not the sole reason for the voyage — there were scientific concerns as well as commercial ones and it was partly commercial concerns that brought them back to New Zealand, with instructions to examine localities in Cook Strait suitable for French whalers. This time two vessels were to sail: the *Astrolabe* under d'Urville and the *Zelée* under Captain Charles Jacquinot. The naturalists were Jacques-Bernard Hombron and Honoré Jacquinot, brother of the *Zelée*'s captain. As was the case on the previous two voyages, one of the naturalists, Jacquinot, possessed sufficient artistic skills to paint fresh specimens before they succumbed to the rigours of preservation.

Their instructions relating to zoology were that the emphasis was to be on invertebrates, with molluscs, lamp shells and starfish amongst those singled out. There was also the kiwi. The French still did not have a specimen of the bird and yet were aware that a second specimen, that sent by the Reverend Yate from Waimate, had already reached England.[75]

The expedition sailed on 7 September, 1837, but it was to be nearly 2½ years before they reached New Zealand, during which time they skirmished with the Antarctic ice pack and suffered the deprivations of scurvy before recovering their health in the warmer climates of the Pacific, only to lose it again — predictably enough — after a stay at Java. Decades after Cook, one wonders why more effort and thought was not expended on avoiding dietary and infectious diseases. The scientific programme of the voyage did not appear to suffer however, and even in moments of tragedy its objectives were not overlooked — consider the unfortunate Tongan who died at sea and was preserved in a barrel of brandy, later to form part of the collections of the Muséum d'Histoire Naturelle.[76]

In January 1840 Adélie Land was discovered. On 11 March, 1840, they arrived at the Auckland Islands from Hobart, observing a vessel of the United States Exploring Expedition in the process of leaving.[77] The Islands were already showing effects of visits by sealers and whalers — rats from the ships had come ashore and there was evidence of their burrowing in the ground.[78] The birds they saw were similar to those from New Zealand, including the Red and Yellow-crowned Parakeets and the Tui. A penguin with a yellow crest, the Yellow-eyed Penguin and the Auckland Island Merganser[79] were the

72. Dupetit-Thouars, 1841, v.3, p.469.

73. Catalogue des Mammifères et des Oiseaux, 5.10.1839, Muséum National d'Histoire Naturelle, (Ornithologie), Paris.

74. Dunmore, 1969, p.342.

75. De Blainville, quoted by Hombron and Jacquinot, 1846, v.1, p.34.

76. Dumont d'Urville, 1845, v.8, pp.13-14.

77. Wright, 1955, p.8, Dunmore, 1969, p.378.

78. Wright, 1955, p.10.

The Kahawai (*Arripis trutta*) collected by J.-B. Hombron and H. Jacquinot on the second voyage of the *Astrolabe* to New Zealand; a hand-coloured engraving from Dumont d'Urville's *Voyage au Pôle Sud . . . Atlas . . . Zoologie*, 1853, *Poissons* (pl. 4). Published by permission of the Alexander Turnbull Library. *(see below)*

only new species of birds they were to collect from the New Zealand region. Fish were plentiful, but oddly enough all or most were parasitised by worms which, although they did not affect the taste, were offensive enough in appearance to prevent the sailors from eating them.[80] New shells were found, including a land snail and a marine mollusc, and several crabs, two new, and some insects (all beetles — about seven of them) were collected here by Hombron and Tardy de Montravel.[81]

From the Auckland Islands the expedition headed for the South Island of New Zealand. Off the Otago Peninsula, the *Zélee* moved in towards the coast trailing fishing lines baited with scraps of red muslin. They caught a large number of 'mackerel' — probably Kahawai. This fish was described and illustrated in the voyage records, although they could not recall whether the locality was Tasmania or New Zealand.[82] The ships anchored in

80. Wright, 1955, p.10, 11. They might have been tapeworm plerocercoids or even larval anisakid roundworms, the adults of which are found in marine mammals, e.g. seals and dolphins.

81. See note 79.

82. Hombron and Jacquinot, 1853, v.3, p.40; described as *Centropristes tasmanicus*.

79. A selection of species collected by Hombron, Jacquinot and others on Dumont d'Urville's third voyage to New Zealand:

AUCKLAND ISLANDS
Birds:
• † Yellow-eyed Penguin *(Megadyptes antipodes)*,
• † Auckland Island Merganser *(Mergus australis)*.
Molluscs:
• *Thelassohelix antipoda* (a land snail)
• *Margarella antipoda* (marine snail)
Crustacea:
• † *Jacquinotia edwardsi*
• *Leptomithrax australis.*

Insects:
Seven coleoptera listed in Catalogue des Animaux articulées, Crustacés, Arachnides, Insectes . . . v.3, 1841, Entomologie, Muséum National d'Histoire Naturelle, Paris; but ten species were described and

illustrated in the Atlas, of which nine remain valid: *Loxomerus brevis, L. nebrioides, Calathosoma rubromarginatum, Oopterus clivinoides, O. plicatollis, Geraniella nitidofuscus, Pseudhelops tuberculatus, Oclandius cinereus,* and *Gromilus insularis insularis.*

NEW ZEALAND
Birds:
• North Island Brown Kiwi (2) *(Apteryx australis mantelli),* (collected and put in spirits by d'Urville and Hombron).
• Black Stilt *(Himantopus novaezealandiae)* (It came to Gould from some early Wellington settlers, Fleming, 1982a, p.266).
Many other birds were credited to Hombron and Jacquinot from New Zealand. But they were frequently recorded species (New Zealand Falcon, Fantails, etc. etc.). Listed in Catalogue des Mammifères

et Oiseaux, 1841, Muséum National d'Histoire Naturelle (Ornithologie), Paris.

Fish:
• (possibly) Kahawai, *Arripis trutta.*

Crustacea:
•† *Notomithrax peronii*
• † *Cancer novaezelandiae*

Molluscs:
•† Trophon Shell *(Paratrophon patens),*
• Wheel Shell *(Umbonium zelandicum).*
• Arabic Volute *(Alcithoe arabica)*

Insects:
Coleoptera (the majority from Akaroa) and Orthoptera (three from Otago) were collected by Jacquinot. Specific identification is not immediately clear from list in Catalogue des Animaux articulées etc. v.3, 1841,

Muséum National d'Histoire Naturelle (Entomologie) Paris. But the following were among those believed to have come from this expedition.

Coleoptera
Otago:
† • *Mecodema sculpturatum,*
Cilibe granulosum,
Akaroa:
Agonochila binotata (Two Spotted Ground Beetles),
• *Omalius elongatus,*
• *Omalius sylvaticus,*
• *Argenta pantomelius,*
† • *Oopterus plicaticollus,*
• *Psepholax sulcatus* (Shotgun Weevils). Some 17 previously described species of Coleoptera were illustrated in the Atlas.

(† new species • illustrated in voyage Atlas.)

Peint par Oudart.

Dirigé par Borromée.

Gravé par Cain.

Gide Editeur.

Imp.r de Bougeard.

1. PLUVIANELLE SOCIABLE. (Nob)   2. ÉCHASSE NOIRE. (Nob) Fem.

The Black Stilt (*Himantopus novaezealandiae*) collected by J.-B. Hombron and H. Jacquinot on the second voyage of the *Astrolabe* to New Zealand; a hand-coloured engraving from Dumont d'Urville's *Voyage au Pôle Sud . . . Atlas . . . Zoologie*, 1853, *Oiseaux* (pl. 30). Published by permission of the Alexander Turnbull Library. *(see p.97)*

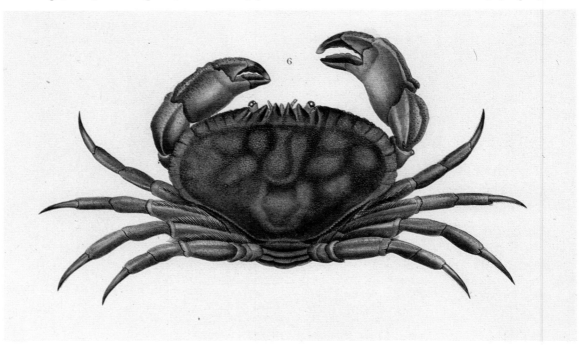

The New Zealand Cancer Crab (*Cancer novaezelandiae*) collected by J.-B. Hombron and H. Jacquinot on the second voyage of the *Astrolabe* to New Zealand; a hand-coloured engraving from Dumont d'Urville's *Voyage au Pôle Sud . . . Atlas . . . Zoologie*, 1853, *Crustacés* (pl. 3). Published by permission of the Alexander Turnbull Library. *(see p.97)*

Hooper's Inlet on the Otago Peninsula on 30 March, 1840.

The well-spaced visits by the French had enabled d'Urville and others to monitor change in the New Zealand scene, so the sight that greeted their eyes on the Otago Peninsula was a particularly depressing one. If evidence was needed that a culture is a fragile thing, unable to resist powerful external influences, then they needed to look no further than the Maori community they saw there. The debasing influence of the Europeans was all too apparent and the French paused to reflect regretfully on the losses to Maori culture resulting from European contact.[83] It would take time for visiting Europeans to extend such feelings to the wider realms of animal and plant ecology, although there were occasional signs that such thoughts did enter the heads of the earlier explorers. Such thinking would develop further and become more openly expressed, but there would be little immediate action taken to minimise the rapidly escalating influence of Europeans on New Zealand's fauna, flora, landscape and indigenous people.

The expedition left Hooper's Inlet to call at Akaroa, where they spent more than a week in the company of French whaling vessels before proceeding at some speed to the Bay of Islands, pausing only briefly at Poverty Bay to pick up supplies. The Bay of Islands behind them on 4 May, 1840, they reached France again on 6 November.

From the New Zealand and zoological point of view, this was not a particularly interesting voyage. The acquisition of a pair of kiwis at the Bay of Islands was a belated reward for their persistence. These went back in spirits to France, one in the form of a skin, the other as a skeleton.[84] Many more birds returned with them, largely duplicates of earlier collections, and one, the Black Stilt, was being described by John Gould in England as the *Astrolabe* and *Zelée* were on their way home.[85] Honoré Jacquinot collected some 24 beetles and some crickets from the Otago Peninsula and Akaroa, a few of which were new, and they added two crabs from Akaroa to the two already discovered from the Auckland Islands. There were a few new shells — the Trophon Shell and the Wheel Shell are examples, but there was nothing like the accumulation of New Zealand molluscs from the previous voyage.[86] Although their New Zealand collections were overshadowed by earlier efforts, the collections from the voyage as a whole were well received.[87] It had been a long voyage and a hard one; there had been deaths, and some sick were left behind at various ports of call. But their achievements in hydrography, 'natural history' and exploration in general were considerable, and although it was not the last time that the French or other nations would appear in the area, it helped bring to an end a major chapter in Pacific exploration. Even before they could finish writing up the work there was the concluding and tragic irony — the death of d'Urville in a fire following a train crash on 8 May, 1842.

Captain Jacquinot of the *Zelée* saw the scientific and other accounts of the voyage through to publication; his brother Honoré and Hombron wrote up most of the zoology, and the entomologist Emile Blanchard looked after the insects. The zoological volumes appeared between 1846 and 1854 and the atlas in 1842, with engravings of the same high quality as before. The kiwi and its skeleton appeared triumphantly on plates all to themselves — lesser animals had to share. Virtually all the animals described from New Zealand, even the invertebrates, were illustrated in the atlas. And yet these specimens, collected by Hombron, Jacquinot and the others, were not the last from New Zealand to find their way to the basements and cabinets of the Muséum d'Histoire Naturelle. Apart from the presence of French missionaries, French interests in New Zealand continued to be maintained by the whale fishery and its protecting vessels and the small French community at Akaroa.

83. Wright, 1955, p.22.

84. Two illustrations, one of which was a skeleton, appear in the *Atlas of Zoology*, Pl.24, 25.

85. Catalogue des Mammifères et des Oiseaux, Fevrier et Mars, 1841, Muséum National d'Histoire Naturelle (Ornothologie), Paris. See note 79 this chapter. p.266.

86. See Note 79 for a selection of species collected; Jacquinot's insects are named in the Catalogue des Animaux articulées, Crustacés, Insectes ... v.3, 1841, Muséum National d'Histoire Naturelle (Entomologie), Paris.

87. Brosse, 1983, p.193.

Leptomithrax australis (upper) collected by J.-B. Hombron and H. Jacquinot from the Auckland Islands (the crab below is not from New Zealand waters); on the second voyage of the *Astrolabe* to New Zealand; from Dumont D'Urville's *Voyage au Pôle Sud . . . Atlas . . . Zoologie*, 1853, Crustacés (pl. 2). Published by permission of the Alexander Turnbull Library. *(see p.95)*

The history of the Nanto-Bordelaise Company, Captain Langlois, the whaling captain who 'acquired' land, and the arrival of the French immigrants at Akaroa is a story told in detail elsewhere.[88] The settlers arrived in the *Compte de Paris*, under Langlois, disembarking on 19 August, 1840. Only nine days earlier, Captain Owen Stanley on the authority of Lieutenant-Governor Hobson had proclaimed British Sovereignty over the South Island, but a compromise was worked out and the French remained. It was only a few months after d'Urville's departure. Arriving complete with a door and a window, indicating a lengthy stay, was the leader of the colonists Sainte Croix de Belligny, a naturalist from the Jardin des Plantes in Paris. Appointed 'travelling correspondent' to that institution he was asked to send specimens back to France when his duties allowed him. He obliged by collecting insects and more than 130 birds including New Zealand Pigeons, Keas, South Island Kokako, fantails, ducks and shags for his colleagues at home.[89] These were probably consigned to France on a visiting whaler or on one of the French naval vessels appointed to the Akaroa station.

In the early-to-mid-1840s came a succession of French vessels with their surgeons and other officer-naturalists who collected New Zealand plants and animals while they watched over the dwindling spheres of French influence — the Catholic missions, the whaling vessels and the tiny colony at Akaroa. The corvette *L'Allier*, with Lavaud as commander and Raoul as surgeon; Ensign Jaeger Schmidt of the *Héroïne*; Arnoux, surgeon of *Le Rhin* — all took back birds and some insects from Akaroa. A few insects were also sent from the Bay of Islands.[90] Much of this material went to the Muséum d'Histoire Naturelle, either directly or through dealers like Maison Verreaux who were large suppliers to the Museum and collected with Government support. A member of the firm, Jules Verreaux, is believed to have visited New Zealand at about this time. It is likely that he was on *Le Rhin* and brought back insects from Akaroa and the Bay of Islands.[91]

What can one say about the French voyages in summary and in retrospect? In certain respects they invite comparison with Cook's: there were three of them, and each carried naturalists; in general they restricted themselves to coastal surveying; and they were characterized by outstanding seamanship and organization. Both groups were well served by their naturalists, the British initially using civilians while the French relied completely on naval officers. This represents a process of evolution rather than a fundamental difference of approach, but what is interesting is the depth of natural history talent to be found on a French naval vessel. In most cases collecting was by fortuitous encounter and would seesaw qualitatively with the enthusiasm of the respective naturalists, methodical scientific surveys not being a feature of the period.

Cook's instructions regarding natural history were less precise, couched more in terms of the practical value of what they might find, whereas the French instructions were more specific with particular animal groups being singled out for attention. This, too, reflects a difference in the times and in attitudes, the French having the benefit of an accumulation of knowledge from the earlier voyages and more muted colonial and commercial concerns. Both were under contract to a single respository — Banks in one case and the Muséum d'Histoire Naturelle in the other. But there is the impression of a more clamorous private market in England and Europe earlier on, which had settled down somewhat by the time Duperrey and d'Urville made their returns.[92] The nature of the repositories and the degree to which private collectors demanded specimens had its effects on the dispersal of the collections in the years following the voyages and here the English suffered more than the French. On the question of post-voyage publication, the French are clearly well ahead. Here, once again, the passage of time had helped,

88. Buick, 1928.

89. Catalogue des Mammifères et des Oiseaux, 9.2.1843, Ornithologie Muséum National d'Histoire Naturelle, Paris. See also Simpson, 1984, pp.68, 89. He also collected plants.

90. *Ibid.*, 14.12.1843; ?Avril 1844, 19.7.1844. Catalogue des Animaux articulées, Crustacés, Arachnides Insects ... v.3, 1846, 1847, Muséum National d'Histoire Naturelle (Entomologie), Paris. Raoul was also prominent in New Zealand botany.

91. Catalogue des Animaux articulées, Crustacés, Arachnides. Insectes ... v.3, 1847, Muséum National d'Histoire Naturelle (Entomologie), Paris. Farber, 1982, p.93.

92. The increasing availability of specimens through the sale rooms and the mobility of professional collectors assisted this process.

with the private funding of the Banks era standing in comparison with the institutional and state support for publication by the French.

Finally, how did each advance zoological science? In concept and outcome both sets of voyages were firmly in the mould of 'natural history'. By their very nature they could not be part of the developing experimental sciences, but beyond classification and description one might look for signs of interest in comparative anatomy, life history and behaviour, as well as a move towards theoretical biology and the sciences of zoogeography and animal and plant evolution, subjects in which exploration had a fundamental role to play. In fact, both series of voyages helped more immediately to fuel the growing science of comparative anatomy. The Cook voyages supplied John Hunter and the Royal College of Surgeons while the French school of comparative anatomy, which was already well established under Cuvier when Duperrey and d'Urville set sail, was provided with a bounty of specimens. Among them we can recall the Kiwi whose skeleton graced the pages of the Zoological Atlas of the third voyage. Although all the voyages — British and French — contributed to the empirical base for the science of animal and plant distribution, their published observations offer little more than a tentative delving into biogeography. We can finish this brief comparison by noting that intrinsically both series of voyages were possessed of roughly equal abilities and potential in terms of natural history, but in terms of outcome each was a product of its times.

Chapter Seven

MISSIONARIES AND TRAVELLERS

ONE OF THE LIMITATIONS OF the exploratory voyages of those early years was that their itineraries restricted them to selected bits of the coast, so we have natural histories of the convenient parts. Although a great deal was achieved this way, it meant that much of the coastline and hinterland remained virtually unexplored for some time. A proportion of the coastal gaps was filled in by the sealers, whalers and flax gatherers, although Dusky Sound, Queen Charlotte Sound and the Bay of Islands in particular were still favoured ports of call. Exploration of New Zealand's green and enigmatic interior needed someone with time and incentive, someone whose instructions were not urging them to resupply, weigh anchor, and proceed at all possible speed to the next country on the list. Further, it needed someone who was educated, preferably with a background in natural history, and who could write about what was seen or collected.

Such people would be drawn from the missionaries and travellers who visited New Zealand in the opening decades of the 19th century. The missionaries in particular were appropriately educated and their calling ensured that they were among the earliest explorers of New Zealand's interior, but it was to be some years before one of them would show more than a casual interest in natural history. It is to the early travellers and visitors that we must turn to usher in the next phase of New Zealand's zoological exploration.

This group, driven by curiosity, a sense of adventure, or sometimes by circumstance to cross the Tasman Sea, later recorded their impressions. Few were naturalists and not many enjoyed collecting plants or animals. But their time in New Zealand was sufficient to learn the language, hear stories and make observations that might have escaped their predecessors. They, along with the later missionaries, produced a number of popular accounts of New Zealand and its natural history that would be read by a far wider public than the learned writings of the voyage naturalists. The first of their number was John Liddiard Nicholas.

Nicholas was not a naturalist. He had arrived in Australia to set up in business, where an unsatisfactory partnership left his plans frustrated. Now with time at his disposal, and a professed interest in studying other cultures, he looked at New Zealand as an inviting prospect. Nicholas had formed a friendship with Samuel Marsden, the Government Chaplain of New South Wales, then planning a mission station in New Zealand, and with the consolidation of these plans it was agreed that Nicholas should go with them. Together with Marsden, three other missionaries with their wives and families, sundry passengers and some livestock, he left Port Jackson on the brig *Active*, bound for the Bay of Islands.[1]

The Tasman Sea lived up to its reputation and proved an early obstacle with high seas making most of those on board seasick. Nicholas recalls '. . . this disagreeable complaint had a strange effect on poor Mr Kendall; it made him forget for the moment that he had a wig upon his head; which falling off, in his endeavours to relieve his stomach, dropped overboard, and left him under the necessity of tying a red handkerchief round his temples, which, with the death-like paleness of his face and the grim languor of his

1. Including: cattle, sheep, horses, turkeys, geese, ducks and fowl. John Savage, the Assistant Surgeon of N.S.W., had earlier visited the Bay of Islands in September and October, 1805. His account of natural history was disappointingly brief, referring to the edibility of the few animals observed and the variety of 'rare and beautiful shells for the cabinets of the curious'. (Savage, 1807, p11).

2. Nicholas, 1817, v.1, p.47. Kendall was one of the missionaries. Banquo's ghost from *MacBeth*.

3. Marsden was more interested in agriculture.

4. 'Snapper', 'bream', 'parrot fish', 'herring', and 'flounder': Nicholas, 1817, v.2, p.259. He notes few birds.

5. *Ibid.*, v.2, pp.256-257. He also encountered mosquitoes.

6. *Ibid.*, v.2, p.43.

7. A large lizard.

8. *Ibid.*, v.2, pp.124-125. The Tuatara, *Sphenodon punctatus*, was discovered much later. They might have been referring to the mysterious Giant Gecko. Nicholas was convinced of the presence of some kind of large lizard (*Ibid.*, v.2, p.254).

9. *Ibid.*, v.2, p.126.

10. His book was background reading for the French, e.g. de Blosseville *in* Sharpe, 1971, p.115.

11. Cruise, 1823, p.316.

eyes, made him appear so complete a spectre, that he forcibly reminded me of Banquo's ghost'.[2] It must have been with relief that they anchored at the Bay of Islands at Rangihoua on 22 December, 1814.

In the two months that followed, Nicholas accompanied Marsden on the *Active* as far south as Thames in the Hauraki Gulf, with visits to points in between. The missionaries did not venture into the interior of the country as they would do later and their preoccupation with establishing the mission meant that there was little time for leisure interests such as natural history, although the group gave little sign of enthusiasm for the subject.[3] Nicholas however had time to record his impressions, and following his return to Australia with Marsden, and subsequently to England, he published a detailed account of his travels, with brief references to the New Zealand fauna.

He was able to corroborate the presence of the kiwi in New Zealand, although at the time his description of it as a species of cassowary with feathers like that of an emu would not have immediately connected it with the bird described from southern New Zealand by Shaw. A few fish and shellfish he recognized by their common names[4] and insects he could identify only crudely: 'the butterfly, the beetle, the flesh-fly, the common fly, and a small sand-fly . . .'. He said 'this latter insect was excessively troublesome to us, by its settling on the instep and about the ancles [sic], and biting those parts in such a manner as to prove very painful when we were warm in bed' — sandflies never failed to be mentioned by visitors.[5] The louse, 'cootoo' (kutu), was equally familiar. The Maoris were fascinated by their European visitors and followed closely their every move, including their toilet. Such close contact resulted in the inevitable sharing of the inhabitants of their cloaks and mats and only frequent changes of clothing kept them from permanent infestation.[6]

One day while travelling on foot Nicholas noticed a hole at the foot of a tree, perhaps the burrow of a four-footed animal. His Maori guide described something to him that gave the impression of a 'Guana'[7] and when Nicholas asked the guide to poke a stick down the hole the Maori indicated that he was very frightened at the prospect of finding the animal there. Possibly this was a hint of the existence of the Tuatara which was not discovered by Europeans until some years later; at this distance it is very hard to tell.[8] He was rightly sceptical of the alligator which was supposed to live in the interior of the country.[9]

The limitations of these few observations might have excused the serious naturalist from reading Nicholas's work, but there is some evidence that he was considered useful background reading for those about to explore New Zealand.[10]

Three years further on another account was published, that of Major Richard Cruise, who had spent ten months in New Zealand while in charge of a detachment of marines on board the *Dromedary*, the ship that brought Marsden on his third visit. Cruise's story contains very little natural history; there is only a brief reference to the kiwi and the first published account of its method of capture.[11] Once again, his reference to an 'emu' would throw all but the most discerning of ornithologists off the scent and it is only in retrospect that we can recognize his description for what it is.

In the decade following Cruise's visit there was a fair amount of coming and going across the Tasman, while the missionaries and others extended the range of their activities in New Zealand, probing a little deeper into the country as they did so. Other visitors came and went, some like Augustus Earle writing of their experiences and here and there making a casual remark that gives us a clue to the changing zoological picture of New Zealand. Earle for example tells us of the increasing populations of European dogs and goats which, with wild pigs and rats, had demonstrated that they were on their way to becoming pests.[12] But overall very little natural history shows in the published record of the period until all else in this field is overshadowed by the work of the French expeditions. Although the hiatus in natural history can be explained by external influences, it must be remembered that New Zealand, and the Bay of Islands in particular, would not, on the basis of what was then known, be an obvious magnet for visiting naturalists.

As we have observed before, there was a traditional interest in natural history among the British clergy, so it was more or less inevitable that sooner or later one would arrive in New Zealand who would take an interest in the natural history of his parish and perhaps collect and describe plants and animals. Fortunately timing his arrival in New Zealand to coincide with an increasingly effective focus for the national collections in England was the Reverend William Yate.

Yate arrived at the Bay of Islands on 19 January, 1828, for a stay that was to last some six-and-a-half years before he returned to England and the chaplaincy of a sailors home at Dover. During his term in New Zealand the missionaries penetrated south to Thames, the Waikato and Rotorua, and south-east to the Bay of Plenty and the East Coast. Yate himself was stationed at Waimate near the Bay of Islands. His recollection of the country survives in the form of book published by the Church Missionary Society after his return to England, in which he devoted a chapter to the natural history of New Zealand.[13] It was in fact one of the first popular accounts of this subject to be written.

Yate lists 33 birds under their Maori names, and for some describes their food, feeding, song, behaviour and breeding. There are several interesting records, at least five of which were undescribed species, including the Stitchbird and the Huia.[14] He could not have seen the latter, as its range was south of the extent of his travels, but Maoris from the southern districts were known to send the highly prized feathers as presents to tribes in the north.[15] These feathers were used to adorn the head on ceremonial occasions, and the esteem in which they were held, coupled with the restricted distribution and ease of catching the bird, helped hasten its extinction. Other birds recorded by Yate were North Island subspecies of Kokako and the Fernbird, already described in the South Island,[16] and yet others were species of still wider geographic distribution that earlier naturalists had overlooked. His greatest contribution to ornithology was the acquisition of a North Island Brown Kiwi, which was only the second specimen to reach England.[17]

A lizard, about six inches (15 cm) in length, is described as having variable colours and a pale belly, and he mentions a rich supply of fish along with the common names of ten or so, but notes only eels from fresh water.[18] Crayfish, oysters, shrimp, prawns, cockles and mussels were also considered abundant. Like his predecessors, he thought the insects lacked variety, but mentions locusts (cicadas) and the noise they make in summer, grasshoppers, mosquitoes, many kinds of butterflies and moths, sandflies, dragonflies, scorpion flies, beetles and forest bugs, which make an unpleasant smell when approached.[19] Spiders were abundant. Shells were to be Yate's other major contribution, although he said they would 'scarcely repay the naturalist for any long search'. He also claimed that conchologists regarded shells from that part of the world as curiosities, and it must have been the latter thought which encouraged him to accumulate a respectable shell collection, many of which were gathered on his excursion to the East Coast in 1834. This collection was given to the British Museum through the Church Missionary Society in London, and a note in the Museum's *Book of Presents* for 11 April, 1835, records 30 shells received.[20] These were subsequently examined by the Keeper of Zoology, Dr J. E. Gray, who wrote to Yate in July 1835, telling him what he had discovered.[21] The

12. Earle, 1832, p.130.

13. Yate's Diary (M.S. Yate, 1834-1845, Alexander Turnbull Library) contains little of natural history interest.

14. Yate, 1835, pp.52-70. The names he uses sometimes differ from those used by tribes from other districts (Fleming, 1982a, p.246), and some of his descriptions are not accurate. Some of his bird species were: Australian Bittern (*Botaurus stellaris poiciloptilus*), Banded Rail (*Rallus philippensis assimilis*), Spotless Crake (*Porzana taubensis plumbea*), Pukeko (*Porphyrio porphyrio melanotus*), Pacific Golden Plover (*Pluvialis fulva*), Eastern Bar-tailed Godwit (*Limosa lapponica baueri*), Long-tailed Cuckoo (*Eudynamis taitensis*), *North Island Fernbird (*Bowdleria punctata vealei*), *Stitchbird (*Notiomystis cincta*), *Huia (*Heteralocha acutirostris*), *North Island Kokako (*Callaeas cinerea wilsoni*), North Island Brown Kiwi (*Apteryx australis mantelli*). (* these would later be described as new species or subspecies.)

15. *Ibid.*, p.61.

16. *Ibid.*, pp.58, 64.

17. See Chapter 5 on the Kiwi.

18. Yate, 1835, p.71.

19. *Ibid.*, pp.72-73.

20. J. E. Gray, in Dieffenbach, 1843, v.2, p.180, records only 29 species.

21. Yate, 1835, pp.73, 307.

collection was found to include ten new species, eight of which are still valid today including the Frilled Venus Shell, a Dosinia, Grey's Sunset Shell and the Horse Mussel, which is a large delicate shell commonly found washed up on ocean beaches all round New Zealand.[22]

22. Respectively: *Bassinia yatei, Dosinia subrosea, Gari lineolata, Atrina zelandica.*

Yate's interest in New Zealand natural history was rather a lonely one. William Williams, who had arrived at the mission two years before Yate, was medically trained; with this advantage he might have been expected to take more of an interest in his natural surroundings than he apparently did, although he was later helpful to d'Urville and would play a valuable role in some palaeontological discoveries of great significance. Otherwise there was little missionary interest, potential or otherwise, suggesting that the work of the mission was their priority and they had their hands more than full. Yate did however meet Baron Karl von Hügel, an Austrian scientist travelling on the *Alligator* which visited the Bay of Islands in March 1834 and was to become well known for its rescue of the Guard family from the Taranaki Maoris later that year. Those on board were known to have collected natural history specimens.[23]

23. Von Hügel was escaping the memory of a broken romance but although he probably collected a little while here, he did not note anything of consequence of New Zealand natural history. King, 1981, p.24. Captain Lambert collected a Spotted Shag which he later gave to Gould who had it drawn by Edward Lear, for the *Birds of Australia and the Adjacent Islands* (see Andrews, 1982, p.117).

His career as a missionary in the South Pacific came to an untimely end on Yate's return to Australia. Following a furlough in England, he was accused of homosexual activities while on board ship; worse still, it was alleged that he engaged in similar conduct with the Maoris in New Zealand, for whose moral and spiritual welfare he had some charge. In those days this was no mere trifle and for Marsden, already concerned about the corrupting influence of the sealers and whalers on the Maoris, it was an outrage. Yate was given little chance to defend himself and accompanied by his sister was swiftly despatched to England.[24]

24. Some years later a Miss Yate?(s) presented some New Zealand molluscs to the British Museum — it was possibly his sister. Accessions Register, Mollusca, 29.3.1842.

In 1831 there arrived in New Zealand a colourful figure whose path crossed that of the Reverend Yate's; his name was Joel Samuel Polack, a traveller whose family in England were comfortably off, and who had followed a variety of occupations in the different countries he visited. Polack was not a naturalist; he was not even a collector in any noticeable way, but he was a keen observer, able to get on well with the Maoris in whose company he spent a lot of time. It was this attribute which allowed him to make some notable zoological discoveries. One of these was the Tuatara — a reptile that had been struggling to free itself from the myths that surrounded it. He saw no specimens, but was told that the 'gigantic lizard, or *guana*, exists principally in the island of Victoria (the South Island). Some are found in the isles of the Bay of Plenty'.[25] Apart from the description of a skull received by J. E. Gray in 1831, this is the first certain reference to this reptile which had been regarded as a 'living fossil', belonging to an order that dates back nearly 200 million years. The islands of the Bay of Plenty would later give up the first specimens of the Tuatara, but in the meantime this brief and important record of Polack's remained ignored by zoologists. His second find, which was to foreshadow even greater discoveries, came about when shipwreck forced him ashore on the North Island's East Coast. There he was shown fossil bones of a very large bird, but for a time this discovery too left the zoological world unmoved. The rest of this story will be told later.

25. Polack, 1838, v.1, p.317.

His other observations included the less significant but still interesting records of now familiar household animals such as the cockroach introduced by the whaling fleets, and the Puhihi (Maori for 'pussy-cat'), the domestic cat, introduced for some 25 years according to Polack and alleged to be very nutritious! He reports on the successful domestic animals such as cattle and horses, and the less successful sheep.[26] There is a bare hint that he may have been aware of New Zealand's rare and unusual native frog when he talks of

26. *Ibid.*, v.1, pp.314-315, 320.

the noise made by the 'frogs and toads' that croaked just before rain. However the New Zealand frog is not noted for the stridency of its call and Polack may have been mistaken, hearing instead the noise of crickets.

Crickets form part of an extensive list of New Zealand animals published in a two-volume account of his experiences after his return to England. Apart from those animals already mentioned, and the more exotic record of three sea snakes washed up in Hokianga Harbour, the list was of long-familiar species.[27] The birds were a curious mélange of Maori names, Latin binomials and names of birds from Europe such as the Lark, Nightingale, and 'Poor cock Robin', but Polack did include notes on the behaviour of some species. His lengthy list of fish with Maori names and their Northern Hemisphere equivalents was no advance over anything written before, but he did try to cover the marine mammals and sharks, including the Sperm and Right Whales and the Hammer-head Shark, and also mentions the whale fishery. Many invertebrates are registered by their common names and can only be vaguely identified. And what is one to make of: ' . . . The forms of the gelabrous molluscae, respiring through their bronchiae, are as various in form as the most inventive mind can well conceive'?[28]

His chapter on the flora and fauna was a bouquet of animals, a glorious mixture, but with the mention in his book of the Tuatara and fossil birds there was a sign here that some of the last and most important bricks were about to go into the wall, although at the time of publication neither he nor anyone else realized the impetus that would result from some of the discoveries of which he gave advance warning. Consider the foresight of his final claim: ' . . . doubtless, the future ornithologist will be surprised by discovery, among the hidden mountain-gorges and wilds of the Island of Victoria [the South Island], many birds at present supposed to be no longer in existence'.[29] There were not to be many, but there would certainly be evident of some!

Contemporary with Polack, although arriving a few years later almost at the end of 1834, the missionary-printer William Colenso can be described as New Zealand's first serious resident naturalist. His was a significant and timely figure, as people and events across the Tasman would soon link up with the fragile focus of natural science in New Zealand represented by Colenso and his ilk. Here was no opportunistic collector, seduced by the wonders of a new country; Colenso had been a staunch and active member of the Penzance Natural History Society, with whom he left a departing promise to send specimens from New Zealand.[30] His interests were strongly directed towards botany, but like other botanists who had visited New Zealand he would allow some room for zoology too.

Apart from his own drives and enthusiasms, Colenso was to be greatly influenced by visitors to the Bay of Islands: the first and most important of these was Allan Cunningham, the Colonial Botanist of New South Wales.[31] Cunningham's first arrival in New Zealand was on a whaler in August, 1826, to stay for three months at the Bay of Islands — the first visit by an Australian-based naturalist, and the forerunner of a future scientific link between the two countries. He returned again on the *L'Héroïne* in April, 1838, the visit on which he collected the kiwi referred to earlier, and on this occasion he met Colenso. An indefatigable explorer and botanist he was a model for his new pupil and a valuable working relationship was quickly established. Cunningham died in 1839, not long after his return to Australia, but his legacy to New Zealand was Colenso and the lifting of the study of natural history in that country to a new and higher level.

There would be other important influences: the connection with Tasmania and its patrons of science, Sir John and Lady Franklin, together with their Tasmanian Society

27. *Ibid.*, v.1, pp.295-326.

28. *Ibid.*, v.1, p.325.

29. *Ibid.*, v.1, p.307.

30. Bagnall and Petersen, 1948, p.27.

31. *Ibid.*, p.72. His brother Richard also collected plants and visited New Zealand.

and its publication *The Tasmanian Journal of Natural Science, Agriculture, Statistics, etc.*, which began in 1841. This was a significant intellectual forum for the tiny scientific community in New Zealand, a nucleus for a fragile but self-contained and functional natural history unit in the Antipodes. For Colenso there was the additional stimulus of visiting naturalists from the voyages of d'Urville and the United States Exploring Expedition under Wilkes and, significantly, the lengthy partnership which resulted from a meeting with Joseph Dalton Hooker, the botanist of the *Erebus* and *Terror* expedition.

Initially, at least, zoology in New Zealand was to benefit more perhaps from the atmosphere created by Colenso's activities than it was by his direct contribution. Although more concerned with his plants he did collect animals: Hooker noted on their first meeting, '. . . Of shells Mr Colenso has 150 species with many insects and minerals . . .'. But he had a philosophical and practical objection to collecting bird skins and he declined a request from Lord Derby for specimens.[32] This objection had a lot to do with his recognition of the increasing rarity or extinction of some species, although he later relented and sent four skins to Derby, including one of a White Heron. Much later in life he would publish several articles on a variety of New Zealand animals, including insects, reptiles and birds, but his greatest achievement in this field was undoubtedly his independent description of a giant flightless bird that had once lived in New Zealand, a story told in greater detail in the next chapter.

32. *Ibid.*, pp.55, 270. He once released a live specimen of a bird thought to be new. (Taylor, 1959, p.42).

One thing we can learn from Colenso is the stamina required of a field naturalist, missionary, or explorer of uncommon zeal who wished to traverse the interior of New Zealand. To venture any distance into this interior one had to face up to the prospect of finding a way through the bush and forest which clothed the hills and ranges, steeply dissected by streams and fast-flowing rivers, in weather that could be by turn glorious and frighteningly bad. Europeans since the time of Forster had complained of the difficulties of walking in the New Zealand bush: the tree roots, moss-covered fallen logs rotten to the touch, vines, the wetness that soaked the clothes and made the way slippery underfoot. The absence of quadruped mammals meant there were no animal tracks, although by the time of Colenso the increasing populations of wild pigs must have begun to take effect in some locations. The Maoris had a network of tracks but these presented problems of their own as Augustus Earle, the artist-explorer who visited in 1827, discovered: 'The New Zealanders always travel on foot, one after the other, or in Indian-file. Their pathways are not more than a foot wide, which to a European is most painful; but as the natives invariably walk with the feet turned in, or pigeon-toed, they feel no inconvenience from the narrowness. When a traveller is once on the path, it is impossible for him to go astray. No other animal, except man, ever traverses this country, and *his* track cannot be mistaken since none ever deviate from the beaten footpath, which was in consequence in some places (where the soil was light) worn so deep as to resemble a gutter more than a road.'[33] Europeans including Colenso stumbled along these paths in the company of their Maori porters and guides, knowing that the untracked alternative was far worse.

33. Earle, 1832, p.72.

Sometimes botanical necessity demanded that Colenso leave the path, although even while on it he found he could make collections of plants and animals: '. . . insects often carry their strong and glutinous webs across the pathway; with which, if you happen to be at the head of a file, your face coming into contact, causes you to suddenly halt, to the detriment of your heels, and the disarrangement of the whole line of march. The largest *Cicadae* and *Libellulae* (dragonflies) are often entangled in those webs, and seized by their ruthless and powerful enemy. I have also secured various species . . . doubtless new to science'.[34] Indeed many of his animals were new to science, such as a fresh-water mussel from Lake Waikaremoana, but the time taken in the field and delays in publication — and there were considerable delays even when he published in the *Tasmanian Journal* — meant that frequently others gained the credit for the descriptions. However the British Museum, through the agency of J. D. Hooker, benefitted from Colenso's collection of insects.[35]

Like others of the period Colenso took a tent, using poles chopped from the bush where he camped, but sometimes fatigue or the unsuitability of the site forced him to bivouac. The days were long, starting early in the morning; the explorers travelled well into the evening, or until they found water and a suitable campsite, their diet largely kumara (sweet potatoes) or ordinary potatoes roasted in a fire. A weka, or if near the coast fish or crayfish, might help break the monotony, but sometimes gnawing a few roasted cabbage trees leaves before retiring for the night was all they could manage. Water was a problem, either too much or too little: '. . . we continued our march, through want of water, until long after sunset. Fortunately I had succeeded in finding some, by the side of which, in the open wilderness, we bivouacked, all too fatigued to care much about anything save rest. Gained nothing new in botany in the whole of this melting day's horrid march — fern, fern, nothing but dry, dusty fern all around'. And on the other hand; 'The morning was most gloomy; the rain still incessantly poured, and our cold, wet, lonely, and starving situation was anything but pleasant; . . .'[36] Little wonder they greeted with such enthusiasm the accommodation and food provided by fellow missionaries, Maoris, or one of the growing number of colonists.

Under the circumstances it is no surprise to find that many a treasured specimen was sacrificed to expediency: 'I have often been surprised at the great carelessness which I have shown to rare natural productions when either over-fatigued or ravenously hungry; at such times, botanical, geological and other specimens — which I had eagerly and with much pleasure collected, and carefully carried for many a weary mile — have become quite a burden, and have been one by one abandoned, to be, however, invariably regretted afterwards.'[37]

Colenso was characteristic of his times. Previous generations of naturalists had also made sacrifices and suffered considerable discomfort in pursuit of their studies — think of what the Forsters had to endure. But after the 1830s the mood of romanticism that had cloaked the natural history movement was fading, to be replaced by a new middle class robustness, manifested by feats of intellectual and physical stamina.[38] Work was to be enjoyed for its own sake, an idle brain and body being the Devil's playground. It was in spite of this ethic (or perhaps because it) that Colenso was driven to temptations that eventually lost him his family, his church, and, for a period his science. Nonetheless he recovered, and in terms of publications, his later years were his most productive. His science '. . . like his faith . . . if broad, yet militant'[39] he pursued almost to the end of a century in which he, more than any other New Zealand naturalist, had seen substantial change.

The number of missionary naturalists based at the Bay of Islands increased to two with the arrival of Richard Taylor in March 1839. Shortly after his arrival, Taylor embarked on a journey to the East Coast with William Williams where the pair made an early encounter with fossil bird bones, a subject of common interest for many of the missionaries who had an involvement with natural history and later, as we shall see, the cause of strong competition among them.

After a period at the Mission at Waimate near the Bay of Islands, Richard Taylor was

34. Taylor, 1959, p.9

35. Bagnall and Petersen, 1948, p.137: the mussel was probably *Hydridella menziesi*. There was an hiatus in Colenso's publishing from the time the *Tasmanian Journal* . . . ceased and the *Transactions and Proceedings of the New Zealand Institute* began. Possibly this may have been a reluctance on his part to publish in English journals, but more likely had to do with his personal circumstances — his dismissal from the Church and separation from his family (see Bagnall and Petersen, 1948).

36. Taylor, 1959, pp.21, 27.

37. *Ibid.*, p.34.

38. Allen, 1976, p.78.

39. Bagnall and Petersen, 1948, p.421.

The Garfish (*Hyporhamphus ihi*), a watercolour from the sketchbook of the Reverend Richard Taylor. Published by permission of the Alexander Turnbull Library (E 296). *(see below)*

40. Mead, 1966, p.104.

41. The chapter 'Natural History' refers only to zoology; the other elements are considered separately. Many of the observations came from his journal.

42. *Aenetus virescens*. In the 2nd edition there were fewer illustrations and they were colour-printed by chromolithography, the first New Zealand natural history book to use this process. His account of the Bulrush Caterpillar reflected a fascination with the 'vegetable caterpillar' which was later revealed as a caterpillar of *Wiseana* spp. infected with a fungus *(Cordyceps robertsi)* which transformed it internally.

43. Accessions Register, Mollusca and Radiata, 7.7.1850, 10.7.1850, British Museum (Natural History). Swainson to Taylor, 1.11.1846, in Parkinson, 1984, p.62. *Essay on The Zoology of New Zealand*, Taylor, MS, Alexander Turnbull Library.

appointed to Wanganui, which was to be the base for his numerous travels in the following years. On foot, horseback and by canoe, Taylor penetrated the North Island from the west coast, as Colenso had done from the east, performing the functions of preacher, physician and dentist as he went.[40] Along the way he kept a detailed record of his activities and observations — of services and ceremonies, of the Maoris, and of plants and animals — interspersing the text with the occasional sketch. His accumulated knowledge led first towards a useful little publication called *A leaf from the Natural history of New Zealand* in 1848 and then to his major work, written while sailing to England on furlough in 1855, *Te Ika-a-Maui*.[41] Although much of this latter work deals with Maori culture, a significant part was devoted to natural history in the broadest sense, including geology, botany and zoology. Taylor's sketches and watercolours were transformed into attractive hand-coloured lithographs and some of New Zealand's less well known animals, such as the giant green Puriri Moth[42] were illustrated for the first time. This work was the most substantial popular account of natural history to date, acquainting its readers with recent discoveries such as the Tuatara, but it provided very little in the way of new zoological knowledge. Further, there were inaccuracies in text and illustration that might have been tolerated earlier, but which were not now acceptable in a serious scientific work.

Taylor's book was preceded by the surveys of Dieffenbach and Stokes, whose material, along with that from expedition and private sources, was beginning to flood the British Museum and to be published. One had to have more than a hobby interest in order to break new ground at this time and the continuing demands from his missionary work had left him with very few spare hours. Unlike Colenso, he did not publish widely in scientific periodicals — only a few articles appear in the *Transactions and Proceedings of the New Zealand Institute,* and in the *Annals and Magazine of National History, 1866,* there was one on the Silvereye. He did however make a substantial collection of shells and in July 1850, the British Museum received 247 specimens, and Swainson evidently thought enough of him to correspond on the subject of molluscs. Also his undated manuscript *Essay on the Zoology of New Zealand,* with its speculation on New Zealand's ancient connections with a large southern land mass, gives an impression of someone whose abilities and potential may have lain beyond merely observing the creatures of his parish.[43] While Taylor's best-remembered service to New Zealand zoology lies in the discovery of an important sub-fossil bird site on the North Island's west coast, his very presence in New Zealand was important, giving weight to the increasing momentum of indigenous science. By virtue of the fact that both remained in New Zealand, Colenso and Taylor

Species of moths including (top – centre) the Puriri Moth (*Aenetus virescens*), top right, the Magpie moth (*Nyctemera annulata*), top left, an introduced Australian species The Moon Moth (*Dasypodia selenophora*) and others; hand-coloured lithographs from R. Taylor's *Te Ika-a-Maui* 1855, (pl. 1). *(see p.108)*

can be isolated from the other missionaries and travellers; they were an important link between this period and the later one of resident naturalists and the growth of scientific institutions.

Through the activities of the missionaries the interior of New Zealand, the North Island in particular, was now becoming better known. But if colonisation were to proceed, the resources of the country including the flora and fauna needed a still closer examination. The person chosen to carry out that survey was yet another German who had made his way to England pursued by adversity at home. Few naturalists who arrived on New Zealand shores can have had such a stormy start to their careers as Ernst Dieffenbach, forced to flee his home town of Giessen and take refuge in Zürich for his part in some revolutionary and political escapades around 1832-1833. Dieffenbach was to be a wanted man for some years.[44] When he completed his medical degree in Zürich, his recent past caught up with him and he was deported from Switzerland to France, arriving in London some time before 1837.

His progress towards appointment as naturalist to the New Zealand Company is not well documented[45] but as was the case in those days, medical men found it easy to make the transition between their calling and natural history. Thus it was not for the first time that a German fugitive with interests in natural history found himself in London and eventually on a boat to New Zealand. His appointment in April, 1839, unsalaried but with the Company undertaking to meet his expenses, was followed by the departure of the *Tory* on 3 May with Dieffenbach on board. New Zealand was sighted on 16 August, 1839, and on the 17th they were anchored in Ship Cove in Queen Charlotte Sound.

Dieffenbach's account of his arrival could equally have come from the pen of Banks, Anderson or Forster with its familiar observations of landforms, people, flora and fauna. The sea birds seen before reaching New Zealand prompted him to speculate on their migration and their feeding relationships. He remarks: 'The [life] history of one bird is closely allied to that of some other animal and so on in a chain the links of which are intimately connected. The same is the case with other animals . . .'[46] At a time when ideas on ecology and animal behaviour were still being forged, this statement comes remarkably close to describing something we now know as the 'food chain'.

In the hills at the back of Ship Cove he was soon observing and shooting examples of the local bird fauna, reminding us that it has been some time in our story since we have heard the sound of guns, the missionaries preferring less robust techniques of collecting. He visited the Tory Channel whalers at Entry Island and from them he was able to determine the types of whale that constituted their catch. He endeavoured to dry out part of the skull of a calf whale that had already been processed by the whalers by putting it on the roof of a nearby hut, but found the following morning that the rats and dogs had all but destroyed it during the night.[47]

The expedition left the Sounds to go to Port Nicholson (Wellington) where they anchored just to the north of Somes Island. They were now not far from the habitat of the Huia, and Dieffenbach was anxious to obtain a specimen of this increasingly rare bird. In the company of a Maori who had been recommended as one who could decoy the Huia and together with Charles Heaphy, the artist, he entered the hills near Lowry Bay on the eastern side of Wellington Harbour, making for the Orongorongo River. But it was not until they were homeward bound next day that their guide's imitation of the bird call, 'uia uia uia,' brought results. Four birds[48] quickly but silently landed on the lower branches of a nearby tree. Dieffenbach and Heaphy shot a bird each and on autopsying them he discovered that they fed on Hinau berries and on insects. The

44. Bell, 1976, pp.19, 20.

45. *Ibid.*, pp.28-29.

46. Dieffenbach, 1843, v.1, p.22.

47. *Ibid.*, v.1, p.50. Charles Heaphy, the draughtsman on the *Tory* accurately sketched the Southern Right Whale (facing p.177, v.2).

48. Only two according to Heaphy's account; (Heaphy, 1880, p.84). Dieffenbach's (v.1, pp.50, 51) and Heaphy's accounts differ in other respects, the latter's appearing the more reliable.

birds, a male and a female, were sent to England, enabling John Gould to include a fine illustration of them in his *Birds of Australia* in 1841; this series was featuring a growing number of New Zealand species. Gould himself had described the Huia some four years before.

Following an exploration of the Hutt Valley and then Kapiti and Mana Islands, the expedition revisited Queen Charlotte Sound before returning to the north, sailing via the Manawatu coast to Wanganui. Continuing further north, the *Tory* deposited them at Sugarloaf Point (now New Plymouth) within sight of the snow-covered peak of Mt Egmont (Taranaki). Unclimbed by Europeans and the subject of Maori superstition[49] the mountain was a challenge to Dieffenbach; after an initial unsuccessful attempt he conquered it, much to the consternation of their Maori guides. Dieffenbach recalled: 'to [the Maori] the mountains are peopled with mysterious and misshapen animals; the black points, which he sees from afar in the dazzling snow, are fierce and monstrous birds; a supernatural spirit breathes on him in the evening breeze, or is heard in the rolling of a loose stone'.[50] Unfortunately no such animals enlivened their climb and they found only a rat skeleton near the summit, from which they descended safely.

Returning to Port Nicholson, in early May Dieffenbach sailed for the Chatham Islands on the barque *Cuba* and remained there until 26 July. He took a specimen of the now extinct Dieffenbach's Rail, even then doomed by the combined attentions of the dogs and cats which preyed on it and the Maoris who snared it, and noted some other birds such as the Chatham Islands Pigeon and Bellbird which were subspecies of those on the mainland. As well, he studied the geology and botany of the Chathams in some detail – the first comprehensive scientific examination these islands had received.[51]

Dieffenbach then left New Zealand for a period in New South Wales, returning in October, 1840, to the Bay of Islands from which base he explored the north of the country. At Mangonui in Doubtless Bay he found a small settlement of timber workers and storekeepers and met an Australian who gave him a damaged specimen of a kiwi that had been killed by a dog. It was only a few months previously that d'Urville had obtained his specimens from the Bay of Islands, and Dieffenbach, thinking that the kiwi was about to become extinct, was keen to find out more about the bird. He questioned the Maoris on its habits and its egg laying and was able to elaborate a little on the earlier findings of the French regarding the capture of kiwis. He ascertained that mats or cloaks made from skins were now extremely rare. Kiwis were seen in the possession of the missionaries Taylor and Colenso; at the Bay of Islands Dieffenbach was a temporary neighbour of the latter who supplied him with numerous examples of shells, insects, ferns and geological specimens.[52]

Dieffenbach now made for the Hauraki Gulf and Auckland, by then a well-developed settlement, and for the thermal regions of Rotorua and Lake Taupo. He found the fauna of the latter as diminished as elsewhere, commenting only on a few birds and on the lake fauna, which accounted for three species including an undescribed eel and a native freshwater crayfish.[53] If he was not overwhelmed by the variety of the fauna, he was struck by the beauty of the lake and thermal areas of Taupo and Lake Rotomahana where the pink and white terraces were still exposed. When he reached Rotorua the Maoris told him about a Totara tree near which the reputed last Moa was killed.

Diffenbach was nearing the end of his time in New Zealand when he left Rotorua for Tauranga in mid June, 1841. While there he attempted to reach Karewa Island north of Mt Maunganui, but a heavy swell turned them back.[54] He knew the island was a refuge for the unusual reptile, the Tuatara, but he had to wait until a specimen was given

49. *Ibid*, v.1, p.140; Moas and Crocodiles guarded it.

50. *Ibid.*, v.1, p.156.

51. Dieffenbach, 1841, p.207. The Rail was *Rallus philippensis dieffenbachi*. Dieffenbach noted that the Chatham Islands Bellbird was larger than the mainland form.

52. Bell, 1976, p.77. A kiwi, probably the one referred to, was sketched in Taylor's Journal and reproduced in *Te Ika a Maui*. Dieffenbach's reference to the kiwi egg being the size of a turkey's was incorrect: It is much larger (v.1, p.232). From the presence of kiwis on Little Barrier Island (p.233) he concluded that it had once been joined with the mainland.

53. *Ibid*. v.1, p.362.

54. *Ibid.*, v.1, p.405.

The Huia (*Heteralocha acutirostris*) drawn and lithographed by John and Elizabeth Gould for Gould's *Birds of Australia*, 1841, (v. 4, pl. 19). The work is typical of Gould, showing a pair of birds (the female with the longer beak) one above the other on the plate. Published by permission of the Alexander Turnbull Library. *(see p.110)*

55. *Ibid.*, v.2, p.205. At about the same time 'guanas' (= Tuatara) were collected by William Wade, a missionary, from the Alderman Islands off Mercury Bay (Wade, 1842, p.178).

to him by Dr Johnson, the colonial surgeon, who had obtained it from the Reverend W. Stack of Tauranga.[55] It was this specimen that Dieffenbach took back to England and gave to J. E. Gray of the British Museum for description. A younger specimen, obtained on the *Erebus* and *Terror* expedition from about the same time, was deposited in the Museum of Haslar Hospital, Portsmouth, the headquarters of the explorer-naturalist Sir John Richardson. Gray had already examined a Tuatara, or rather part of one, over ten years earlier when he received a skull for which he described a new genus, *Sphenodon*.[56] But he did not make the connection between it and the whole animal now presented by Dieffenbach and which he named *Hatteria punctata*. It was not until 1867 that the two descriptions were reconciled by Günther, who succeeded Gray at the British Museum, and the reptile became *Sphenodon punctatus*. Although the Tuatara did not attract much scientific or public attention in the same way as the kiwi and other New Zealand animals, it was a major discovery, a reptile of great antiquity whose relatives were known only from fossil material elsewhere. The distribution of the Tuatara on the offshore islands of New Zealand caused Dieffenbach to speculate later on possible connections with the mainland, and the forces of extinction which had stranded the Tuatara on these islands.[57]

56. Accessions Records for reptiles at the British Museum do not mention this bone or it origin. Also, see Accessions Register, Zoology, 1.3.1842, 19.8.1842, for *Sphenodon* and lizards presented.

57. Dieffenbach, 1843, v.1, p.406.

The Tuatara *(Sphenodon punctatus)* lithographed by J. Ford for J. E. Gray's *The Lizards of Australia and New Zealand*, 1857 (pl. 20). The Tuatara had earlier been sketched by the Reverend Richard Taylor and reproduced in the 1855 edition of *Te Ika a Maui* (pl. 18). *(see p.111)*

He returned to England in early 1842 having surveyed a large part of the North Island of New Zealand but very little of the South Island.[58] His two-volume account of his travels, published in 1843, was the vehicle for a report on the natural history of the country and the customs and language of the Maori. The zoology in *Travels in New Zealand* was particularly detailed, being virtually a book within a book authored by J. E. Gray and John Richardson and other qualified naturalists. As a 'Fauna of New Zealand' it was a model for future work of its kind and it had an authority greater than anything that had preceded it, containing references to species previously discovered as well as those now added by Dieffenbach and others. A lengthy introductory note by Gray

58. Bell, 1976, pp.80-83.

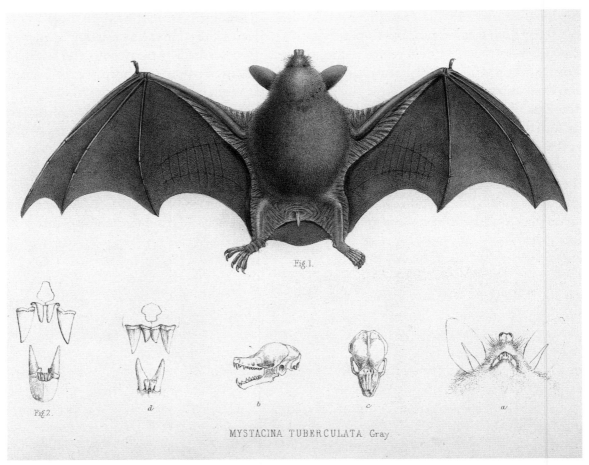

The Short-tailed Bat (*Mystacina tuberculata*), a hand-coloured lithograph by G. H. Ford for R. F. Tomes' paper *On two species of Bats . . .* in the *Proceedings of the Zoological Society of London*, 1857 (v. 25, pl. 51). This was the first published illustration of this animal. *(see p.115)*

59. *Bulla quoyi*, Gray, in Dieffenbach, 1843, v.2, p.243.

60. *Pterodroma cookii, Rallus philippensis dieffenbachi*, and *R. philippensis assimilis*.

61. Dieffenbach, 1843, v.2, p.194.

62. Gray, 1845, in *Genera of Birds*, pt.xvii, but not fully described until Gray, 1847, p.61-62. *Strigops habroptilus*

63. The lizards: *Leiolopisma zelandicum, Cyclodina ornata, Naultinus elegans*, and *Hoplodactylus pacifica*. The fish: *Anguilla dieffenbachi, Galaxius fasciatus, Gobiomorphus basalis*, and *Rhombosolea plebeia*. Lizard donations in Accessions Register, Reptiles, British Museum (Natural History).

describes the steps in the discovery of the New Zealand fauna to date, and contains more than a hint of his discomfort at there being so few New Zealand specimens in the British Museum, apart from those contributed by recent collectors. He does not disguise his envy of the French and too hastily dismisses the considerable work of Quoy and Gaimard, although names a shell after the former in the book.[59]

Directly, or through the New Zealand Company, Dieffenbach contributed 38 bird specimens, five reptiles, three fish, 58 shells, 20 insects and six crustaceans. Among the birds Cook's Petrel was a new species, and a further two, Dieffenbach's Rail and the Banded Rail, were new subspecies.[60] It is from Dieffenbach that we first learn of the Kakapo, the rare ground parrot, which he knew only from some green tailfeathers found in 'the interior'.[61] These were probably found in the North Island, indicating how recently this bird had disappeared from there, and from all but a few localities in the South Island of New Zealand; a disappearance which, in Dieffenbach's view, was hastened by dogs and cats. The Kakapo was described from a skin by G. R. Gray two years after this announcement.[62]

The reptiles, so frequently overlooked in previous explorations of New Zealand, were now receiving attention. The Brown and Ornate Skinks, Common Green Gecko and Pacific Gecko were described by Gray from Dieffenbach's material, collected in the Cook Strait area and further north. Four new fish, the Long-finned Eel, Banded Kokopu, Cran's Bully and the Sand Flounder were all collected by him.[63] Richard Owen had commissioned Dieffenbach to do some collecting for him and the Sand Flounder and some of the lizards were sent to the Royal College of Surgeons. Later Owen donated the lizards to the British Museum. Among the shells described by Gray were the familiar Toheroa, picked up by Dieffenbach on a west coast beach of the southern North Island, and the freshwater Mussel from Lake Taupo.[64] Invertebrates other than shells were described in this work by Gray's assistants, Adam White and Edward Doubleday; they included a few that Dieffenbach had collected, notably the Puriri Moth, the Giant Weta and the freshwater crayfish (Koura).[65] Although it is not clear who presented it, a new species of bat was described as a footnote to this work. The Short-tailed Bat was only the second native terrestrial mammal to be named, and the last to be discovered.[66]

In spite of Dieffenbach's contributions, many of the specimens that went to form this work came from other collectors, most notably Dr Andrew Sinclair, who supplied the new New Zealand Dabchick as well as most of the insects including the Giant Dragonfly, and other invertebrate species.[67] There were contributions also from James Busby, the British Resident, who had a shell named after him[68] and Rebecca Stone, as well as the earlier collections of Yate, Stanger and others. Just as the year 1840 was a turning point in New Zealand history, witnessing the arrival of the New Zealand Company ships, the signing of the Treaty of Waitangi, and the proclamation of sovereignty over the whole of New Zealand by the British Crown, so it signalled a change in the rate of outflow of zoological material, most of which went to the British Museum.[69] In the early 1840s representatives of a rapidly growing population donated or sold specimens to the Museum, resulting in a manifold increase on the previous decade. A trickle of molluscs was swelled by a rush of specimens from many different vertebrate and invertebrate groups.[70] The species named in Dieffenbach's publication represented part of that contribution.

With the limitations of his contribution to the zoology of the *Travels*, it would be easy to dismiss Dieffenbach as just another observant collector.[71] But in his writing — as we have already observed in his comments about the food chain — there is a scrutiny and an analysis of the fauna, flora and geology that runs deeper than casual observation. He subscribed to the view of a fluctuating organic world in which introductions of humans, fauna or flora could cause a 'chain of alterations', even when only a single species was the catalyst. Geological and climatic causes might have contributed to the decline of the Moas, introduced animals to the loss or decline of the New Zealand Quail and the Tuatara. He was aware of faunal and floral relationships between New Zealand, Australia and (South) America, and that 'different regions of the globe are endowed with peculiar forms of animal and vegetable life'.[72] Each plant or animal 'has its natural boundaries, within which it can live, and thrive and attain its fullest vigour and beauty'; and this crude ecological concept could be applied to colonisation (and this after all was the purpose of Dieffenbach's survey) where transplanting of a human population to an analagous clime would favour the continued development of that population. New Zealand, in that context, was found very suitable.[73]

The collective work of Dieffenbach, Gray and his colleagues was thus a valuable one, but by its omission of most of the South Island the survey was not complete; the gaps in this work might well have been filled had Dieffenbach's wish to remain in New Zealand been granted. Nor did the *Travels* win universal approval. Back in New Zealand Swainson predictably attacked Gray,[74] and Colenso, never overlavish with praise for contemporary biologists, annotated the margins of his copy of the book with remarks critical of his former neighbour, such as: 'Stuff', '!!!', and 'this is not your opinion but one you obtained

64. *Paphies ventricosa* and *Hydridella menziesi*.

65. Respectively *Aenetus virescens, Deinacrida heteracantha, Paranephrops planifrons*.

66. Gray, in Dieffenbach, 1843, v.2, p.296; *Mystacina tuberculata*. Two specimens were donated by Dr F. Knox of Wellington, in 1844, Accessions Register, 1844, Zoology, British Museum (Natural History).

67. The dabchick, *Podiceps rufopectus*; dragonfly, *Uropetala carovei*.

68. *Paryphanta busbyi*.

69. A lesser amount to Paris from Akaroa.

70. As seen from examinations of the British Museum (Natural History) Accessions Registers.

71. The detailed zoology of his report was written by others, but the sizes of the growing fields of zoology, botany and geology were such that they were increasingly difficult to fit within the compass of one person's abilities.

72. Dieffenbach, 1843, v.1, p.421.

73. *Ibid.*, pp.2-3.

74. Parkinson, 1984, p.26.

75. Colenso's copy of Dieffenbach's *Travels* is in the Alexander Turnbull Library. Bagnall and Petersen, 1948, p.102.

76. Relatively little has been written about him. He became a friend of Dr E. Shortland, Protector of Aborigines for the Colonial Government. (Shortland, 1851, pp.124, 165-168, 187) and almost certainly knew Johnny Jones of Waikouaiti; George French Angas, an artist and a collector in a small way, recalled pleasant discussions with Earl when he passed through Wellington (Angas, 1847, p.243). Earl took a role similar to that of Mathew Friend, an earlier visitor to Australia in 1830; perhaps another indication of 'metropolitan primacy' – maintenance of the London societies as a dominant influence in science (see McLeod, 1982, p.10.).

77. *New Zealand Gazette*, 5.10.1842, p.2.

78. Accessions Registers for Zoology, Mollusca, Insects, of the British Museum (Natural History), contain several examples.

79. *Sceloglaux albifacies*; others were the Fiordland Crested Penguin *(Eudyptes pachyrhynchus)*, Stewart Island Shag *(Leucocarbo carunculatus chalconotus)*, Brown Teal *(Anas aucklandica chlorotis)*, Marsh Crake *(Porzana pusilla affinis)*, Black-fronted Tern *(Sterna albostriata)*. There were several others that were not new species, (Fleming, 1982a, refers to these).

80. Earl to Owen, 24.2.1845, MS. 275-h-7, Library of the Royal College of Surgeons. Earl wrote from 29 Henrietta Street, Covent Garden, but efforts to discover more about him through this address failed.

in New Zealand'. Colenso it seems felt injured by the lack of acknowledgement for specimens and information passed to Dieffenbach during his stay in New Zealand.[75]

Hard on the heels of Dieffenbach came another visitor from England with a commission to collect rather than to survey, although in those days there was less difference between these activities than the terms might suggest. The figure of Percy Earl is an enigmatic one,[76] descending on the New Zealand scene in October 1842, to leave it two years later with one of the more substantial and important natural history collections made from that country. On his arrival on the brig *Nelson* on 23 September, 1842, at Wellington, Earl was introduced as '. . . a Naturalist of the very first order, in the prime of life, and possessed of all the requisite qualifications suiting him for the important commissions which, we were aware, had in the course of 1841 been entrusted to him. Mr Earl comes to the colony at the request of the Medico-Botanical and other learned societies, with the express purpose of collecting specimens of Natural History . . . to enable such information of undoubted authenticity, as will enable scientific and practical men in the Mother-country to direct their attention to those parts of natural history which shall be of practical utility to the Colony.'[77] This was not just a simple collector funded by a few enthusiasts, but someone who would reinforce the influence and position of the London-based institutions and the individuals within them.

But there was another purpose in his coming. It was seen as a favourable opportunity to establish 'a Society . . . As shall combine the comparatively few Naturalists in the Colony'. This very early attempt to institutionalize science in New Zealand did not advance, even though it was contemporaneous with the development of the Tasmanian Philosophical Society, but it was a sign that a more corporate approach to science within the colony was not too far away.

Earl's sponsors were obviously not prepared to wait on the findings of the New Zealand Company's naturalist, as the arrangements to send Earl were probably under way before Dieffenbach had a chance to present his report in England. Apart from the concern expressed by Gray at the standard of the colonial collections at the British Museum, there must have been a growing interest in the recently announced discoveries of the fossil remains of giant birds, and rather than wait on the uncertain efforts of early colonists or missionaries, a collector on the spot would make a great deal of sense. Whatever the reasons, Earl soon got down to business beginning with the molluscs of Port Nicholson.

Most of his collecting it seems took place in the south of the South Island, where he accumulated molluscs, insects, birds, and fossil bird bones. In October 1844, he left New Zealand for England taking his specimens with him, where shortly after his arrival he offered them for sale to the British Museum and the Royal College of Surgeons. The accessions registers of the British Museum list several lots of shells, insects and birds from this collector, mostly sold but one or two donated to the Museum.[78] His valuable bird collection, some 32 species, caused G. R. Gray, who was preparing the ornithology of the *Erebus* and *Terror* expedition, to recast his report to include Earl's material, which contained six new birds. The most notable of these was the Laughing Owl, now probably extinct. Others were birds seen since the time of Cook but never collected or described.[79]

Of still greater importance was his collection of fossil bird bones from Waikouaiti. In a letter dated 24 February, 1845, to Richard Owen of the Royal College of Surgeons, Earl offered him first choice of the bones, adding that he would dispose of the whole collection for £250/-/-. A similar offer was made to the British Museum.[80] It was indeed a valuable collection and it was later found to contain the remains of several species of giant birds. Having disposed of his recent haul, it appears that Earl headed once again

Wolf del et lith.

Printed by Hullmandel & Walton.

ATHENE ALBIFACIES. G.R.Gray.

The Laughing Owl (*Sceloglaux albifacies albifacies*), a hand-coloured lithograph by Joseph Wolf from G. R. Gray in Richardson's *Zoology of the Voyage of H.M.S. Erebus and Terror* . . ., 1844 *Birds* (pl. 1). One of relatively few New Zealand works by this prolific natural history artist. Published by permission of the Alexander Turnbull Library. *(see p.116)*

81. No connection with the French vessel. *Wellington Independent*, 2.9.1846, p.2. By a curious coincidence, another Earl (George, S.W.) an explorer and naturalist was in the same area a year later and also was later in Singapore (the destination of the *Heroine*'s survivors). These and a few other coincidences warrant further investigation. See Anon, 1958, p.324.

82. Among the better known: Sir George Grey, Lt Col Bolton, Dr Sinclair, and the English shell collector G. B. Sowerby. Less well known were Rev J. Medway, Mrs Mauger, Mrs Dunn, Dr Andrew Smith; there were others.

83. Particularly during 1846, 1847, 1854.

84. Bagnall and Petersen, 1948, p.86.

85. Dieffenbach, 1843, v.2, pp.228-296; Accessions Registers, British Museum (Natural History), e.g. Mollusca, November, December, 1842; Zoology, June, 1845, etc; Fitzroy to Sinclair, 6.12.1843, Sinclair papers No. 21, Alexander Turnbull Library.

for the South Pacific, for we next hear of him sailing to the north coast of Australia, where tragedy struck. His ship the *Heroine* hit a reef in Torres Strait and Earl was among the passengers who drowned.[81]

It was now becoming increasingly difficult to obtain material of interest to institutions such as the British Museum or private collections; this drove collectors to more remote and dangerous parts of the world in pursuit of their stock. Both Gilbert and Strange who collected for Gould in Australia also died in the course of their work. Percy Earl can be regarded as the first commercial natural history collector to visit New Zealand and stay for any length of time, and the value of people like himself and Strange, who came not long after, can be seen in the quality of their specimens. Their knowledge of the business allowed them to be more selective than amateur collectors and although they charged for their work, often quite heavily, the recipient could be reasonably assured of obtaining something unusual or new.

The recipients of this material in England were now clearly institutionally based or had strong institutional connections. White, Doubleday, J. E. and G. R. Gray and others were part of a growing scientific staff at the British Museum. Richard Owen was based at the Royal College of Surgeons, and even John Gould who ran a private establishment had strong connections with the Zoological Society of London and published many of his papers in its *Transactions* and *Proceedings*. Thus with amateur and professional collectors abroad, and professional zoologists at home, this phase of colonial zoology reached its zenith as far as New Zealand zoology was concerned in the 1840s and 1850s. The surpremacy of the French in this area was now being seriously challenged.

Leaving aside the major expeditions, there were many other collectors whose names appear in the British Museum accession records of the period.[82] There were prominent visitors such as the British naval captain Sir James Everard Home, son of Sir Everard Home of the Royal College of Surgeons who had written on the anatomy of the New Zealand Hagfish some years before. A commander of British naval vessels of the Sydney station, the younger Everard Home and his ship were available to provide a show of strength at various trouble-spots in New Zealand that began to show in the early 1840s, and it was on these patrols that he was able to pick up specimens of molluscs and birds that were later deposited in the British Museum.[83] Also from this period is George French Angas, the artist and traveller, who donated fish and insects; there were in addition many other collectors who were less well known.

One of these arrived as a visitor but later became resident in New Zealand. He was Andrew Sinclair, a Scottish naval surgeon who first visited New Zealand in 1841, when he was at the Bay of Islands at the same time as J. D. Hooker and members of the *Erebus* and *Terror* expedition.[84] During this period he made a substantial collection of molluscs, insects and other invertebrates, many of which were named and described in Dieffenbach's *Travels in New Zealand*. With Governor FitzRoy's support he returned to New Zealand in late 1843, to act in a medical capacity and as an explorer of the country's resources and in January the following year was appointed Colonial Secretary.[85] He continued to send plant and animal material to the British Museum and to Hooker, and thus played an important part in the beginnings of indigenous science in New Zealand, the subject of a later chapter.

Disquiet over animal introductions and the effects of human activity on native plants and wildlife had been expressed by naturalists visiting New Zealand well before Dieffenbach. But now, as the effects of these introductions were making themselves more obvious, so the problem was increasingly raised in published works. In 1859 another

Pencil and watercolour sketches of a weka *(Gallirallus australis)*, the Huia *(Heteralocha acutirostris)* and a fantail *(Rhipidura fuliginosa)* by G. F. Angas. The Huia was later used in *The New Zealanders Illustrated*, 1847 (pl. 34). Published by permission of the Alexander Turnbull Library (A 20). *(see p.118)*

voice was added to the chorus, that of A. S. Thomson, an Army surgeon who spent eleven years in New Zealand.[86] Thomson said: 'It is high time some good collections of the birds of New Zealand were made, as some species have entirely disappeared, and others are decreasing . . . This decay may spring from Nature's laws; but the introduction of man, dogs, cats, rats, pigs, and sheep into the country must have proved destructive to birds without wings, or to birds that fly with difficulty, and more particularly, to birds not instinctively aware (of such enemies) . . . It is possible that some birds may increase when grains are extensively cultivated.'[87] The idea of making collections had some rough common sense. The task of conservation appeared hopeless and the prevailing attitude was that as some species were going to become extinct anyway, people might at least remind themselves of what they looked like.

Apart from these thoughts, Thomson made another contribution which led to the description of the last major element of the terrestrial New Zealand vertebrate fauna, the native frog, until this time only hinted at by previous writers. On the eastern side of the Hauraki Gulf is the Coromandel Peninsula, a range of hills which in 1852 was thickly covered by trees shading the numerous streams draining into the Gulf. It was an area rich in minerals, and in November, 1852, some goldminers were washing the gravel of one of the streams that emptied into Coromandel Harbour when one of them dislodged a large quartz boulder and found a small frog underneath. The animal was such an unusual sight that a bottle was found and the frog placed in it with some water and the bottle then sealed with a cork. In this airless and probably overheated environment the frog died. It was then taken out of its container and put on the trunk of a nearby

86. Thomson, 1859. Scholefield, v.2, 1940, p.381.

87. Thomson, 1859, p.28.

88. Thomson, 1853, pp.67-69.

89. Fitzinger, 1861, p.217.

90. Schmarda, 1861, pp.183-229. He published a work on invertebrates (Schmarda, 1859) that included New Zealand species, e.g. polychaetes. The 'vegetable caterpillar' is discussed in Note 42 above.

The so-called "vegetable caterpillar"; these are usually caterpillars of the Porina Moth (*Wiseana* spp.) infected by a fungus, causing its unusual appearance; from R. Taylor's *Te-Ika-a-Maui*, 1855, p. 425.

91. Von Lendenfeld in his spare time explored the Mount Cook region and conquered the Hochstetter Dom, (Bagnall, 1980 v.1, p.581).

92. Von Lendenfeld, 1889, pp.1-2.

93. King, 1981.

94. Stresemann, 1975, pp.226-230; Whitley, 1959, p.112.

95. Dawson and Dawson, 1958, pp.39-49. (These authors also note the visit of G. Thilenius from Hamburg); Ramsay, 1979, p.5. See also Fleming, 1982b, pp.25-32 for a record of German visitors to New Zealand.

Kauri so as to preserve it by drying. Subsequently the miners were visited by Major Thomson who, on being given the specimen, showed it to some nearby Maoris who at the time were holding a conference with Lt Governor Wynyard. These Maoris and others who were later shown the frog were amazed by it, could not recollect having seen one before, and furthermore did not seem to be able to offer any convincing stories that would suggest that the frog and the Maoris had some past association. Thomson described the frog without giving it a scientific name and sent off the description and the story of how he came by the specimen to the *Edinburgh New Philosophical Journal*, where it was published in 1853.[88]

Ferdinand von Hochstetter, who arrived in New Zealand as part of the *Novara* expedition, discovered a frog under circumstances rather similar to those experienced by Thomson. He sent this specimen back to his native Austria, where it was described by Fitzinger in 1861 and named *Leiopelma hochstetteri*. Later the frog was pronounced to be of a very primitive type.[89] Hochstetter's place in New Zealand science was a special one and will be examined in greater detail in pages to come. His visit to New Zealand with the *Novara* was only a few years away from the date of Thomson's discovery, but future Austrian interest in New Zealand was heralded by the private visit to Auckland of Ludwig Schmarda, a traveller and naturalist who had a reputation as a zoogeographer. Schmarda arrived at Auckland on 30 September, 1854, on board the *William Dennie* from Sydney, and stayed for five weeks. He collected a few invertebrate species among which was a presentation from Andrew Sinclair, the Colonial Secretary, of a 'vegetable caterpillar'.[90]

Many years after Schmarda and Hochstetter, Robert von Lendenfeld came to New Zealand in 1883. From the University of Graz, he had also worked at the Zoological Station at Trieste, then an Austrian port. He was given room at the Canterbury Museum by Haast and collected marine invertebrates by dredging at Lyttelton and Sumner and further afield at Timaru.[91] The culmination of his visit to Australia and New Zealand was a monograph of the horny sponges, published in 1889. A few New Zealand species are described.[92] Von Lendenfeld coincided with another Austrian who came to New Zealand to work with Haast, but remained for far longer than he intended. Styled as 'The Collector', Andreas Reischek has a place in New Zealand zoology which will be discussed later.[93]

The lineage of German and Austrian naturalists which began with Forster continued and divaricated throughout the late 19th century. In the spirit of the Hanseatic city where they successively held the directorship of the Bremen Museum, first Otto Finsch and then Professor H. H. Schauinsland came out to New Zealand to observe, collect, and consult with local naturalists. Finsch, an ornithologist already known to Buller, Haast, Hector, Hutton and others, arrived in October 1881, explored widely while in New Zealand, and returned laden with specimens in 1882.[94] Schauinsland came in 1896-1897, accumulating a substantial collection of New Zealand fishes, insects, reptiles and crustaceans. On a visit to the Chatham Islands he procured a specimen of the Chatham Islands Red Admiral, and Stephens Island supplied him with the Banded Rail, Yellow crowned Parakeet, South Island Kaka, Morepork and New Zealand Falcon as well as numerous sea birds. Both localities supplied him with wetas.[95] The small commensal crab he discovered in mussels from French Pass bears his name — *Pinnotheres schauinslandi*.

Interestingly enough, like several other naturalists from their part of the world, Finsch and Schauinsland were not part of an expedition but used commercial vessels to carry themselves and their specimens. Major German expeditions were less frequent, the most

RALLUS ASSIMILIS or STRIPED LANDRAIL.

The Banded Dotterel (*Charadrius bicinctus*); (right) a juvenile of this species or possibly a New Zealand Dotterel (*Charadrius obscurus*), and (lower) a Banded Rail (*Rallus philippensis assimilis*); original watercolour illustrations from Richard Laishley's *Notes and drawings . . .* (n.d.) (pl. 39). Published by permission of the Trustees of the British Museum. *(see p.122)*

significant scientific voyage to the area being the German expedition to the Auckland Islands in 1874-5.

What was it that drove these Teutonic naturalists to the far side of the world in their ones and twos, instead of as members of nationally mounted expeditions? Adverse circumstances at home might have formed part of the answer for the Forsters or

Dieffenbach, but surely that did not apply to them all? Some explanation might reside in the old tradition of wandering scholars who would study in several places before completing their education, moving among the universities in the different states that existed before the German-speaking peoples united themselves on a national basis.[96] Lacking national cohesiveness and institutional bases early on, the Germans were unable to visit New Zealand as the members of national expeditions until quite late in the piece. But even the Austrians (and we can include them here) demonstrated that same spirit of enterprise and individualism, such as when Hochstetter left his ship in New Zealand to embark on a lengthy and solitary exploration, only to encounter a German-speaking colleague who had arrived there on his own account. It was these qualities and traditions that led to a characteristic and unique contribution to New Zealand's natural history which, along with the 18th century German work on molluscan shells, demonstrates a marked difference in style from that of French and British counterparts.

This chapter should not be left without brief mention of a representative from that group of early New Zealand naturalists which was so prominent at its beginning — the ministers of religion. The Reverend Richard Laishley was not a collector of note, nor a publisher of traveller's journals or frequent scientific articles. But his contribution lends perspective and colour to the history of New Zealand zoology, if it does not leave us with much in the way of valuable scientific information.[97]

A Congregationalist minister, Laishley arrived with his family at Auckland on 12 October, 1860, settling in Onehunga at the end of that year. He passed the next seven years there before moving temporarily to Melbourne, and then returning to a parish in Thames. But it is at Onehunga that most of his observations and drawings of plants and animals were made. From his journal a picture has been created of a man absorbed in his surroundings, exploring the reaches of the Manukau Harbour on horseback and rambling in the nearby hills in the company of his sons, who shared their father's love of natural history.[98] He made notes on many of the species observed and was particularly taken with birds and insects[99] and like Colenso and Taylor before him, kept a kiwi in captivity. Most striking however, are his illustrations.

Laishley had trained as an artist before deciding to become a minister, and for ten years or so attended the Royal Academy Schools.[100] The artistic traditions of the time did not perhaps sit altogether comfortably with natural history illustration — some of Laishley's animals are posed in formidable backgrounds and one searches for an allegory — but his subjects and he are at greater ease when he reduces his scale and concentrates on the detail, or on creating the impression of detail, of his subject. Thus his insects, birds' nests and eggs are more true to life than a rather vulnerable-looking Little Blue Penguin struggling with a dramatic landscape. Overall, however, his work is most attractive, rather better than Taylor's, and we can only regret that a popular natural history written and illustrated by Richard Laishley never saw the light of day.

The man and his work make a point that is repeated throughout this history, and is true of many other 19th century naturalist-artists, Swainson, Gould and Keulemans amongst them: that 'Art and science are not opposed but are complementary ways of exploring the world and man's place in it'.[101] With his pencil and brush, Laishley did for the creatures of his parish what Gilbert White of Selborne had done with his pen.

96. Russell, 1983, p.85.

97. He has been the subject of short biographies by Sibson, 1983, and Whitehead, 1984.

98. Whitehead, 1984, p.110.

99. Laishley (n.d.) MS. *Gleanings ... etc*, (351 fols.); Laishley (n.d.) MS. *Notes and drawings etc ...*, Zoology Library, British Museum (Natural History).

100. Whitehead, 1984, p.156.

101. *Ibid.*, p.101.

Chapter Eight

A GIANT
STRUTHIOUS BIRD

IT WAS ONLY GRADUALLY THAT Europeans became aware that a giant flightless bird had once roamed New Zealand. None of these birds had appeared either in myth or reality to any of the members of Cook's expeditions. And despite the accounts of huge reptiles and other mythological monsters, decades of visitors would pass through the country before word was heard of the Moas. Even the secretive kiwi had been part of the scientific literature since 1813, but at that date the first report of a Moa was some distance away. The 1830s were reached and there were still no myths or bones of giant birds to excite the curiosity of Europeans until an accident in the form of a shipwreck stranded the trader Joel Polack on the shores of the North Island East Coast.

Sailing from Thames in the cutter *Emma* in July 1834, Polack struck a series of four gales that badly damaged the vessel, forcing it to put into Tolaga Bay for repairs. But the damage was too great and the *Emma* was to remain there a wreck.[1] During his extended stay on the East Coast the Maoris presented Polack with 'the petrifications of the bones of large birds', and thus, after returning to England to write his memoirs, he became the first European to publish on Moa bones and the mysterious tales associated with them.[2] He concluded: 'That a species of emu, or a bird of the genus Struthio, formerly existed in the latter island, I feel well assured, as several large fossil ossifications were shewn to me when I was residing in the vicinity of the East Cape, said to have been found at the base of the inland mountain of Ikorangi (Hikurangi). The natives added that, in times long past, they received the tradition, that very large birds had existed, but the scarcity of animal food, as well as the easy method of entrapping them, had caused their extermination'.[3]

In the light of subsequent discoveries and knowledge of the Maori myths relating to the Moa, there is no doubting the accuracy of Polack's account. He was not making any claim to fame by virtue of his discovery and the reference to the Moa in his book initially almost died for want of scientific attention. But such would be the attachment that some scholars later felt towards the bird and its remains and the prestige associated with its discovery, that indifference to his findings would turn into a charge against Polack that his claim to have seen Moa bones was a fabrication. This was made by William Colenso, jealous of Polack's deserved prominence amongst the early Moa writers, who many years later in 1894 denounced his account, saying it was concocted by '. . . a Jew of the lowest grade and type. I have often been in his rum store on the Kororareka Beach . . .'.[4] He virtually accused Polack of lying, and of having his book written by another hand, but we can dismiss this outburst as the intemperate grumblings of a man nearing the end of a long, hard road, disappointed that his was not the first written European record of the Moa's existence.

Polack's first sojourn in New Zealand was coming to an end when in the same part of the country another more significant encounter with a Moa bone was taking place. This time the bone was a small piece, an ambiguous fragment that was ultimately to come before a competent authority who would eventually recognize it for what it was and so bring New Zealand and its hitherto interesting but unremarkable fauna on to centre stage.[5]

1. Polack to Busby, 8.1.1835. National Archives, New Zealand. Later accounts (including Polacks' own) have the date wrong.

2. Polack, 1838, v.1, p.345.

3. *Ibid.*, v.1, p.303.

4. Colenso to Mantell, 22.8.1894, *in* Buick, 1936, p.45. He displayed an anti-semitism fairly typical of the period.

5. Buick, 1936 researched this story in detail, producing a lengthy but interesting account which is followed in part here.

First in this chain of events was John Williams Harris. As the agent of a Sydney firm, he had been sent to New Zealand to arrange trade in flax and other commodities and had settled on the North Island's East Coast at Turanga (Poverty Bay) where he established a shore whaling station. Domiciled for several years on the East Coast, he had probably become aware of Moa bones some time before Polack had visited, but in this isolated spot he had no-one to whom he could report his find. The nature of his work required him periodically to visit Sydney; on one such trip aboard the brig *Martha* Harris carried with him a piece of bone, a greenstone mere and other bits and pieces, arriving in Sydney on 23 January, 1837.[6]

6. *Ibid*, p.75.

During his visits to Sydney, Harris made it his practice to call on his uncle by marriage, Dr John Rule, a retired English naval surgeon who had come out to New South Wales in 1833. This particular January the two had been unable to meet, Rule being out of town and Harris, discovering that the schooner *Currency Lass* was to return to the Bay of Islands somewhat earlier than expected, was unable to wait for his return. Harris sat down and wrote a note to Rule on 28 February, explaining his hasty departure and noting that he was leaving a few items for the doctor in the care of a Mrs Pike, including a bone from an extinct bird and a greenstone club. He added that the bird 'was of the eagle kind', was extinct, and was known to the Maoris as 'A Movie'.[7] How Harris had come by this term is hard to imagine as it bears very little resemblance to any other name given to the Moa by the Maoris, even allowing for the variations that Europeans contrived. It is at this point that John Harris bows out of the story, appearing briefly later on to acquaint the missionaries William Williams and Richard Taylor with the Moa. To this largely unrecognized collector and his concern to pass on his discoveries we owe much of the early progress in this field.

7. Rule to Harris, 28.2.1837, *Ibid*., p.107-109.

Dr Rule received the letter, collected the bits and pieces from Mrs Pike, and after some time had elapsed he returned with them to England in 1838 or in 1839. Driven by curiosity and a shortage of money he took the bone around various museum collections to make comparisons that would further convince him of its claimed origin and facilitate its sale. Here great credit must go to him for his persistence, as in appearance it was as unpromising a specimen as had ever come this far and he was clearly going to have difficulty in persuading someone to buy it. Rule first offered it to the British Museum where in spite of the ornithologist G. R. Gray's support it was turned down. He then took the plunge and offered it to the Royal College of Surgeons for 10 guineas, encountering for the first time that undoubted authority on all matters to do with bones and comparative anatomy, Richard Owen, Hunterian Professor and Professor of Anatomy and Physiology at the College, already familiar with New Zealand flightless birds through his classical series of studies on the anatomy of the kiwi. Who better to evaluate Dr Rule's bone?

8. Rule to Owen, 18.10.1839, v.22, Owen Correspondence, British Museum (Natural History), *in Ibid*., p.111-113.

With the backing of his studies and comparisons in the museums, Rule wrote to Owen on 18 October, 1839, stating his views on the bone's origin.[8] What he had to say in his letter was no great advance on the original claims of Harris and he perpetuated the erroneous supposition that the bird was able to fly. He also had a lot to say about the veracity of Maori traditions which owed as much to mythology as they did to accurate recollection. Hard on the heels of his letter, he visited Owen in person, only to find the eminent man about to be engaged in a lecture to his students. This was an unlucky coincidence as Owen had only 24 such lectures a year (although he gave numerous others) and in concentrating his mind before the occasion he was probably in no mood to entertain a stranger with a piece of bone in a small paper parcel. The delicate links in the chain of discovery were about to be broken then and there as Owen hastily and

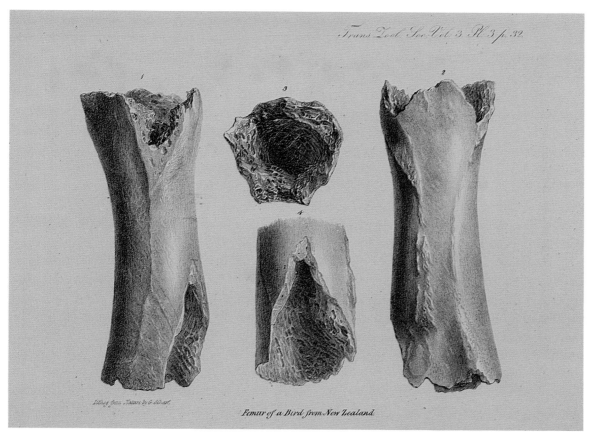

The piece of Moa bone given to Dr Rule by J. W. Harris, and later described by Owen; lithographed by G. Scharf for Owen's *Nature of a Fragment of the Femur . . .* in *Transactions of the Zoological Society of London*, 1842 (pl. 3). *(see p.124)*

probably disdainfully dismissed the unspectacular piece as a marrowbone. But Rule was stubborn as well as soundly prepared. He restrained the Professor in his flight to the lecture room by demonstrating the typically avian appearance of the internal wall of the bone, and in support of the story of its having come from New Zealand, he produced the greenstone club. At last Owen paused to reconsider, indicating that he would look at the fragment more closely if Rule would like to return the next day to hear his judgment.[9]

His lecture over, Owen began to compare Rule's specimen with specimens in the Hunterian Museum. He began with an ox and then proceeded to other quadrupeds that might have been introduced to New Zealand and left their remains there. None of these bore any likeness at all to the one before him so he was compelled to move on to the Ostrich, and it was here that some points of similarity began to emerge. His eventual conclusion was that the piece came from the femur (thigh bone) of a bird that was probably larger than the Ostrich, but whose bones were not pneumatic and were possibly filled with marrow. It was later recalled by Owen's friend Broderip that Owen took that very first fragment and made a projection of what the whole bone must have looked like on a piece of paper. When a perfect bone subsequently arrived it was laid on the paper and fitted exactly the outline that Owen had drawn for it.[10]

On the basis of his findings, Owen decided to recommend purchase of the bone to the Museum Committee at the price asked, but unfortunately they declined. He was however able to find Rule a purchaser, a Mr Richard Bright, whose family collected natural history specimens and whose collections later passed to the British Museum, but the price had now dropped to £3.[11] Before the bone went to its new owners Owen

9. *Ibid.*, p.117.

10. Owen, 1894, v.1, p.151.

11. Mantell, 1940, p.225. See Buick, 1936, pp.124-135 for a detailed history of the Bright family collection.

had some drawings made, apparently without Rule's knowledge, and these with his description and conclusions were placed before the Zoological Society of London on November 12, 1839.[12]

The publication committee of the Society was in turn as sceptical as Owen himself had been when the bone was first shown to them. Their chief difficulty lay with the fact that it was not a fossil, and the notion that the bird had only recently become extinct did not seem to occur to them. The inference that it must therefore have been a living species caused them to question how such a large bird had escaped observation in a small country which accomplished naturalists had visited and explored.

There was also the problem that too much was apparently being deduced from a single fragment. Out of deference to Owen's ability, but with obvious reluctance, the Committee agreed to publish the paper in the Society's *Transactions* with the rider that responsibility for the paper rested 'exclusively with the author'.[13]

The first announcement of the discovery was published in the Society's *Proceedings* in March 1840. The more detailed description of the bone accompanied by illustrations appeared in the *Transactions of the Zoological Society of London* in June 1842. One hundred extra copies of one of these papers were printed and sent to New Zealand in the hope that further information or specimens would be supplied.[14] The dates of publication of the papers, as we shall see later, are important.

While that first piece of bone was on its way to England, a sequence of events was taking place in New Zealand that would incorporate another group of personalities in the story of Moa discoveries. These were the missionaries, who for pastoral reasons were showing a great deal of interest in the East Coast. It was almost inevitable, should one

12. Mantell, 1940, p.225.

13. Owen, 1879, v.1, pp.iv, v.

14. There is confusion as to which paper had extra copies. Most evidence points to the second paper, but it seems strange that Owen was prepared to accept two years delay before his need for more specimens was made public.

Professor Owen exhibited the bone of an unknown struthious bird of large size, presumed to be extinct, which had been placed in his hands for examination by Mr. Rule, with the statement that it was found in New Zealand, where the natives have a tradition that it belonged to a bird of the Eagle kind, but which has become extinct, and to which they give the name "Movie." Similar bones it is said are found buried in the banks of the rivers.

The following is an abstract of Professor Owen's paper upon this bone :—

"The fragment is the shaft of a femur, with both extremities broken off. The length of the fragment is six inches, and its smallest circumference is five inches and a half. The exterior surface of the bone is not perfectly smooth, but is sculptured with very shallow reticulate indentations : it also presents several intermuscular ridges. One of these extends down the middle of the anterior surface of the shaft to about one-third from the lower end, where it bifurcates ; two other ridges or lineæ asperæ traverse longitudinally the posterior concave side of the shaft ; one of them is broad and rugged, the other is a mere linear rising.

No. LXXXIII.—PROCEEDINGS OF THE ZOOLOGICAL SOCIETY.

The first published notice of Owen's identification of Rule's fragment of Moa bone; published in the *Proceedings of the Zoological Society of London*, 1840. *(see above)*

of them be interested in natural history, that they would encounter stories of the Moa's existence and even the bones themselves. William Williams and William Colenso from the Bay of Islands were rewarded for their exhausting work on the production of the New Testament in Maori with a 'vacation' which, at the instruction of the Committee of Missionaries was to take place in the East Coast district where the Mission was yet to have influence.[15] They sailed on the *Columbine* in January 1838, leaving the vessel at Hicks Bay to walk overland to Waiapu where the Maoris told Colenso about a 'monstrous animal' that they called a Moa. Bones were referred to, but not seeing them himself he was forced to discount the whole story as 'fabulous'.[16] At the end of the following summer a second missionary party made the same journey to look for sites for another mission. This time a new arrival, the Reverend Richard Taylor, was in the company of William Williams.

The two sailed from Paihia in March 1839 on the cutter *Aquila*. Their accommodation on board was not quite what the reverend gentlemen had been used to. Taylor remarked '. . . for what with bugs and fleas we had not much rest and the filthy way in which the meals were prepared quite destroyed what little appetite we had.'[17] But they survived the journey and left the cutter at Hicks Bay, near East Cape, from which point they tramped overland, first to Waiapu and then further south. It was at Waiapu, according to Taylor's later and rather unreliable recollection, that while resting in a hut belonging to the chief Rukuata he noticed a bone fragment jammed in the thatch of the ceiling.[18] It appeared to be part of a toe or claw, although Williams apparently scoffed at the idea of its being a bird bone. These were Taylor's later recollections but his journal of the day makes no mention of this discovery. Rather, it records that on 21 April, 1839, when much further south at Poverty Bay, they were invited to dine at the home of a Mr Harris, the same John Williams Harris who had given Dr Rule a piece of bone two years earlier. Harris gave Taylor a bone too and told him that he had found others 'larger even than an ostrich', and the Maoris there told them the same stories heard by previous visitors to the region.[19] William Williams also recalled being shown some bones on this occasion. On his return journey Taylor found another bone in a riverbed and concluded that the bird must have been of a large size.

Colenso revisited the East Cape area in the summer of 1841-1842; for one who would gain great pleasure from the esteem granted pioneers of Moa discovery, the outcome of this expedition would ultimately prove something of a disappointment. At Waiapu the Maoris presented him with seven Moa bones obtained from Whakapunake, a mountain reputed to be inhabited by the bird.[20] There were five femora, 9 tibia, and one bone undetermined, and although the specimens were not perfect, there was enough in the way of characters to convince Colenso of their origin. He travelled down the coast to Turanga to find that William Williams, now resident there, had also some luck and was in possession of a huge tibia. Colenso left two bones from his collection with William Williams, who was going to send them to Oxford University, and went on his way, the two having agreed to offer rewards to the Maoris for any further finds.

Colenso returned to the Bay of Islands, only to be shortly notified by Williams that their offers of reward were having good effect and some huge examples were arriving. Colenso felt that he now had sufficient information to draw conclusions and sat down to write a paper entitled 'An Account of Some Enormous Fossil Bones of an Unknown Species of the Class Aves, later Discovered in New Zealand'. It was completed on 1 May, 1842 and on 4 November of that year he forwarded it to Tasmania for publication in the *Tasmanian Journal of Natural Science.* . . .[21] Unfortunately all was not well with the

15. Bagnall and Petersen, 1948, p.65.

16. Colenso, 1880, p.65.

17. Taylor's Journal MS. 953 papers, 1839-1868, Alexander Turnbull Library; *Aquila* also spelled *Aquilla* in some works.

18. Taylor, 1873, p.97.

19. Taylor's Journal, MS. 953, p.156 (p.111 typed transcript), Alexander Turnbull Library.

20. Colenso, 1880, p.66.

21. Bagnall and Petersen, 1948, pp.142-143.

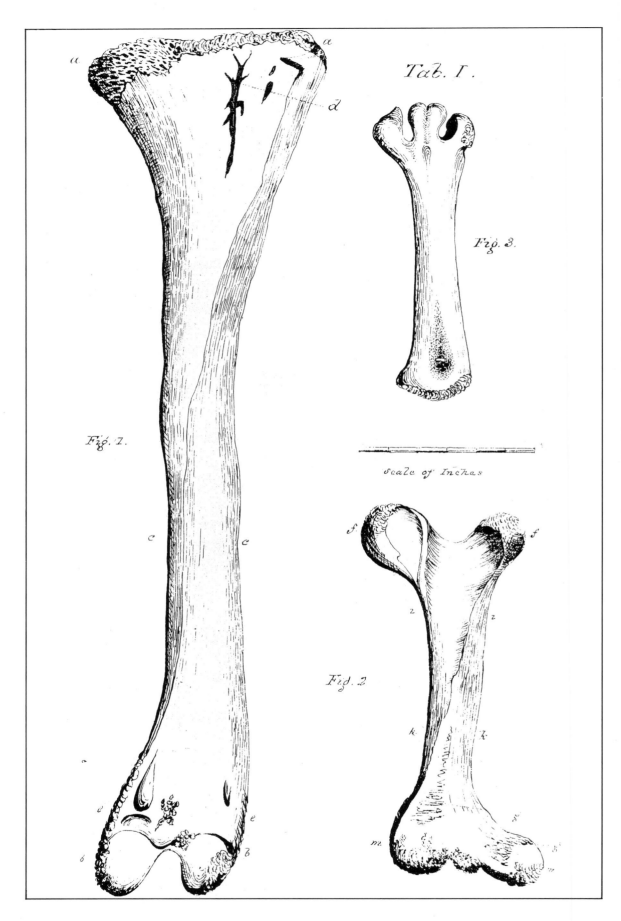

Tab. I.

Fig. 1.

Fig. 2.

Fig. 3.

Scale of Inches

Sketches of Moa bones made by William Colenso and published in his paper *An account of some enormous Fossil bones* . . . in the *Tasmanian Journal of Natural Sciences* . . . ,1846 (pl. 1). Published by permission of the Alexander Turnbull Library. *(see p.127)*

Tasmanian Society, as the new Governor, Sir John Eardley-Wilmot, had been responsible for the inauguration of a rival group, causing the remnants of the Tasmanian Society to move to Launceston where they continued publishing the journal. Although received at a meeting of the Society on 17 May, 1843 these moves had the effect of delaying this and other publications of Colenso's until 1846.[22]

Earlier, in England, Owen had published his conclusions on Rule's bone, but caution had prevented him from scientifically naming the bird, until July 1843 when he described it as *Dinornis novaezealandiae*.[23] It has long been claimed that Colenso described but did not formally name his specimen, although he gives a list of the genera in the Family Struthionidae that clearly includes 'No. 7 *Moa*', having already used *Moa* throughout the text. It appears Colenso (perhaps unintentionally) employed *Moa* as a generic name which, had it been published when originally intended, might have had priority over Owen's *Dinornis*. As it was, Owen magnanimously complimented Colenso on his conclusions and arranged publication of the latter's paper in the *Annals and Magazine of Natural History*, 1844 where it appeared with *Moa* removed from the list of genera.

Many years later Colenso found it necessary to defend himself against a charge of plagiarism with respect to his discovery and subsequent publication.[24] His attacker claimed that Colenso had prior knowledge of the Moa as a result of the distribution of one of Owen's first two reports published in 1840 and 1842 respectively. It is nowhere absolutely clear which of these papers was distributed to the colony and both Owen[25] and Colenso's accuser later appear confused about the relevant dates of publication. However, if as has been widely assumed it was the 1842 paper with its illustration, then it would have been impossible for Colenso to have read it before writing his own. It is possible, but unlikely, that Colenso saw Owen's first note published in the *Proceedings* in March 1840, as this account, or knowledge of it, could have reached New Zealand (perhaps with Sinclair) before Colenso sent his manuscript to Tasmania. With something less than wholehearted emphasis, Colenso denied that Sinclair or Hooker knew of the paper when in the Bay of Islands, so in the absence of any evidence the charges must be dismissed.[26]

While Colenso was contemplating the prospect of his first paper, William Williams, as we have already discovered, was receiving Moa bones from all sides as the Maoris discovered the extent of the missionaries' interest and the value they placed on them. In the end he estimated he had the bones of some 30 birds some of which were massive. A tibia was 2′ 10″ [85 cm] in length, and there were others almost as long; a femur at its narrowest point was 8″ [20 cm] in circumference. He decided to send this collection to the Reverend Dr Buckland, a leading cleric and geologist and one of those influential English scientists who were part of the booming interest in palaeontology; Buckland was known to William Williams from his days at Oxford. On the last day of February 1842, Williams sat down to write a letter to Buckland, describing the circumstances of his acquisitions and passing on the unsubstantiated rumour that two Americans had seen a live bird near Cloudy Bay. At this time, aided by a training in surgery and by comparisons with the bones of a domestic fowl, Williams was in no doubt that he was handling the remnants of a giant bird, estimated in his letter at 14-16 feet [4.3-4.9 metres] in height.[27] Here, while acknowledging that great credit must be granted to Williams and Colenso for their interpretation of the bones, the endeavours of John Williams Harris should not be forgotten. He too was well aware of their origins, possibly sharing his views with the missionaries, and only the lack of a direct contact with the English zoological establishment prevented him from becoming more publicly associated with the discoveries.

22. Oliver, 1949, p.4, claims August, 1843 as Colenso's date of publication, but see Hoare, 1969, pp.205-207 for a history of the Tasmanian Society's difficulties and the dates of its activities.

23. Owen, 1843, pp.8-10, 1844, pp.144-146.

24. Colenso, 1892, pp.468-478.

25. Owen, 1879, v.1, pp.iv, v. Owen implies that the extra 100 copies were made of the *Transactions* paper (with plate), but then goes on to give the date as '1838'. As his first note was published in 1840, and the *Transactions* paper in 1842, he is astray with the date — extra copies could have been made of either paper, but see Note 14 above.

26. Colenso, 1892, p.476, says he is 'pretty sure' that Sinclair, with whom he had contact, had not seen Owen's first paper.

27. Williams to Buckland, 28.2.1842, No. 191, Additional MS. 38091, British Library.

Bearing in mind the dangers of sending valuable property by sea, Williams resolved to divide his collection into two consignments which were put separately on a boat to Port Nicholson and from there went to Sydney and London. They were received by Dr Buckland at Oxford who transmitted them almost immediately to Richard Owen. He later respected Williams's wishes concerning their ultimate use, and asked Owen that casts be made for distribution to an assortment of museums in England and France, including his own at Oxford.[28]

Since his first announcement on the Moa and his appeal for more specimens, Richard Owen had established himself as the logical ultimate receiver of Moa bones. Insofar as he was destined to become even more involved with fossil New Zealand birds, it is time to take a closer look at the man and his times, for they have a great deal to do with New Zealand zoology.[29]

For some twenty years Owen held undisputed hegomony over the natural sciences in Britain. Narrower in outlook than Banks, and vastly different in style, he nevertheless adopted the same central position, part of the 'hub' towards which exotic zoological and palaeontological collections must flow. He was at once hated, feared and admired, a man who had colleagues and correspondents by the score but few friends. If one practised as a serious amateur or a professional in the realms of natural history, an encounter with Owen was almost unavoidable. Some, like Gideon Mantell, found that everywhere they turned there was Owen, aloof, deprecatory, and all-powerful in the higher councils of British science.

Owen began his career with a training in medicine, but shortly thereafter was persuaded to become an assistant to William Clift, the Conservator of the Museum at the Royal College of Surgeons, where he began by cataloguing the Hunterian collections. From there he embarked on a lifetime study of the anatomy of vertebrate and invertebrate animals, in the process becoming Hunterian Professor and then succeeding Clift as Conservator, before becoming Superintendent of the Natural History Department of the British Museum. This department later became a museum in its own right as a result of his efforts.

The science of comparative anatomy had been well developed by the French under Cuvier,[30] but England had no rival figure until Owen started publishing in 1830. Two things fortified his career and reputation: the first was the developing science of palaeontology and the second the increasing supply of zoological specimens, fossil and living, from the British colonies — marsupials from Australia, the fossil reptiles from South Africa, and the kiwi and fossil birds from New Zealand. Palaeontology itself was enjoying unprecedented popularity during the middle decades of the 19th century; visions of large and fearsome monsters from the past fired the public imagination and authorities such as Gideon Mantell were in great demand for soirées, *conversaziones* and public lectures. For naturalists who were social climbers and craved public attention it was a godsend, their intriguing specimens providing them with an entrance to the salons of the aristocracy. The extinct New Zealand birds came fortuitously right in the middle of this period, their discovery coinciding with the presence of Owen and heightened enthusiasm for palaeontology and natural history in general.[31]

One or two decades earlier, Rule's bone might have been totally ignored. Two decades later and the discoveries would have been overshadowed by Darwin and the debate on evolution. Instead, there was to be a chain of papers on the extinct birds, ending in a two-volume monograph; they would be looked upon as the 'greatest zoological discovery of our time'.[32] By the time of the appearance of his concluding work on Moas in 1886,

28. Buckland to President, Royal College of Surgeons, 22.2.1844, MS. 275-h-5, Royal College of Surgeons Library.

29. *The Life of Richard Owen* (2v) written by his grandson, the Rev. Richard Owen and published in 1894, is a fairly laudatory account with a concluding essay on Owen by Huxley. The essay was not written with a 'light heart' and Huxley struggles to find praise for his subject.

30. Along with other French anatomists like Geoffroy St Hilaire, on the foundation developed by Buffon and Daubenton. Owen comes fairly late into the field.

31. Desmond, 1982, pp.4, 5.

32. Broderip to Buckley, 20.1.1843, Additional MS. 30891, British Library.

Owen's authority had already been undermined by the discussion on evolution and by other smaller defeats. But for a time he had the power to bring into the limelight an animal, a fauna and a country. Owen, Gray and the Hookers were among the imperial masters of New Zealand natural history and it was part of Owen's skill that when something like Rule's bone turned up, he would quickly recognize its potential and move to make the field 'peculiarly his own'.[33]

Owen had to exercise considerable patience before he began to receive any confirmation of his first analysis of Rule's bone. This finally arrived in a letter on 10 January, 1843, from the Reverend William Cotton,[34] who confirmed Owen's findings and told him that he had seen the bones collected by Williams (which were in fact specimens collected following Williams's original consignment to England). Hard on the heels of this letter the first box of bones sent by Williams arrived on Buckland's doorstep. The Prince Regent was notified, His Royal Highness expressing great interest in 'this feathered monster'[35] and a desire to be kept informed. Buckland sent the consignment smartly on to Owen. On 19 January the box was opened in the presence of William Broderip[36] and Owen himself. Fascinated, they watched as the huge bones came out one by one — a pelvis, vertebrae, femur — becoming so preoccupied that the time for their evening meal came and went. 'Yesternight we supped upon the mysterious moa' recalled Broderip in a letter to Buckland the next day.[37]

The second box arrived as safely as the first in October that year, and again Broderip was present at the Owen house for the opening.[38] Following notices in the *Proceedings*, Owen described and figured this material in the *Transactions of the Zoological Society of London*, 1844, and Williams was given credit by Owen for his interpretation of the bones: 'wholly unaware that its more immediate affinities had been determined in England'. Owen's attempt to attract specimens by sending copies of his early publication out to New Zealand eventually had effect, although after some time had elapsed. Colonel William Wakefield received a letter from the New Zealand Company in London and a copy of Owen's paper late in 1843 and took action immediately to supply specimens.[39] While all this was going on another sizeable discovery was about to take place in New Zealand, this time in another part of the country and by a missionary who was already familiar with the fossil birds.

In May 1843, the Reverend Richard Taylor arrived at Wanganui which was to be the headquarters of his large parish for some years to come. A few weeks after his arrival, in July, he was returning along the coast from a trip to South Taranaki when he stopped at Waingongoro to baptize two children. He paused for a while on the shore near the mouth of the Waingongoro River and noticed a fragment of bone lying in the sand. He was reminded of the bone seen on the East Coast and asked the Maoris what it was. 'A Moa's bone,' was the reply 'what else? Look around and you will see plenty of them.' To his astonishment he saw that the sand flat was heaped up into mounds full of Moa skeletons, 'a regular necropolis'.[40] But many of the bones when examined were found to be fragile and disintegrated when touched. He emptied out his box of supplies and packed a few handy specimens, intending to return at some later time to make a more comprehensive selection. Unfortunately for him, the site was visited in the intervening period by Walter Mantell, who made a large collection of these sub-fossil bird bones to send to his father Gideon Mantell in England. It was a great loss for Taylor, although he had well over three years in which to make his return, years in which he placed his pastoral duties ahead of this particular interest in natural history. Mantell's collection from Waingongoro was a valuable one, containing species other than Moas and of

33. Spokes, 1927, p.204.

34. Owen, 1879, v.1, p.74. Cotton was Bishop Selwyn's chaplain.

35. Owen, 1894, v.1, p.208.

36. Broderip was a fellow naturalist and one of the founders of the Zoological Society of London.

37. Broderip to Buckland, 20.1.1843, Additional MS. 38091, British Library.

38. Broderip to Buckland, 11.10.1843, Additional MS. 38091, British Library.

39. Williams's contribution resulted in the description of *Anomalopteryx didiformis*, *Dinornis giganteus* and *Dinornis struthioides*. Colenso, 1891, p.475.

40. Taylor, 1873, pp.97-101.

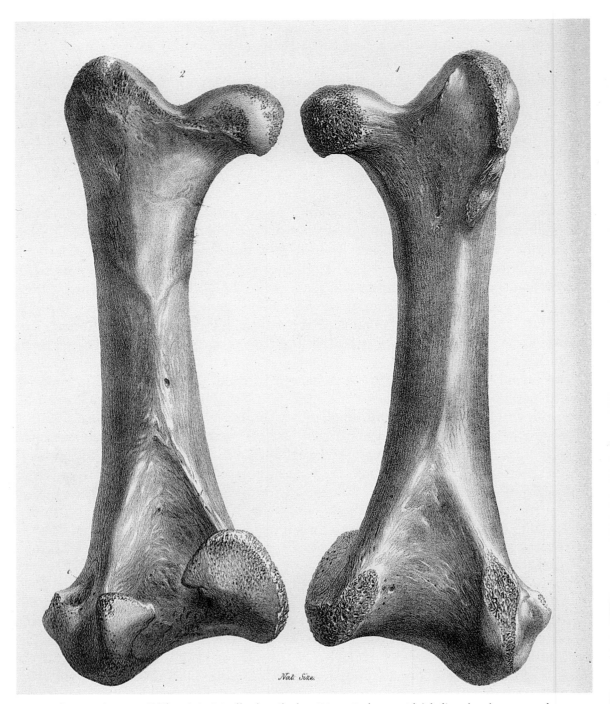

Bones of *Anomalopteryx didiformis* (originally described as *Dinornis dromaeoides*) believed to be among those sent by the Reverend William Williams to Dr Buckland and described by Richard Owen; lithographed by G. Scharf for Owen's *On Dinornis an extinct genus . . .* in *Transactions of the Zoological Society of London*, 1844 (pl. 22). *(see p.131)*

sufficient importance to make them the subject of a separate story to be told later on.

Over the years Taylor continued to find Moa skeletons on his coastal travels around Wanganui and was once brought a near-perfect specimen, lacking only the skull. Eager to find the missing portion he rode off at once to the site, where with great care he unearthed the delicate skull bones. Wrapped up and placed in the crown of his hat, the bones seemed safe until Taylor's horse bucked, sending him sprawling on the ground and shattering the bones. He hobbled around salvaging what he could for Owen, but once again the full measure of exciting discovery had slipped through his fingers.[41]

41. *Ibid.*, p.99.

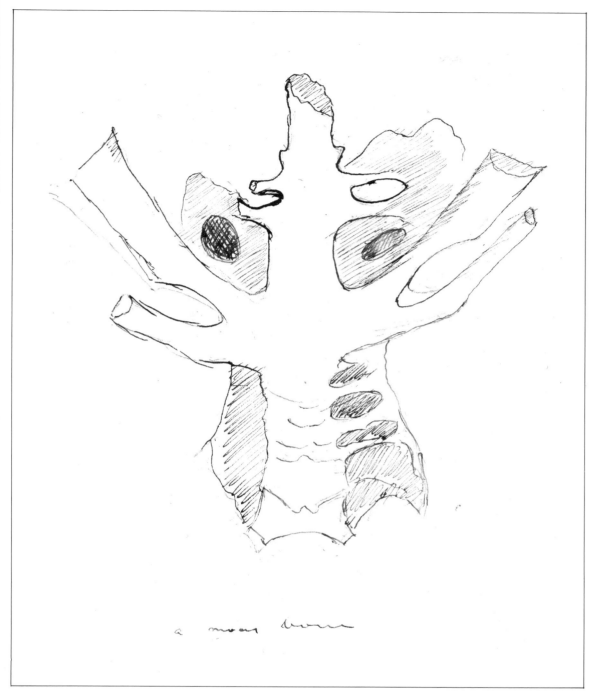

A sketch of a Moa bone by the Reverend Richard Taylor, thought to have been done at some time in 1840's. Published by permission of the Alexander Turnbull Library (E 296). *(see p.131)*

While Taylor was making his find at Waingongoro, new Moa sites were being exposed in the South Island where a discovery was made at the mouth of the Waikouaiti River north of Dunedin. The first to view this site was possibly Johnny Jones, a whaler and land dealer who established a station at Waikouaiti early in 1840. The site was visited during 1843-1844 by the collector Percy Earl, who took his collection back with him to England to offer it for sale to Owen in February 1845.[42] Two new species of Moa came from this collection and they were described by Owen in 1846. Material from the site was also aquired at about the same time by a Dr MacKellar, who sent the bones to Edinburgh University — a departure from the normal course of events. It appears

42. Earl to Owen, 24.2.1845, MS. 275-h-7, Royal College of Surgeons Library. Not 1846, (as in Oliver, 1949, p.5). According to Burdon, 1941, pp.112, 115, the spot was a 'Golgotha' covered in whale-bones, smelly and with few natural attractions.

A watercolour of Moa egg shell fragments from Waingongoro; by an unknown hand, but part of the collection of Mantell papers (MS 83/422) in the Alexander Turnbull Library. Published by permission of the Alexander Turnbull Library. *(see p.131)*

43. Bennett to Owen, 20.2.1844, Additional MS. 38091, British Library. McKellar acquired his specimens at about the same time as Earl.

44. Spokes, 1927, p.204.

that MacKellar was from Sydney and had been given the bones by 'a sailor' (probably one from the whaling station) during a visit to New Zealand.[43]

If Owen did not receive Moa bones directly, such was his position of authority and control over the publication of the fossil bird papers that he received material from even the most reluctant of sources. Gideon Mantell has already been mentioned in connection with his son's collections from the Waingongoro site, and it was the result of this and later efforts by Walter Mantell in New Zealand that placed the elder Mantell firmly among the followers of the Moa. But he was always a planet to Owen's sun, feeling he had the right to lecture on his son's discoveries but not to describe them: 'I determined to forego the pride and pleasure of describing these new acquisitions, and allow him (Owen) to have the use of all the novelties my son has collected'.[44] If Owen felt he could be generous in appreciation and praise for the work and interest taken by the missionaries and others in the colony, a rival lecturer on the subject on his very doorstep was a different matter, and Mantell was often the victim of bruising encounters in spite of his generosity to Owen in passing on specimens. The products of middle class backgrounds, both men

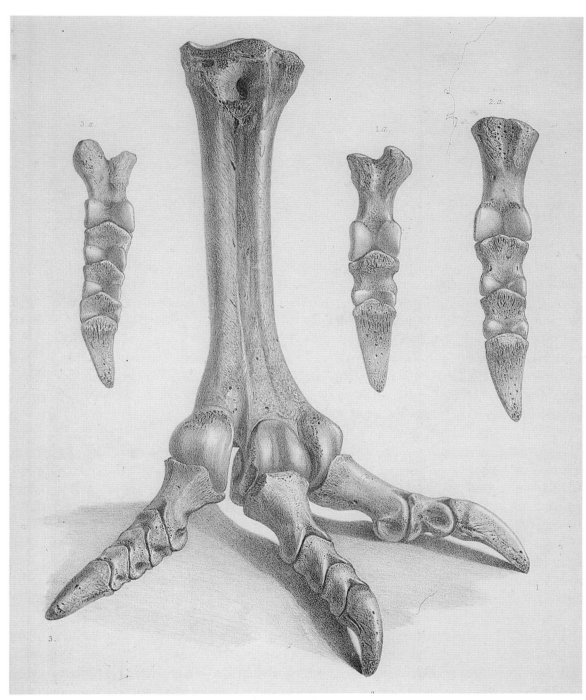

A lithograph of one of a pair of Moa feet collected at Waikouaiti by Tommy Chaseland and given to Walter Mantell who sent them to his father. The illustration made the frontispiece for Gideon Mantell's *Pictorial Atlas of Fossil Remains* . . . 1850, in which it was hand-coloured. Several uncoloured versions were circulated, some possibly by Mantell to his son. Published by permission of the Alexander Turnbull Library. *(see p.136)*

were snobs and courted the attention of the titled and famous. But Mantell was at a severe disadvantage in that the need to keep up his medical practice conflicted with his palaeontological interests, whereas Owen was a professional in this field, albeit a busy one. Mantell suffered from a further handicap in the form of a persistent and agonizing spinal condition which was no help in preserving his magnanimity and good humour at the time of Owen's attacks.

The brief truce which established itself when Mantell first offered Walter's collection

Gizzard stones of the Moa; a hand-coloured lithograph by J. Smit for G. D. Rowley's *Ornithological Miscellany*, 1876-1878, (v. 3, pl. 113). Published by permission of the Alexander Turnbull Library. *(see below)*

to Owen was soon broken and the enmity between them reached a peak three years later following a second consignment of bones from New Zealand. Earlier in 1846 Walter had also visited the Waikouaiti site, where the outgoing tide had exposed a pair of Moa tibia and feet standing in the mud. A whaler, Tommy Chaseland, carefully dug them out and gave them to Mantell who despatched them to his father.[45] The two feet were articulated and on 27 February, 1850, Gideon Mantell read a paper on his son's recent collections. Owen arrived at the same meeting, also loaded with Moa bones, and 'commented in his usual deprecating manner', declaring that the feet in Mantell's possession were imperfect, lacking the hind toes, and that he himself possessed a perfect foot which he had described the night before at a meeting of the Zoological Society. 'Poor envious man,' wrote Mantell, but in fact Owen was infuriatingly right, although the hind toe has long been a contentious and difficult problem, only recently resolved.[46] The following March saw Owen read a paper on *Dinornis* that again earned Mantell's contempt, but Owen was virtually unassailable, and he went on to publish more than thirty papers on Moas and other flightless New Zealand birds. Were it not for this work, the premium zoological journals of the day would have been largely devoid of New Zealand species.

Bones were not the only remains of the Moa that were discovered: it was said that an ancient taiaha given to Captain Cook was embellished with a Moa feather and feathers were found at Alexandra buried under river silt.[47] Egg fragments were among the earliest discoveries so it was to be expected that a perfect egg would be a highly prized discovery. One was eventually discovered, in a grave, clasped in the hands of the human occupant, and when the natural history dealer J. C. Stevens put this one up for auction in London in 1865 a bid of £105 was entered although the egg was passed in at £200. In 1899 a really perfect egg was discovered in the most unusual circumstances in the Clutha River where a gold dredge had undermined a bank allowing the imprisoned egg to drop down into the water where it floated downstream into the well of the dredge where it was recovered by one of the workers. This discovery earned the dredgehand £50 and the egg went to the Otago Museum. Earlier, Walter Mantell had remarked of Moa eggs: 'Roughly guessing I should think an ordinary hat would serve as an egg cup for it — what a loss to the breakfast table!'[48]

There were also other signs of the Moa's presence. In 1866 David Miller, who ran the ferry across the Taruheru River (near Gisborne) noticed some curious impressions in the soft sandstone rocks near his landing place. They were too evenly spaced to be there by chance, so one day he drew the attention of Archdeacon William Leonard Williams to them. Williams almost immediately recognized them as Moa footprints, and more were found later in other parts of the country.[49] Even more remarkable were

45. Buick, 1937, p.108. Some Waikouaiti bones sold to Philadelphia by Gideon Mantell. Mantell G. to Mantell, W. 14.11.1850, MS. Papers 83, Alexander Turnbull Library.

46. Mantell, 1940, p.250; P. Millener, *pers. comm.*

47. Buick, 1937, p.340. Gizzard stones were often encountered.

48. *Ibid.*, p.255. The first was probably the huge *Dinornis maximus* egg (253 × 178 mm) collected by Fyffe in 1859 and eventually purchased by the British ornithologist G. Rowley (Oliver, 1949, p.43). The second egg from *Pachyornis elephantopus* (Benham, 1902, pp.149-150); the identification of eggs with particular species of Moa has never been absolutely certain. Mantell, W., to Mantell G., 20.6.1848, MS. Papers 83, Alexander Turnbull Library.

49. Williams, 1872, p.124-127. The son of William Williams.

the mummified remains found in caves and fissures in the south of the South Island. The lower leg of a *Dinornis novaezealandiae*, with much of its ligaments, muscles, and skin, must have looked almost fresh to its discoverer who found it in a rock fissure in 1874, while the dried head of a *Megalapteryx didinus* found in a cave in Queenstown presented a most fearsome aspect.[50] No wonder the latter-day Moa hunters were ever-hopeful of catching a glimpse of a huge feathered shape striding through the bush. The Moas had left a distinct presence behind them and there were parts of this uninhabited and untracked country that gave the watcher the feeling of a recently vacated room.

The new discoveries accumulated at a slow but steady pace for the remainder of the century and on into the next. From localities such as Awamoa, Glenmark, Hamilton (Otago), and Pataua (North Auckland) came further new species.[51] With these later discoveries came also a significant change: the Moas were able to become part of indigenous scientific activity in New Zealand, even to the extent of being described in locally produced scientific journals. Interest reached as far as the Governor, although Sir George Grey's first forays into Moa bone collecting met with disaster when Government House burnt down in Auckland on 23 June, 1848. 'I lost,' wrote Sir George in a letter to Richard Owen, 'all my plate, china, linen, wine, and the most valuable of my books, besides curiosities, native songs of different countries, and objects of natural history, which I had been many years in collecting. In your department I lost a magnificent collection of moa bones, including a complete skeleton of the largest moa which had ever been found. I had three complete moa heads of different species . . .'; he went on to describe more of his collection, which included many Kakapo.[52] Grey's enthusiasm remained undampened and he indicated he would take the field again. Some years later, in 1866, he visited the Waingongoro site with the Reverend Richard Taylor, its original discoverer. Taylor recalled 'it was amusing to see his Excellency grubbing up the old ashes and carefully selecting what he thought was worth carrying away'. A large cloth spread on the ground was soon covered with an assortment of specimens: fragments of eggshell, chert flakes (of which more shortly), polished implements and other artifacts.[53]

But the person who most clearly put his stamp on Moa discovery and description during this period was a recent arrival from Germany, Julius Haast. In the company of his friend and colleague, Ferdinand von Hochstetter, he became involved in the investigation of the remains found in the caves of the Aorere River, near Collingwood in north-west Nelson.[54] This was in 1859, and nine years later, in the same year as he was appointed Director of the Canterbury Museum, he became associated with the Glenmark site from which he acquired specimens of the largest of the known Moas, *Dinornis giganteus*, which he described in the *Transactions and Proceedings of the New Zealand Institute*.[55] It was perhaps fitting that Haast would also describe one of the smallest of the New Zealand Moas and in so doing make a point that Owen himself could not make: he named the bird *Anomalopteryx oweni*. Haast did not succeed or eclipse Owen but rather worked in parallel, publishing some of his papers in New Zealand and others in England, while Owen, now in the twilight of his career, laboured on with his anatomical researches. Owen's last Moa was described in 1883. Of the 12 or so species currently believed to have existed he had described eight.

With the establishment of scientific societies and other institutions throughout the New Zealand provinces, it was inevitable that others would soon join Haast in studying the remains of the giant birds; in the wake of the discoveries would come increasing speculation as to the origins of the birds and the reasons for their extinction. Such surmise became fruitful ground for disagreements which flared into disputes as the combatants

50. Oliver, 1949, p.10.

51. *Pachyornis elephantopus, Dinornis maximus, Emeus huttoni* and *Anomalopteryx oweni.*

52. Owen, 1894, v.1, pp.319-320.

53. Taylor, 1873, p.100.

54. Oliver, 1949, p.6.

55. It was, appropriately, in the first volume of the *Transactions* (Haast, 1869, p.87).

56. Burdon, 1950, p.182.
His paper (pp.178-192)
summarizes the battle
between Haast and his
adversaries.

An engraving of the Moa and a kiwi from F. von Hochstetter's *New Zealand* . . . . . 1867, (pl. 176). *(see p.137)*

indulged in the 'polite censures and veiled sarcasms upon which the progress of science thrives'.[56]

The most significant and most divisive of these disputes might be said to have had its origins in the paper written by Colenso and published in 1846, in which he concluded that the Moa had died out prior to, or contemporaneously with, human arrival in New Zealand.[57] His argument stressed the importance of the very sketchy and often fanciful recollections that the Maoris had of these birds. Much later these ideas were developed at great length by Haast in a series of addresses to the Canterbury Philosophical Institute, the first of which was on 1 March, 1871.[58] Haast recognized that Moa bones had been discovered by Mantell and others in kitchen middens associated with stone implements and human remains, but he stated that these should not be interpreted as a sign that the Moas were hunted and eaten by the Maoris. Instead, he claimed that an earlier Polynesian culture had hunted the Moas to extinction, and that this race of Moa hunters were superseded by the Maoris who arrived in New Zealand after the Moas had died out. The substantial part of Haast's argument rested on the artifacts found in the middens where, mixed in with the polished stone tools of probable Maori origin, were primitively worked fragments reminiscent of a stone-age culture.

The following month Haast read a supplement reinforcing his first paper, attempting to iron out some of the contradictory statements and quoting other work in support of his argument. An answer was not long in coming, and it arrived in the formidable person of James Hector, Director of the Colonial Museum and the Geological Survey, a man of unrivalled power in the New Zealand scientific establishment. On 16 September, 1871, Hector read a paper before the rival Philosophical Institute of Otago in which he firmly refuted Haast's theory.[59] The primitive stone implements, he explained, were simply the result of pouring water onto heated stones in the Maori ovens. The stones fractured into sharp flakes, conveniently available for a rough-and-ready refashioning into knives, spearheads, or what-have-you. They may even have been used to cut slices of cooked Moa from the steaming joints in the oven. Hector was supported in his argument by another speaker, W. D. Murison, who reached similar conclusions: it was the forefathers of the Maoris who had hunted the Moa, not the 'Moa-hunters'.[60] Haast retaliated the following December, when he turned aside the arguments of Hector and Murison and firmly held his ground.[61]

With battle lines drawn, others such as the Reverend Stack and F. W. Hutton joined the fray. But the issue really came to a head as a result of some excavations made at Moa-bone Point Cave near Sumner, Christchurch. Here under the supervision of Haast worked his two assistants, one of whom was Alexander McKay, an ex-miner whose curiosity about geology caused Haast to take him under his wing.[62] Amongst the methodically exposed bones, pieces of eggshell and Maori artifacts were some stone chips of the type referred to earlier and, significantly, a polished basalt adze — clear evidence that the bones and the chips were contemporary with a culture well capable of producing finished tools. This of course did not suit Haast's theory at all, and although he assimilated the information revealed by his two assistants he did not publish it until some time later, claiming that the business of directing the Canterbury Museum kept him from giving this particular collection a proper examination.[63]

In the interim, Alexander McKay moved to Wellington to work under James Hector, taking with him his notes on the discoveries made at Moa-bone Point. In view of the continuing silence from Canterbury, he decided to read them to the Wellington Philosophical Society at a meeting on 8 August, 1874. Haast was furious, perhaps less at the content of the paper — for some parts of it paralleled his own thinking — than at McKay's effrontery in taking it upon himself to broadcast what Haast regarded as his intellectual property. Worse, Haast read about this perfidious deed in the newspaper.[64]

57. Colenso, 1846, pp.81-107.

58. Haast, 1872, pp.66-90.

59. Hector, 1872, pp.110-120.

60. Murison, 1872, pp.120-124.

61. Haast, 1872, pp.94-107.

62. Buick, 1937, p.53

63. Von Haast, 1948, p.722.

64. Buick, 1937, p.76.

His response was to disinter his notes and specimens of two years earlier and read a paper to the Canterbury Philosophical Institute the following month. There he explained away the presence of polished implements at the site by consigning them to a period of greater antiquity than earlier supposed, and by advancing the state of civilization he had previously attributed to the 'Moa-hunters'. But at least he now acknowledged that the chipped and polished stone implements were part of the same culture.

McKay came almost to the point of regarding the Moa-hunters and the Maoris as one and the same. But in the end he drew back from a definite conclusion, saying that the Moas were either exterminated by a race different from the Maoris, or that the Maoris around 1350 years before (ie 520 AD) had set about exterminating the birds. Haast sought to have McKay's paper banned from publication in the *Transactions*, but the Journal published the two papers in the same issue, ensuring that the protagonists fell out completely.[65]

65. Haast, 1875a, p.54-85; McKay, 1875, pp.98-105.

Echoes of the dispute continued to reverberate around the Wellington and Canterbury branches of the New Zealand Institute until eventually the President of the Royal Society of London, Dr Joseph Hooker, the botanist, was called upon to adjudicate. He found that McKay had a right to publish although it was thought inconsiderate of him to do so, and there the matter more or less ended. Haast patched up his relationships with his colleagues — all, that is, except McKay.[66] Although he had already resigned his presidency of the Canterbury Philosophical Institute on 6 May, 1875, he was not finished with Moa disputes. There was more to come.

66. Von Haast, 1948, p.747.

Three months after he read his paper on the Moa-bone Point material, Haast was making equally forceful pronouncements with respect to the middens found at Shag River; these too he thought to be of considerable antiquity.[67] The following year, F. W. Hutton, who was Professor of Natural Science at Otago University, countered Haast in a paper read to the Otago Institute on 24 August, 1875, in which he found not the slightest evidence to back Haast's claims.[68] That year there were seven papers on Moas in Volume VIII of the *Transactions* — discoveries at Pataua River and Poverty Bay fuelled further discussion on the Moas and their relationship to the Maoris. And then of course in 1892 there was Colenso's injured response to critics and usurpers of his claim to be amongst the foremost pioneers of Moa discovery in New Zealand. The same issue of the *Transactions* contained Hutton's lengthy review of the New Zealand Moas, and although by no means the last word on the subject, it provided a fitting conclusion to more than forty years of discovery and lively discussion.[69]

67. Haast, 1875b, p.98.

68. Hutton, 1876a, p.107.

69. Colenso, 1892; Hutton 1892.

For a time Moas were *the* issue in New Zealand science, a match for the debate on evolution, and if figures like Colenso, Taylor, Haast, Hutton and the others bridged the gap between imperial and indigenous colonial science, then the Moa was one of the major platforms on which this was carried out.

Chapter Nine

A SPECIES NEW TO SCIENCE

THE CONVENIENT PACING OF ORNITHOLOGICAL and palaeontological discovery in New Zealand was to continue. The kiwis, followed by the moas, provided the British naturalists and public with a sustained regimen of flightless birds — some extant, others of magnificent size but extinct. Now, with appetites whetted, they would be shown that a bird once believed to be extinct was now alive.

In the last few days of January 1847, Walter Mantell searched for Moa bones on the coast south of the volcano the Maoris called Taranaki (Mt Egmont). He had spent the day of the 26th walking to the Kaupokonui Beach, arriving at the Methodist Mission at Heretoa where he relaxed in the company of the missionary William Woon, revelling in a good wash and the unaccustomed luxury of mutton chops, bread and butter.[1] He farewelled his hosts at sunset and left with his Maori assistants for the mouth of the Waingongoro River, some miles to the south. They went by the beach and it was dark when they reached the mouth; they passed over the undulating flat known as Te Rangatapu and clambered up to the Ohawetokutoku pa where they were welcomed by the Maoris and passed the night. The next few days were fine and were spent digging in the surrounding area, with sufficient success to distract Mantell from the numerous tribulations of palaeontological field work, including his flea-ridden accommodation, the sandflies and a painful boil.[2]

The open, undulating surface of Te Rangatapu was believed by the Maoris to have been inhabited by their ancestors. Mantell thought it had been picked over by Richard Taylor, so turned instead to a site under the cliffs at Ohawetokutoku where he had spent the night. It turned out to be an excellent choice. His digging exposed perfect examples of the bones of the giant birds, but so soft they virtually turned to clay when firmly grasped. His joy at their discovery was short-lived as the Maoris, attracted by the activity, came down in droves — men, women and children — to help in the digging. They trampled on the bones laid out to dry and competed for each new one as it emerged from the sand, with scant regard for its fragile nature. Mantell was reduced to despair: 'You can imagine how exasperated I must have been to see specimens destroyed before my eyes'.[3] Miraculously some survived intact, and although he appeared unaware of it at the time, amongst the Moa bones, eggshells and artifacts were the skull and a few other skeletal bones of a bird that on later examination was to be declared new to science — the *Notornis*. The place of this discovery, Mantell was to recall later, was the camping site of the shore party from *HMS Alligator*, which had rescued the Guard family from the Taranaki Maoris and had rested there unaware of what lay beneath them.[4]

The bones that survived the excavations and the attention of his overzealous helpers were packed in readiness for the overland journey that was to follow, and the remainder of his time was spent in further exploration and a last call to Heretoa to see the Woon family. It was February when finally he set out for New Plymouth — an arduous walk through the bush as he had chosen an inland route, to the west of the mountain, rather than risk being thwarted by the tides of the coast. He arrived on the evening of 3 February, 1847, exhausted but with sufficient energy remaining to write a few lines of birthday greetings to his father and to tell him of his exciting discovery.[5]

His laden assistants had left Waingongoro separately and some days passed before they trailed into New Plymouth and the task of sorting and repacking the specimens began.

1. Mantell, W., 1847 *in* Moa Journey, Waingongoro, Jan. 1847-Feb.3.1847, Walter Mantell, MS. 1847, Alexander Turnbull Library.

2. *Ibid.*, 1847.

3. Mantell, W., to Mantell, G., 25.6.1847, MS. Papers 83, Alexander Turnbull Library.

4. *Ibid.*, 25.6.1847; 20.6.1848.

5. *Ibid.*, 3.2.1847/5.4.1847.

A notebook sketch of the Waingongoro River mouth made by Walter Mantell at the time of his Moa discoveries. Published by permission of the Alexander Turnbull Library (MS Man 1847, 1242). *(see p.141)*

At length the work was done and he was able to see the bones safely aboard the *Inflexible*, a schooner bound for Wellington. It would have been quicker and easier to have joined his collection on board, rather than return to Wellington overland, but the prospect of further discoveries lured him back to Waingongoro. He returned in the company of Charles Nairn, who already had a commission from Colonel Wakefield to collect Moa bones for Professor Owen. But the success of the earlier digging was not repeated and the two left the site virtually empty-handed.[6] They carried on their journey to Wellington through Wanganui, which was under martial law, and then down through the Manawatu where signs of civilization, including road gangs of Maoris and soldiers, increased as they went further south. They finally joined the nearly completed road that stretched from the Porirua harbour down to Wellington, eventually arriving there soaked to the skin at 10 p.m. on 29 March — almost a month after the discovery of the bones. The stress and the hard physical demands of the expedition showed in Mantell's face. His friends said later that they hardly recognized him, and that he looked 'uglier and older'.[7]

The *Inflexibile* had arrived in Wellington long since, on 13 February. Within a few days of Mantell's arrival the bones were packed in Toi-toi flowers in a large box and loaded aboard the *Comet* bound for Sydney on 8 April, 1847, where a friend arranged its onward journey to London and his father, Gideon Mantell.[8] In the event, the box took a great deal longer than expected to reach its destination, and anxious letters were exchanged between father and son as the months went by and nothing was seen or heard of Walter's valuable collection. His friend in Sydney had confirmed the safe departure of the box on board the *Eugenia* which was to sail in May, but nearly a year later, in the absence of any favourable news, Walter was to write, 'I tremble for the Moa bones'. But although he did not realize it at the time, all was long since well.[9]

On 13 December, 1847, eight months after its departure from New Zealand, the collection finally arrived in London. Gideon Mantell unpacked the box and spread out its contents in his dining room. He was delighted with the well-preserved specimens, more than 500 of them, including a number that were new to him. News of the arrival soon reached members of the scientific community who converged on the Mantell house at Chester Square. Lyell, the geologist, was to call, and so was the eminent Dr Buckland, Dean of Westminister. All who visited were impressed by what they saw.[10]

Apart from establishing the presence of the collection in London, Gideon saw himself faced with three further courses of action. First, Walter would have to be financially recompensed for his labour and expenses; secondly, the collections should be used to

6. *Ibid.*, 3.2.1847/5.4.1847.

7. *Ibid.*, 3.2.1847/5.4.1847.

8. *Ibid.*, 5.4.1847; ?5.6.1848. The Toi-toi flowers provided a soft packing for the delicate bones — they were ideal.

9. *Ibid.*, 24.4.1848.

10. Mantell, G. to Mantell, W., 18.1.1848, MS. Papers 83, Alexander Turnbull Library.

A sketch of the Notornis (*Notornis mantelli*) skull made by Gideon Mantell in a letter to his son, Walter Mantell, on 18.1.1848. Published by permission of the Alexander Turnbull Library (MS 83). *(see below)*

enhance his son's standing with the New Zealand authorities; and thirdly, the bones had to be examined and described by a competent authority, and it was this last matter which probably exercised Gideon the most. Gideon was a vertebrate palaeontologist of some consequence, and although birds were not his speciality, he might have been excused for wanting the collection for himself. It was greatly to his credit, then, that he recognized that there was another in the field whose reputation in fossil ornithology, particularly in relation to New Zealand birds, was already well established, one who 'had made the subject his own'.[11] This was of course the now familiar figure from the Royal College of Surgeons — the redoubtable Richard Owen. The bones were placed at Owen's disposal, a noble gesture considering their dislike of each other, and on Christmas Day Owen replied to Gideon with a courteous letter thanking him for his liberality. This seasonal good humour was shortlived however, for 1848 had barely started when Owen received a consignment of bones from Colonel Wakefield in New Zealand which he did not allow Mantell to see.

Owen worked hard on the collection and on Tuesday, 11 January read descriptions of some of the skulls to a meeting of the Zoological Society. One skull was a *Dinornis*, the second a *Palapteryx*. The third he placed in the family Rallidae, saying that it was closely related to the genus *Porphyrio* (including the Pukeko), which still existed in New Zealand. To this third skull and accompanying skeletal bones he gave the name *Notornis* (meaning 'southern bird'), adding the specific name *mantelli* in honour of its discoverer. Thus the scientific name, *Notornis mantelli*, was established, although the full connection between the Maori names — moho or takahe — and the new one had yet to be drawn.[12]

Meanwhile Gideon made plans to use the bones to promote his son's interests. He proposed giving several lectures on the collection and, capitalizing on the goodwill created by his earlier gesture, he asked Owen to write to Governor Grey in New Zealand and to recommend Walter in the strongest terms.[13] Whether the publicity in England and the pressure brought to bear in New Zealand had any significant effect on his son's career is difficult to establish, but it certainly did him no harm, and in October 1848 he was appointed a Government Commissioner with special responsibilities for Maori land in the Middle (South) Island. (He was previously Superintendent of Military Roads.)

Gideon decided to bide his time on the sale of the collection, thinking that it would be of greater value once properly examined by Owen and catalogued. He intended applying to the British Museum in the first instance, although as they already had a

11. Spokes, 1927, p.204.

12. Mantell, G. to Mantell, W., 18.1.1848. Published by Owen in *Proceedings of the Zoological Society of London*, 1848, pp.1-11, and the Society's *Transactions*, 1848, pp.345-378, with a drawing of the skull. Gideon's letter contained pen sketches of the *Notornis* skull and that of the two other genera recognized by Owen from Walter's collection.

13. *Ibid*, 18.1.1848, 'my great desire is to make this collection a means of promoting your interests . . .'. It was not entirely his father's ambition. Walter, (20.6.1848) wrote: 'too much influence cannot be used at the Colonial Office . . .'.

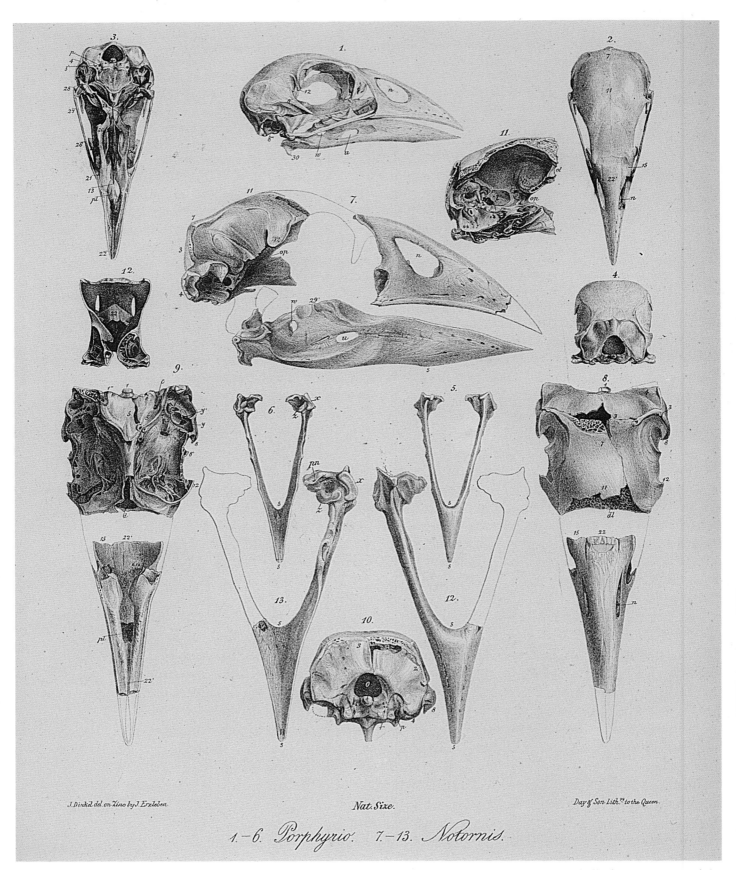

1.–6. Porphyrio. *7.–13.* Notornis.

A lithograph of some of the bird skeletons found by Walter Mantell, including the skull of *Notornis*; part of the original description by Richard Owen in *On Dinornis (Part III): Containing a description of that genus . . . and . . . Notornis . . .* in *Transactions of the Zoological Society of London*, 1848, (pl. 36). *(see p.143)*

large collection of Moa bones they would take only the choicer parts of his offering. The remainder he thought to take to other institutions. In the end however he was able to sell the collection to a Mr Forshall of the British Museum for £200 on 4 September of that year, an occasion that was celebrated by his going into the city and investing in £300 worth of Consols.[14]

Interest in the *Notornis* continued into the following year as a result of the elder Mantell's lectures, although nothing much new could be added in 1849. Late in the year he received another collection from Walter with *Notornis* bones not being mentioned, but already events were taking place in New Zealand that were to lend a new dimension to the earlier discovery.[15] It was one thing to unearth the fossil or sub-fossil bones of a new species, but it was quite another to produce evidence of a living representative of a species hitherto thought extinct. Such a discovery was to raise the reputation of the *Notornis* to the level of that of the Dodo — a bird that had already roused the imagination of the Victorians. Once again, it was Walter Mantell's good fortune to become associated with the discovery.

Later in 1849 he encountered a sealer who had a collection of bird skins, amongst which was one that he must have recognized as belonging to a *Notornis*. He tried to persuade the man to sell it, but this proved a difficult task and in the end he had to buy the whole collection at the price asked.[16] When writing to his father on 19 January in the following year he recounted the circumstances of the acquisition of the skin and the capture of the bird. He said that it had been shot and was one of two seen at Jacob's River (now called the Aparima River) which discharges into Foveaux Strait. A day or so after writing this he spoke again to the man who had sold him the skin. What he learned compelled him to revise his earlier statement with the following addendum, dated 24 January: 'Saw the man who obtained the Takahe. It was caught by a dog up a gulley of the sound behind Resolution Island, Dusky Bay. When caught it made a great noise which he and his brother encouraged thinking to bring up its mate but were afterwards told by the old natives that such noise was sure to drive the others away. As it was snowing and bitterly cold they did not continue the chase. The ground was covered with snow retaining tracks which they noticed and followed long before they saw the bird. Kept it alive on board the schooner for several days then killed and ate it. It was very good and as it was the first of its kind all hands had a taste for curiosity'. Before parting from the man, Walter asked him to forward the next *Notornis* packed in a keg of grog.[17]

The *Notornis* skin was packed in a box along with the skins of several other species including two kiwis, a Kakapo and two Huias, plus some other odds and ends to fill up the space. On 11 February Walter insured the lot for £50 and waited for a suitable boat to take the collection to England. Possibly deterred by his previous experience of sending a valuable cargo via Sydney, he chose to wait for a boat that would sail direct for London. This took some time, but the box finally left New Zealand on Sunday 5 May, 1850, on board the *Woodstock*.[18]

Gideon Mantell waited expectantly for news of his son on each occasion that a boat from New Zealand reached England. He was becoming increasingly anxious at the late arrival of the *Woodstock* when two letters from Walter and some New Zealand newspapers finally reached him on 10 October.[19] The ship must have dropped the mail at one of the Channel ports, for she had not yet reached the London docks. A servant was despatched to Gravesend the next day in search of the *Woodstock* and the box that Gideon knew was on board, but unfavourable winds prevented her docking for another two days.

14. Mantell, 1940, p.288: following earlier agreement by the Trustees of the British Museum (11.3.1848). Forshall was the Museum Secretary.

15. *Ibid.*, p.244.

16. Mantell, W. to Mantell, G., 19.1.1850, MS. Papers 83, Alexander Turnbull Library.

17. *Ibid.*, 19.1.1850.

18. *Ibid.*, 16.5.1850.

19. Mantell, G. to Mantell, W., 29.10.1850, MS. Papers 83, Alexander Turnbull Library.

Trans Zool Soc 4 Pl 25 p 74

Wolf, lith.

M. & N. Hanhart, Imp^t.

Notornis Mantelli. Owen.

The Notornis (*Notornis mantelli*) lithographed by Joseph Wolf; a remarkable illustration as it appears that the bird was not properly mounted at the time the work was carried out; from G. A. Mantell's *Notice of the Discovery . . . . . .* in *Transactions of the Zoological Society of London*, 1852 (pl. 25). *(see below)*

Mantell was to face further frustrations: the customs house delayed release of the box and imposed 11/6d duty on some oil paintings that it contained. Mantell reacted unfavourably to the 'rascally customs house', but he finally got his box with its contents 'mixed up pell mell together'.[20] Nothing was damaged and he was delighted with the bird skins, particularly the skin of the *Notornis*.

News of its arrival spread quickly. John Gould, who had a nose for rarities and a formidable reputation as an illustrator of birds, was soon on Mantell's doorstep, begging for a loan of the bird. He asked if he could make a drawing of it for the supplement to the *Birds of Australia*, a monumental work that had already included several New Zealand species.[21] The *Notornis* — still unstuffed — left with Gould and was returned three days later with the information that he thought the bird belonged to the same genus as the common rail *Porphyrio* (Pukeko) — an opinion he would later revise. Gould had been a central figure in ornithological illustration for the last two decades and his cooperative ventures, first with his wife and later with other artist-lithographers, brought him commercial success and the recognition of the scientific community. Those currently associated with him were Joseph Wolf, a highly regarded animal illustrator who had recently emigrated from Europe, and Henry Richter, who had a hand in much of Gould's work. Both men were to produce coloured plates of *Notornis* quite soon after the arrival of the specimen in England, although the order in which this was done is not quite clear.

The day after he loaned the specimen to Gould (24 October) Mantell attended a Council

20. *Ibid.*, 29.10.1850.

21. *Ibid.*, 29.10.1850.

NOTORNIS MANTELLI. Owen.

The head of the Notornis (*Notornis mantelli*) drawn by H. C. Richter; from Gould's *Remarks on Notornis mantelli* in *Proceedings of the Zoological Society of London*, 1852 (pl. 21); apparently taken from a larger plate by Richter and Gould and used in Part 1 of the supplement to the *Birds of Australia*, 1851, (pl. 76). *(see p.146)*

meeting of the Royal Society, where Owen congratulated him on his acquisition. But he was to discover to his annoyance that the wily Owen had been allowed by Gould to have a surreptitious look at the bird, and had later reported his viewing to the Philosophy Club.[22] Matters were made even worse by Owen's apparent failure to disclose the part of either Mantell in its procurement. However from Gideon's point of view some good was to come of it all as Owen pronounced the bird distinct from *Porphyrio* and belonging to the genus *Notornis*, already described by him from fossil bones. This was really exciting news — a bird once thought extinct was now authoritatively declared a living species. The *Times* carried the story in its issue of Saturday 26 October under the heading 'Interesting ornithological discovery', and the ornithologist Sir William Jardine wrote requesting that anticipatory notice of the discovery be given in some ornithological papers he was preparing.

Following the initial excitement, Mantell's mind turned once again to the more practical problem of obtaining credit for his son for the acquisition, and also some recompense for the time, trouble and not inconsiderable costs incurred. He first offered the bird to Gray, the ornithologist at the British Museum, for £25, but this was declined with a counter-offer of £20.[23] Although Gray was later to agree to the higher figure, Gideon decided to keep the bird for the time being in his drawing room. As he said, he could have got the money abroad (for example at the Philadelphia Museum) but the thought of sending it out of the country cannot have appealed to him.

22. *Ibid.*, 1.11.1850.

23. *Ibid.*, 29.10.1850; 1.11.1850.

24. *Ibid.*, 14.11. 1850.

25. *Ibid.*, 31.12.1850. This letter was sent with a tracing of the woodcut from the *Illustrated London News*.

26. Mantell, 1940, p.263.

27. It was to be the first of several — most coming from the 'shop' of John Gould. Gould had been a central figure in ornithological illustration for the last two decades and his cooperative ventures, first with his wife and later with other artist-lithographers, brought him commercial success and the recognition of the scientific community. (Sauer, 1982, provides an excellent Gould bibliography and chronological history.) The two artist-lithographers currently associated with him at the time of the *Notornis* discovery were Joseph Wolf, a highly regarded animal illustrator who had recently emigrated from Europe, and Henry Richter. Both were to produce coloured plates of *Notornis*, quite soon after the arrival of the specimen in England, although the order in which this was done is not clear. At the meeting of the Zoological Society of London, on Tuesday 12 November, 1850, when the discovery was formally announced, Gould displayed a 'full-size' figure of the bird (this could refer to either of the subsequent full-size plates), and a lithograph was thought not far off. It was completed in a fortnight, and Gideon Mantell told his son that he would send some coloured prints of it as soon as possible. It was at this point that Mr Bartlett, the taxidermist, received the skin for stuffing and mounting.

An ink and wash drawing (possibly a tracing by Gideon Mantell) of the *Notornis*, taken from the large plate by Richter and Gould, in the *Birds of Australia*, 1851. Published by permission of the Alexander Turnbull Library. *(see p.149)*

A meeting of the Zoological Society was arranged for Tuesday 12 November; Mantell was to read the first formal account of the discovery and Gould was to follow with a description of the bird's ornithological characters. Owen however was to precede them both with a paper on another fossil flightless bird from New Zealand. Gideon was extremely apprehensive that Owen, with his knowledge of the *Notornis* from the earlier preview, might pre-empt them all. But although he went on at length, Owen kept to his own subject and the evening passed without incident. Gould displayed his drawing of the bird and even Owen joined in praising Walter Mantell for his part in its recovery. Gideon wrote to Walter the following evening, saying that '. . . he [Owen] does not visit the sins of his father upon the child when it would not answer his purpose'.[24]

The evening at the Zoological Society concluded with Gould departing with the bird in order to make a lithograph of it, after which it was to be stuffed and mounted by a Mr Bartlett. Bartlett completed the work in late November and Gideon was able to write to Walter at the year's end to tell him that the bird was 'a noble fellow and capitally stuffed and mounted'.[25] It had been placed in a glass case and some white substance scattered around its feet in imitation of the snow in which it had been captured. The account for all this was to come to £6/2s, which together with some other work done by Mr Bartlett made a total of £15 — a sum which Gideon said he could ill afford.[26] On 4 January, 1851, the *Illustrated London News* informed its readers of the discovery and printed a woodcut — the first illustration of the entire *Notornis* to be published.[27] The bird and Gideon Mantell then began a round of scientific meetings and soirées. On his arrival at one of the latter, hosted by Lord Rosse, the glass case was broken, fortunately without damage to the bird, and a replacement had to be made in a hurry. This was

THE MOIIO; OR, NOTORNIS MANTELLI FROM NEW ZEALAND.

The first published illustration of the *Notornis* which appeared in the *Illustrated London News*, 4 January, 1851. There is some resemblance to the works of Wolf and Richter which appeared later. Published by permission of the Alexander Turnbull Library. *(see p.148)*

done with time to spare and the bird in its new case attended the soirée and contributed to its success.[28] In due course the *Notornis* came to rest in the British Museum, where it remains to this day.

A party of whalers shot a second specimen of the *Notornis* in 1851; again through Walter Mantell's agency this went to the British Museum, although later it was presented to the National Museum in Washington. Some men from the *Acheron* possibly sighted another while the ship was at Milford Sound, but it was not until 1879 that a third was caught, again by a dog. This one was auctioned at Covent Gardens, fetching the considerable sum of £105, its new owner being the Dresden Museum. The next *Notornis*, captured in 1898, also taken by a dog, ended its days in the Otago Museum.[29] These four specimens, distributed almost literally to the four corners of the world, were all that were taken in the 19th century, although there were many attempts to find more. Fortuitously, the later *Notornis* hunts were unsuccessful, otherwise the small colony that exists today might have been wiped out.

However, by the end of that year Gideon was still wondering when either Gould or the Zoological Society would publish the illustration. In the event neither was to have the honour of publishing the first drawing – on 4 January, 1851, a woodcut appeared in the *Illustrated London News*. This unsigned illustration bore a resemblance to Wolf's (which was published in Mantell's paper in the *Transactions* of March 1852), and possibly one of the newspaper's own artists had been sent round to make a sketch from Wolf's original drawing. A large plate featuring two birds by Richter and Gould, intended for the *Birds of Australia*, was published in 1851 (Waterhouse, 1885, p.38) part of which (the head of one) was used by Gould in the *Proceedings of the Zoological Society of London* of 24 January 1852. It seems likely that Joseph Wolf and Henry Richter (the latter probably under Gould's supervision) did their original drawings at about the same time – possibly when Gould first borrowed the bird. Responding to a request from Walter Mantell to see 'the large plate', a lithograph from Gould's *Birds of Australia* was apparently despatched to New Zealand on the *Lord Williams Bentinck* on 31 July, 1851, and reached him on 14 March, 1852 – this was probably a copy of the large double plate by Gould and Richter eventually published in the supplement to *Birds of Australia* in 1869. The presence of an ink and wash copy on tracing paper (C107/1 Alexander Turnbull Library) with the appearance of having been sent in a letter, suggests that it was this

continued overleaf

rather than an actual lithograph that was sent.

28. Mantell, 1940, p.267.

29. Hall-Jones, 1976, pp.136-138.

Notornis. head

A sketch of the Notornis head made by Walter Mantell, from his notes. Published by permission of the Alexander Turnbull Library (C103/116).

Chapter Ten

SUNDRY VOYAGES

DURING THE 1820s NEW ZEALAND had seen little of voyages of discovery other than those of the French, but by the 1830s Britain had again found the need to revise and extend her domain of colonial science, for both scientific and geopolitical reasons.[1] One of the symptoms of this revival was a surveying mission undertaken by a small brig which slipped out of Plymouth on 27 December, 1831. The ship was the *Beagle*, Robert FitzRoy in command. On board was a civilian, a young naturalist who had barely graduated from university. His name was Charles Darwin, and in a world of natural history greatly in need of some new generalizations to be derived from years of exploration and collecting, his were to have enormous impact. FitzRoy himself was well trained, a graduate of the Royal Naval College at Portsmouth; he was destined for Fellowship of the Royal Society and later the Governorship of New Zealand. Members of the *Beagle*'s crew, including Stokes and Chaffers, would later visit New Zealand on other voyages.

Apart from the more general scientific and hydrographical motives for this voyage, there was also the specific and unusual one of returning some Fuegians to their homeland along with an English missionary who would 'extend the benefits of civilisation' to the inhabitants of Tierra del Fuego.[2] It is a trifle ironic that one of science's greatest voyages should have begun under such auspices when other expeditions with motivations perhaps socially less worthy but scientifically more ambitious, laid foundations that were far less substantial. We are reminded as the voyage began of parallels with Banks's voyage with Cook — a personable and intelligent young man about to embark on his first major natural history venture, a ship expensively fitted out, a competent, interested and sympathetic commander, and a diet guaranteed to keep them all healthy.[3]

Their stay in New Zealand from 21-31 December, 1835, was short and not particularly productive, although it would be too parochial to question the value of this voyage on the basis of the contributions made from these islands. First and often quoted impressions of the country were not good, and Darwin particularly found little to attract him although he saw only the Bay of Islands. He was puzzled, however: 'with regard to animals it is a most remarkable fact, that so large an island . . . with varied stations, a fine climate and land of all heights . . . with the exception of a small rat, should not possess one indigenous animal [mammal is probably meant in this context]'.[4]

The expedition collected a few fish, of which only the Inanga, *Galaxias maculatus*, was new. Taken by Darwin in a freshwater stream, it is the most important constituent of whitebait and the best known member of this genus in New Zealand.[5] He also collected a new green tree gecko, similar to the one described earlier by Gray.[6] A Tiger Beetle (no expedition was complete without one) and two chalcid wasps were also taken.[7] If they were unimpressed by the fauna, they came away with warmer feelings towards William Williams and the other missionaries and what they were trying to achieve in New Zealand. FitzRoy in particular entertained ideas of administrative changes that might be beneficially applied to this country, little realizing the future role he would play there.[8]

The *Beagle* returned to England in 1836. Darwin then successfully persuaded the Government to financially support the voyage publication to the extent of £1000 and the first part of the zoology appeared in February 1838. Its authors were Richard Owen, who described the mammals, John Gould, the ornithology, and the Reverend Leonard

1. MacLeod, 1982, p.8.

2. Mellersh, 1968, p.60.

3. *Ibid.*, pp.69-70.

4. Darwin, 1839, p.511.

5. McDowall, 1978, p.69.

6. Darwin's gecko (the Northland Green Gecko) is still thought by some to be a separate species or subspecies (i.e. *Naultinus elegans grayi* or *Naultinus grayi*; see Meads, 1982, pp.321-325).

7. Darwin presented these (*Pteromalus* spp.) and the Tiger Beetle to the British Museum: Accessions Register, Entomology, 1841, 1843, British Museum (Natural History).

8. Mellersh, 1968, pp.167-168.

1. *Gymnodactylus Gaudichaudii*
2. *Naultinus Grayii*

The (Northland) Green Gecko (*Naultinus elegans grayi*) lithographed by B. Waterhouse Hawkins for T. Bell's *Reptiles* in C. Darwin's *The Zoology of the Voyages of H.M.S. Beagle . . . .* 1843 (pl. 13). Published by permission of the Alexander Turnbull Library. *(see p.151)*

Jenyns and Thomas Bell who described the fish and reptiles respectively. Their contributions marked a departure from earlier British expeditions where the collectors wrote up their own work, and this demonstrated something of the growing institutional base of the natural sciences in England. Now their findings both in their authority and sumptuous presentation could stand on an equal footing with the publications of the recent French expeditions to the South Pacific, although the official zoological publication dealt only with vertebrate animals — invertebrates were relegated to the scientific journals.

It was a brief chapter in terms of physical contact with the New Zealand fauna, but the influence of the voyage on scientific thinking in New Zealand and elsewhere would be profound. It was Darwin's journey on the *Beagle* that paved the way for his epoch-making *On The Origin of Species. . .*, published in 1859. This new wave of expeditions abroad was helping to stimulate English minds including those of Darwin, Wallace and their followers to bring in the age of evolution and biogeography. A new order of science and scientists was now coming to the fore, signalling a democratization of science and the beginning of the end of the dominance of imperial zoologists such as Owen.[9]

Just as the first reports from the *Beagle* were appearing, another expedition to the South Seas was preparing to get under way, this time from a country newly entering the scientific exploration of the region: the United States of America. Commercial vessels from this country had been active in the South Pacific since the late 18th century, but as America was still in the process of self-discovery, the need for voyages mounted at public expense was not seen until the middle decades of the 19th century were reached. When the moment finally came it was not a matter of harmonious agreement, and the intrigues and disputes of the occasion were more reminiscent of 18th century European voyages.[10]

The explorer Jeremiah Reynolds first proposed the expedition, which subsequently received Congressional approval, but disagreements between the Executive and the Navy, and among the prospective personnel, resulted in the loss to the expedition of the first appointed commander and then Reynolds himself, despite the entreaties of President

9. MacLeod, 1982, p.10.

10. Reingold, 1966, pp.109-110.

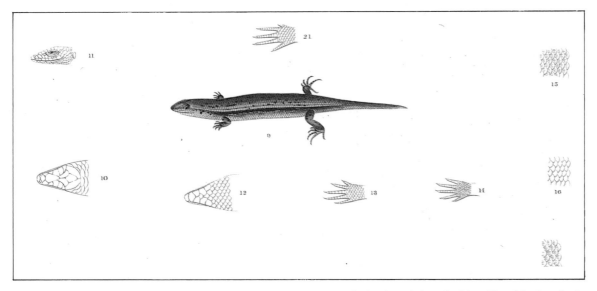

Lizards collected by the U.S. Exploring Expedition from the Bay of Islands and described by Girard in *Reptiles* in Wilke's *United States Exploring Expedition* . . . . 1858; (top) The Brown Skink (*Leiolopisma zelandicum*); (middle) The (Northland) Green Gecko (*Naultinus elegans grayi*); and Copper Skink *Cyclodina aenea*. Published by permission of the Alexander Turnbull Library. *(see p.154)*

11. *Ibid*, p.110.

Andrew Jackson. Conflict on whether the emphasis should be on physical rather than natural sciences joined with disagreement over the appointment of naval over civilian scientists. The military patronage of science, and interservice rivalries, brought to light what was to become a long-standing difficulty; one that has achieved a particular notoriety in the American context.[11] Ultimately a young lieutenant, Charles Wilkes, was chosen to lead the expedition and on 18 August, 1838, two years after Congress had approved it, the *Vincennes*, *Peacock* and four other vessels, with nine scientists and their assistants on board, left Hampton Roads for the Pacific. By the standards of previous Pacific voyages it was an armada, and must have been the subject of nervous speculation by the inhabitants of many of the foreign ports in which they dropped anchor.

12. Wilkes, 1845, v.2, p.1.

Only three of the original fleet — *Vincennes*, with *Porpoise* and *Flying Fish* — arrived at the Bay of Islands on 30 March, 1840, and the pattern of their visit was by now a familiar one: a quick round of collecting; some observations of the local inhabitants and their manners; later generalization on the country's fauna. They left on 6 April, not many weeks before d'Urville's arrival, the following observation later being made by Wilkes: 'I believe that no person in the squadron felt any regret at leaving New Zealand, for there was a want of all means of amusement, as well as of any objects in whose observation we were interested'.[12] This by now familiar complaint came from one whose interests were in the physical sciences, and was not entirely borne out by the work of some of his scientists. They had made a respectable collection of crustaceans which included some new species, and with the help of the Maoris they did better by the New Zealand lizards than many of their predecessors. The draughtsman, Drayton, made coloured sketches of them and one species among the seven collected was new.[13] Shells were collected from the Bay of Islands and the Auckland Islands, among which were some familiar specimens including the Auckland Rock Oyster and the Hairy Mussel and a new subspecies of Paua.[14] Wilkes's return to the United States was marked by the turmoil of his court martial and later controversy, none of which helped the publication of the voyage results. Finally brought out under the auspices of Congress, the zoological parts with their handsomely illustrated atlases appeared between 1848 and 1858.[15] The last parts appeared nearly 14 years after their return and many of the findings had begun to lose their novelty, although ironically it was the natural sciences, less favoured at the voyage's commencement, which fared best at the end.[16] Although the impact of their visit to New Zealand might be considered slight, the Americans, along with the British and French in their contemporary visits, gave momentum to the development of indigenous science in New Zealand during what is regarded as a formative period.[17] Two of these expeditions we have now described; the third was a British one under the command of James Clark Ross.

13. *Cyclodina aenea*. The Copper Skink.

14. Respectively *Crassostrea glomerata*, *Modiolus areolatus*, *Haliotis virginea crispata*.

15. Mollusca, Gould, 1852, accompanying Atlas, 1856; Zoophytes, Dana, 1848, Atlas, 1849; Crustacea, Dana, 1852, Atlas, 1855; Reptiles, Girard, 1858, Atlas, 1858; Mammals and Birds, Cassin, 1858, Atlas, 1858. See Wilkes, 1845.

16. Reingold, 1966, p.110.

17. Hoare, 1977a, p.19.

Ross was already a seasoned explorer and navigator when he was asked by the Admiralty to command and lead a scientific expedition to Antarctic waters, with the object of studying terrestrial magnetism. The tradition of surgeons doubling as naturalists continued with this voyage: Robert McCormick and John Robertson were to collect zoological and geological specimens, Joseph Hooker and David Lyall were responsible for the botany. Other members of the crew were allowed to collect on their own account. With instructions that included 'the remaining winter months may be advantageously engaged in visiting New Zealand and the adjacent Islands',[18] and supplementary instructions from the Royal Society on various aspects of science, including zoology, the *Erebus* and *Terror* sailed on 25 September, 1839.

18. Dodge, 1973, p.188.

While at Hobart, Ross received news of the activities of d'Urville and Wilkes who had preceded him in the high southern latitudes; this news he received with great and rather unjustified indignation.[19] The Auckland Islands were reached on 20 November, 1840; there they discovered that d'Urville and Wilkes had preceded them. Several birds were noted and two new species, the Auckland Island Teal and Auckland Island Snipe, were taken as a result of McCormick's ability with a gun.[20] Keeping to their instructions they skirted the Antarctic before returning to Tasmania and thence to New Zealand, where they arrived at the Bay of Islands on 17 August, 1841. Two observatories were set up on the banks of the Kawakawa River, and it was characteristic of the times that they were flanked by two gin-shops. During their stay the naturalists were given every assistance by two of the resident missionaries, William Williams and Richard Taylor.[21] Although they were to confine themselves to the Bay of Islands, the length of their visit allowed them to move further inland: one day they carried the seine net five or six miles [8-10 km] from Waimate to Lake Omapere. Their first haul produced only roots and limbs of trees—to the amusement of the Maoris — but on the second attempt they were able to catch a few specimens of the Common Smelt and some small mussels.[22] They later visited the falls on the Kerikeri River as had the French before them. Despite these and other excursions there were limits to the extent of their travel as a result of the Maori unrest that had followed the recent signing of the Treaty of Waitangi. But significantly this was the occasion on which Hooker met up with William Colenso and Andrew Sinclair, collected with them and gave them some botanical instruction, creating another of those vital scientific connections between England and one of her colonies.

Numerous species of birds, fish, reptiles, crustaceans, molluscans, insects and other invertebrate life were collected during their three months at the Bay of Islands. Later, the published accounts included species taken from other sources and it is sometimes difficult to distinguish *Erebus* and *Terror* material. New fish from New Zealand waters were the Orange Clingfish, Silver Conger Eel and Butterfly Tuna. Some insects (mainly Coleoptera) were also new and molluscs included a Turrid Shell and Razor Mussel.[23] Otherwise much of what they saw duplicated earlier efforts. They left the Bay of Islands on 23 November, passing the Chatham Islands on their way back to Antarctic waters. Their voyage ended at Folkstone on 4 September, 1843, nearly 4½ years from the time they set off. By all accounts it was a highly successful journey, although they were not exempt from the usual minor disputes and wranglings that accompanied expedition returns. The British Museum benefited greatly by the collections donated by Ross, the Admiralty, and Lieutenant Smith.[24]

The appointed editors of the voyage zoology were Sir John Richardson of Haslar Hospital and John Edward Gray of the British Museum. The Government agreed to grant £2000 towards the costs of illustrating the zoology and the botany, and by 1844 the first parts of the work had started to appear. Various authors were assigned to different zoological groups but financial problems intervened and publication dragged on for years with many of the authors failing to complete their assignments. The work was finally completed in 1878, over 30 years after their return. John Gray was responsible for the mammals and (with Günther) the reptiles; G. R. Gray did the birds, Sir John Richardson the fish, and White made a start on the insects. But it was virtually a new team that finished the work: Arthur Butler the insects, Edward Miers the crustaceans[25] and Edgar Smith the molluscs. Günther concluded the reptiles and R. Bowdler Sharpe finished off the birds, using additional information from Walter Buller's works.

While the work did justice to the New Zealand collections by members of the *Erebus* and *Terror* expedition, there was an attempt over the long haul of publication to provide

19. *Ibid.*, p.194.

20. *Anas aucklandica aucklandica* and *Coenocorypha aucklandica aucklandica* respectively, the latter falling to the gun of Surgeon McCormick (Fleming, 1982a, p.50).

21. Dodge, 1973, p.204. McCormick, 1884, pp.223, 245.

22. Ross, 1847, P.103. The Common Smelt, *Retropinna retropinna*.

23. Respectively, the fish: *Diplocrepis puniceus*, *Gnathophis habenatus*, *Gasterochisma melampus*; the Coleoptera: *Mondella antarctica* and *Philoneis harpoloides*; the Molluscs: *Phenatoma zelandica*, *Solemya parkinsoni*.

24. e.g. Crustaceans and Insects from Lt Smith and Ross of *Erebus*, Accessions Registers, Entomology, 1843, 1844; Mollusca and Radiata 31.5.1843, 20.11.1843; Zoology, 14.9.1843, British Museum (Natural History).

25. Miers also later authored the *Catalogue of the Stalked and Sessile-Eyed Crustacea of New Zealand* (1876). This was one of a series of catalogues produced during this period several of which were tied to the *Erebus* and *Terror* publications: e.g. Butler, the butterflies, 1874, 1880; G. R. Gray, the birds, 1862; J. E. Gray, the reptiles, 1867.

RALLUS DIEFFENBACHII.

C.Hullmandel's Patent Lithotint.

Dieffenbach's Rail (*Rallus philippensis dieffenbachi*), one of several birds illustrated in Sir John Richardson's *The Zoology of the Voyage of H.M.S. Erebus and Terror*, 1854-75 (78); this illustration shows a development of the lithographer's art in the use of printed tints (generally only two colours over black); a process that led to the more complex one of chromolithography. Published by permission of the Alexander Turnbull Library. *(see p.155)*

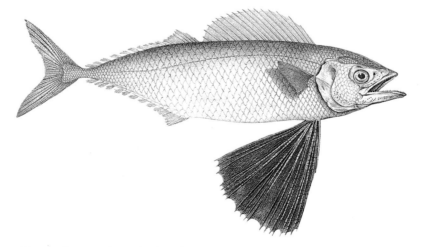

A Butterfly Tuna (*Gasterochisma melampus*); this illustration is of a young fish which differs in several features from the adult; a lithograph by W. Mitchell for Sir John Richardson's *The Zoology of the Voyage of H.M.S. Erebus and Terror . . .*, 1876, (Fish, Vol 12, pl. 37). Published by permission of the Alexander Turnbull Library. *(see p.155)*

a more complete faunal listing; it is thus difficult in places to sort out what was collected on the expedition and what came from additional collections by Percy Earl, Andrew Sinclair and others.[26] The illustrations were lithographs of varying quality, a sign that the high standards of early years in zoological illustration were about to decline. Many of the New Zealand illustrations — which are mostly of birds — are unattributed, although some, including one of a Kakapo, were prepared by notable natural history artists such as Joseph Wolf.[27]

FitzRoy, Wilkes and Ross, with their naturalists, and several of those who came before and after them, made notable contributions to New Zealand zoology, with one limitation — they visited only the Bay of Islands or were otherwise restricted in their ports of call. By then of course local collectors were providing a better coverage, but even they had their limitations. What was now needed was another survey, an extension of Dieffenbach's, particularly one that would take in more of the South Island. Just such a survey was in fact about to take place — in the *Acheron* under Captain Stokes.

The voyage of the *Acheron* was rather different from some of its predecessors. First, although not directly connected with its capacity for scientific exploration, the ship had steam-driven paddle-wheels in addition to her sails. Not the first such vessel to visit New Zealand, she was nonetheless the first to be involved in any lengthy exploration there. In the second place, apart from a delay at Sydney, New Zealand was the sole object of the venture and one that would hopefully plug some of the gaps left by earlier attempts such as that of Dieffenbach and the *Tory* — a golden chance for a comprehensive faunal survey. Unfortunately this did not eventuate, even though the *Acheron* carried two surgeon-naturalists. One, Lyall, was assistant surgeon on the *Terror* when she last visited New Zealand, the other was Dr Charles Forbes. The Captain, J. Lort Stokes, had also earlier visited New Zealand on the *Beagle*. The *Acheron* left Plymouth on 21 January, 1848; they were joined at Sydney by the natural history dealer Frederick Strange and reached Auckland on 7 November, 1848.[28]

They were soon to be despatched to Wellington to lend assistance following the violent earthquakes there between 16 and 18 October. In February 1849 the vessel arrived at Akaroa, where a nautical survey of Banks Peninsula was carried out. Meanwhile Strange had been collecting in the area and rejoined the ship at Port Cooper (Lyttelton) with what appeared to be a teal (probably a Brown Teal), a Banks Peninsula Gecko or Jewel Gecko, and an iridescent beetle, amongst other specimens.[29] From Port Cooper, Stokes allowed Strange leave of absence to make a trip up the Waimakiriri River into the area of the Torlesse Range — a river draining the nearby Lake Lyndon is today called the Acheron River. Strange started on 4 March, 1849, and on his way he encountered several species of birds including a Grebe, the Australasian Harrier (an early report), the New Zealand Quail and what was most likely a Black-backed Gull. He shot a pair of Paradise Shelducks, but some snails he picked up were found to have been destroyed by a fire.[30] His Maori companions caught some rats but these were Norway rats, not the 'native' rats which he suggested would shortly become extinct. He crossed the plains near the foot of the ranges, seeing piles of broken Moa bones, camping at the end of the day at a place where he took two new species of snail and 'some fine insects'; he saw the South Island Kokako, and numerous Tuis.[31] On the 8 March they began to climb, reaching a peak where they could view the precipitous country beyond. He returned to camp with two butterflies and three snails for the day's work, and the following morning began his return journey. On rejoining Stokes he found that they had decided to make an expedition to the Grey Range, a 16-day trip that did not net anything substantial.

26. See Note 79, Chapter 5.

27. These (including the Laughing Owl, the Kakapo, and the White-fronted Tern) were perhaps not his best work and the hand-colouring was of varying quality. Birds of prey were his speciality.

28. Natusch, 1978, p.21.

29. *Anas aucklandica chlorotis*; the Banks Peninsula Gecko or Jewel Gecko. *Naultinus elegans gemmeus*.

30. Either the Southern Crested Grebe, *Podiceps cristatus australis*, or the New Zealand Dabchick, *P. rufopectus*; *Circus approximans gouldi*; *Coturnix novaezelandiae*; *Larus dominicanus*. The Paradise Shelduck; *Tadorna variegata*.

31. Whittell, 1947, p.107.

Following this, the *Acheron* sailed to the east coast of Otago before moving on to Cook Strait and Tasman Bay. Near Nelson, Strange obtained a specimen of the 'native' rat and also a Brown Creeper in Tasman Bay that he gave to Stokes for the British Museum.[32] In Croiselles Harbour they picked up an example of the Imperial Sun Shell, beloved of Cook's period. With the company of a couple of live kiwis, the *Acheron* set sail for Sydney for a refitting, at which point Strange left them.

There were other observations and collections during the survey: Evans, the Master, took some shells and with a party from the ship, 30 to 40 kiwis and many Kakapo; more Moa bones were seen on a journey inland from Otago Harbour; Bradshaw, a midshipman, found a brachiopod; Lyall made some significant observations of the Kakapo in Dusky Sound and a party led by him thought they saw a Takahe (*Notornis*) in the Milford Sound area.[33] They returned to England with ample evidence of their industry. Lyall's collections, passed on to the British Museum by Stokes, were substantial, including well in excess of 500 molluscs; there was a large collection of over 100 insects including many Coleoptera, a few crustaceans, starfish and other invertebrates. Vertebrates included a large number of birds and a few fish, collected mainly in the south of the South Island (including Stewart Island and Fiordland). Two kiwis, the South Island Brown Kiwi and the Little Spotted Kiwi, were collected along with some birds' eggs from Thompson Sound.[34] Strange too threw in some of his specimens with Lyall's: a few insects only, most of his other material probably being sold to other collectors or institutions.[35] Evans took tea with Gideon Mantell; always eager for news from New Zealand, Mantell heard descriptions of the South Island, West Coast and that traditions of Moa and *Notornis* were rife amongst the Maoris and whalers.[36]

Without a major publication to follow, the survey of the *Acheron* could easily be construed as zoologically insignificant, although there was Lyall's paper on Kakapo behaviour and the description of Polyzoans collected by him, published by Busk.[37] But there was no expensively illustrated account of the natural history, and nor could there be. Nevertheless the specimens they collected were a useful contribution to the growing colonial holdings of the British Museum, possibly the last major collection of its kind to be sent to England from New Zealand for some decades. From now on the New Zealand collections there would be augmented largely by contributions from individuals, some of them visitors but most of them residents. The days of the great collecting expeditions were almost over.

For many years voyages of science and exploration had been part of the lifeblood of New Zealand science and from the time of the *Acheron*'s survey until the end of the century, only the voyage of the *Novara* had substantial local impact. However two remaining voyages were to have different but profound influences on the zoological sciences and both touched briefly at New Zealand.

The *Rattlesnake* was returning to England from a hydrographical survey when it was compelled to put in for repairs to the Bay of Islands for a week in May 1850. The surgeon on board was T. H. Huxley, undergoing an informative 'grand tour' in preparation for a notable career in biology and eventual intellectual leadership of the British scientific community. As his biological interests on the voyage were confined to the collection and study of coelenterates, his activities in the Bay of Islands do not survive in detail – his influence on New Zealand science was to be an indirect one, as he rose in stature to lead a new wave of scientists and scientific thought that would be universal in its application.

The paths of Huxley and the New Zealand fauna crossed again in 1869 when he

32. *Ibid.*, p.108.

33. Mantell, 1940, p.208; Natusch, 1978, p.159.

34. The Accessions Registers for Zoology, Mollusca, and Entomology, 1849-1853, British Museum (Natural History) contain several lengthy lists of material donated by Stokes.

35. Strange's insects (rather inappropriately) noted in Accessions Register, Mollusca, 1850, British Museum (Natural History).

36. Mantell, 1940, p.280.

37. Lyall, 1852; Busk, 1852.

described a fossil reptile which had some kinship to the Tuatara. In discussing the two he was compelled to ask: 'What if this present New Zealand fauna, so remarkable and so isolated from all other faunae, should be a remnant, as it were, of a life of the Poikolitic period which has lingered on isolated and therefore undisturbed, down to the present-day?' He also observed the affinities of the native frog and freshwater fishes with South American forms, suggesting some vast alteration of the physical geography of the globe as part of his thesis that new forms of life originated in the spread from the northern hemisphere with the survivors of more ancient kinds finding a refuge in the southern lands.[38]

Huxley was not in fact the official naturalist on board the *Rattlesnake*; the position was filled by MacGillivray, who on his return wrote a lengthy but dull account of the voyage. He gives his brief impressions of the Bay of Islands, mentioning particularly the numbers of molluscan shells he saw — which included abundant terrestrial and freshwater species — and remarking that a collector from Sydney had obtained nearly 100 species of such shells during a nine-month visit to New Zealand. He was undoubtedly referring to Frederick Strange, whom he must have met while in Sydney.[39]

The second voyage was that of the *Challenger*, a steamship which was making a vast and comprehensive survey of the world's oceans and which reached Wellington in June 1874 to stay for ten days. The arrival was tinged with tragedy, as the leadsman had been lost overboard in a rough crossing of Cook Strait.[40] The expedition found Wellington undeveloped compared with Sydney and something of a disappointment. They collected in the Cook Strait region, and although their interest was mainly in ocean life, a lizard was collected while in Wellington.[41] Later, the naturalist H. N. Moseley recalled the Black-backed Gulls in the harbour, the New Zealand Kingfisher perched on the telegraph wires, and the corpses of North Island Kakas hanging in the poulterers' shops. He also remembered the kindly assistance given him by local naturalists such as W. T. L. Travers who came out to the ship to provide him with specimens of planarians (flatworms) and some strange caterpillar-like animals. Their interest in the latter was such that collectors were sent out from the *Challenger* to find more, and they returned with some fifty specimens.

The published output of the *Challenger* expedition was prodigious, running to 50 volumes in which some species encountered from New Zealand waters are described. On his return Moseley had put aside his notes on the caterpillar-like animal, only to resurrect them when he read a paper by F. W. Hutton (then Otago provincial geologist and lecturer in zoology at the university) on a new species then called *Peripatus novaezealandiae* — the same species Moseley had in his possession. Published in the *Annals and Magazine of Natural History* (1876), it was to this journal that Moseley replied with a forceful account of Hutton's errors. A politely heated exchange followed, Hutton thanking Moseley for his 'courtesy' and his '. . . telling me exactly what I saw, what I did not see, and what he imagined I saw . . .' and wishing to use: '. . . my own eyes in preference to looking through his'. This was stern stuff, but Hutton had been a little hasty and was later compelled to conduct a modest retreat. On the other hand, one suspects that Moseley was indignant that someone from the colonies had stolen a march on him by describing this specimen.

The animal they were arguing over, the *Peripatus*, was a representative of a group that became of great interest to zoologists. Distributed in the southern hemisphere, these animals achieved the status of a 'missing link', having features intermediate between the earthworm-like animals and the arthropods, the major group that includes the insects

38. Huxley, 1869, in Foster and Lankester, v.3, pp.388-389 (from a paper read to the Geological Society of London that year).

39. McGillivray, 1852, v.2, pp.87-94. He also noted the scarcity of birds (p.92).

40. Thomson, 1885, pp.472-474; Linklater, 1972, pp.305, 306.

41. The lizard was the Moco Skink *Leiolopisma moco* — Accessions Register (Reptiles), British Museum (Natural History). The caterpillar-like animal, *Peripatoides novaezealandiae* (Hutton, 1876b), Moseley, 1879, pp.278-279. The first issue of the *New Zealand Journal of Science* contains some of their records (e.g. copepods and molluscs) from the Cook Strait region: Brady, 1882, pp.35-59, Boog Watson, 1883, pp.319-321.

42. Thomson, 1885, pp.1053-1072, contains a bibliography relating to the scientific results of the expedition; Hutton, 1876b, p.361; Moseley, 1877 pp.85-91; Hutton, 1877, p.83.

43. Scherzer, 1861, v.1, p.viii.

44. Anon, 1911, v.17, p.924.

45. Medically trained and with interests in science, Purchas took up the priesthood in New Zealand.

46. Scherzer, 1862, v.3, p.147; *Eudynamis taitensis*.

47. Scherzer, 1863, v.3, (Appendix) pp.508-510.

and crustaceans.[42]

The last voyage to be described in any detail here is one that has an importance beyond its natural history collections, an importance that lies in its timing and in the person of one of its participants. The expedition came from Austria, a country relatively new to New Zealand exploration but one that continued to maintain a line of interest in New Zealand from Germanic Europe. The reasons for Austria's belated concern for this part of the world were several. Having temporarily settled disputes in various regions of her unwieldy empire, her rulers were able to look beyond Europe in the same way as the major maritime colonial powers had done before them, and to unfurl 'the imperial flag of Austria in those distant climes where it has never before floated'.[43] There were the expected commercial and scientific objectives for the voyage, and the benefits of sponsorship by the Archduke Maximilian, brother of the autocratic Emperor Franz-Josef, and himself the future Emperor of Mexico. Maximilian had maintained strong interest in the sciences, particularly botany, and was prominent in the development of the Austrian Navy. He was an ideal patron.[44]

The commander of the expedition was Commodore B. von Wüllerstorf-Urbair and among the naturalists on board were the zoologists Johann Zelebor and Georg Ritter von Frauenfeld. But it was the geologist from Vienna, Ferdinand von Hochstetter, whose presence was to have such a far reaching influence on New Zealand science. That their vessel the *Novara* (named for a recent Austrian victory) called at New Zealand at all is the result of their previously visiting Capetown and there meeting the Governor of the Cape Colony, Sir George Grey, recently Governor of New Zealand. Grey persuaded Wüllerstorf-Urbair to visit New Zealand and the *Novara* subsequently arrived at Auckland on 22 December, 1858. Coincidentally, the *Evening Star* had arrived a day earlier with a German immigrant, Julius Haast. With his kindred interests in science, Haast must have been surprised and delighted at the arrival of German-speaking colleagues in this remote corner of the world. It was to be a significant meeting.

Not long after the party from the *Novara* had landed, a request came from the Provincial Government for Hochstetter to be allowed to make an exploration of the coalfields at Drury and Hunua, south of Auckland. Some of the naturalists chose to go with him while the others remained in the Auckland area to visit the kauri forest at Titirangi. Here, an Irishman named Smith ran a farm and sawmill and had built a comfortable home at the edge of the forest where he made the visitors welcome. Wüllerstorf-Urbair accompanied them and manifesting the enjoyment of hunting characteristic of his countrymen shot several birds in the nearby forest, including a Tui and a North Island Robin. Other birds were collected by Zelebor and Frauenfeld; amongst the specimens collected by Zelebor were some kiwi eggs.

The party that went down to Drury was joined by the Reverend Arthur Purchas[45] and Julius Haast. While the botanists and zoologists followed their different inclinations in the forest, Hochstetter examined the coal beds there before moving further south to the banks of the Waikato River where a specimen of the Long-tailed Cuckoo was found: a 'rare example — its capture was greeted with great joy by the zoologists'.[46]

Not long after their return Governor Gore-Brown wrote to Wüllerstorf-Urbair asking him if he would release Hochstetter to carry out further work in New Zealand. The Commodore replied in the most favourable terms, emphasizing the bond that would result from the 'communication between scientific men of both countries'.[47] The only condition placed on Hochstetter's stay was its duration (which in fact he exceeded) and the recognition of his findings and collections as the property of the *Novara* expedition.

On 8 January, 1859, the *Novara* sailed from Auckland, a raft of Austrian language and custom deserting Hochstetter but leaving him only momentarily ill at ease. He was soon hard at work. Although geology and geography headed his brief, he was also charged by his colleagues from the *Novara* with collecting zoological and botanical specimens. To this end, he placed advertisements in the newspapers asking for specimens to be brought to him with duplicates where possible: in the same spirit that promoted his being there in the first place, he wished to donate some to the Auckland Museum. The results of his advertising were overwhelming. The increasing volume of specimens soon put him at risk of being driven from his lodgings, until the Government provided him with a small house close by in which to house his collections.

Hochstetter's stay of nine months in New Zealand was crowded with travel through a substantial part of the North Island, including Auckland, the Coromandel Peninsula and the central New Zealand thermal area. Also visited were the northern part of the South Island from Nelson down to Lake Rotoiti. He was often in the company of Haast, 'the true and trusty German' — they were now good friends and would be lifelong colleagues — and Charles Heaphy.[48] He encountered the commonplace in the form of wekas as well as the exotic, such as the *Dinornis* and *Palapteryx* remains found in the Aorere Valley of north-west Nelson. He observed and collected a wide range of New Zealand animals, mostly previously described but with a few important additions. He visited the western Coromandel Peninsula, mainly to see the gold fields in the area; it was most likely here that some Maoris presented him with a specimen of a primitive and unusual frog — similar to the one that had been procured earlier by Thomson.[49] He collected or was given a large terrestrial snail from the southern Alps that was found to be a new species.[50] The trustees of the Nelson Museum presented him with a fine skeleton of *Dinornis robustus* for the Vienna Museum.

Hochstetter left New Zealand for Austria on 2 October, 1859, leaving behind him a legacy of goodwill and cooperation between scientists of the two countries, an active and growing scientific community that had been the richer for his presence, and the

48. Heaphy had been the draughtsman on the *Tory* and had remained in New Zealand. Von Hochstetter, 1867, p.15.

49. See Chapter 7.

50. *Paryphanta hochstetteri hochstetteri*.

Hochstetter's Frog, the first published illustration of a New Zealand native frog; a hand-coloured lithograph by A. Hartinger for Fitzinger's *Eine neue Batrachier − Gattung* . . . in the *Verhandlung der k.k. zoologische-botanischen Gesellschaft*, 1861 (pl. 6). *(see above)*

foundations for the science of geology in New Zealand, on which he would publish after his return to Austria. The specimens went back to Vienna to the Imperial Museum, one of the great museums of Europe and renowned for its natural history collections, particularly those in geology. The New Zealand material formed only a small part of the vast amount brought back by the *Novara* and Hochstetter; with the exception of the native frog, their small vertebrate collection contained little of any novelty apart from the Moa skeleton already mentioned and the Rock Wren. It was the invertebrates that benefited most from the expedition and its sequel. Hochstetter's lengthy stay had allowed him to collect several species of land and freshwater molluscs, and the large insect collection — there were 65 Coleopteran species alone — included the Long-nosed Kauri Beetle, the Cabbage Tree Moth, the Kowhai Moth;[51] at long last that villain of so many previous expeditions, the sandfly — or more correctly the New Zealand Blackfly — was collected and later described and named.[52]

The official zoological reports began to appear in 1864 and continued to be published for more than a decade, resulting in six illustrated volumes including four on insects. The reports were multi-authored and bibliographically complex.[53] The geological part of the voyage reports (1864) was mainly devoted to New Zealand geology (particularly Auckland and Nelson), and was one of the first lengthy descriptions of New Zealand invertebrate palaeontology. In 1863 Hochstetter's *Neu-Seeland* appeared, a general account of the geology and natural history of the country he had recently visited. In this, his general conclusions on the fauna were similar to his predecessors but he acknowledged that there was still much to be learned about the lower orders of animals.[54] In the closing pages of his chapter on the fauna he talks of the 'struggle for existence' through competition — Darwin's *On the Origin of the Species. . .* was published not long before Hochstetter's return.

It was almost as if New Zealand had been waiting for the likes of Hochstetter to come along and take the natural sciences out of the straightjacket of collecting and systematizing and away from its resident naturalists acting as brokers for remote imperial scientists and institutions. To be sure, the *Novara* scientists including Hochstetter collected for the Imperial Museum in Vienna, but at the same time he was able to widen the perspectives of New Zealand scientists and help them towards self-respect and independence, as well as link his science with practical, economic objectives. For a while he belonged, he was one of them, a pleasant change from the distant and lofty acknowledgements of an Owen or a Gray who would seldom leave London in search of specimens. A paragraph from *Neu-Seeland* is revealing: 'The kind and valuable services rendered me by word and deed,

51. The Rock Wren – *Xenicus gilviventris*. The moths were *Epiphryne verriculata* and *Uresiphila polygonalis maorialis*, respectively.

52. Described as *Austrosimulium australense*, Schiner, 1868, v.2, p.15, in the voyage zoology. Hochstetter also noted the presence of the venomous Katipo spider, *(Latrodectes katipo)* being warned of it by the Maoris (Hochstetter, 1867 p.440). It was not described until 1871 (Powell, 1871, pp.56-59) although there were several accounts since it was first reported by Dr Ralph of Wellington (see Buller, 1871, pp.29-34).

53. Bagnall, 1980, v.1, pp.925-929.

54. Von Hochstetter, 1867, p.197.

The land-snail *Paryphanta hochstetteri hochstetteri* acquired by Hochstetter and illustrated in his work *New Zealand*, 1867, (p. 169). *(see p.161)*

wherever my scientific peregrinations led me; the numerous and attentive audience in Auckland and Nelson at my evening lectures on the geology of New Zealand; the honours and distinctions, with which I was overwhelmed at my departure, – all this imparted to me the soothing consciousness and the cheering certainty, that I had not laboured only for myself or for a few initiated in the science, but happily for a whole nation, who with their lively interest and national energy participated in the results of geological and geographical researches, and most vigorously endeavoured to turn them to account. Perhaps I have also aroused a dormant taste of natural sciences in many a distant friend . . .'.[55] So once again the course of New Zealand science was modified by a series of historical coincidences and encounters abroad: Archduke Maximilian's interest in the navy and in science, the persuasions of an ex-Governor of New Zealand, the presence on the *Novara* of an able and enthusiastic geologist prepared to leave his ship. The sequels too were important: Hochstetter's continuing friendship with Haast,[56] his encouragement of the collector Andreas Reischek, and Austrian visits later in the century. In a country struggling towards independent scientific effort and institutions, he was an important catalyst for change. The time was fast arriving when New Zealand could start to look inward to her own scientific institutions and needs.

55. *Ibid.*, 1867, p.13

56. Von Haast, 1948, p.7.

Chapter Eleven

INDIVIDUALS AND INSTITUTIONS

ISSIONARIES, TRAVELLERS, TRADERS, AND RESIDENTS had scattered themselves about the first four decades of the 19th century, filling in their journals and collecting natural history specimens. Now and then they were charged with enthusiasm by visiting voyage naturalists who came and went, sometimes taking away as little as a fleeting impression of the Bay of Islands and a few plants and animals, sometimes making friendships that would last a lifetime. But for many years there were no museums, only the private collection of enthusiasts like Colenso; nor were there evening meetings with attentive audiences — the few naturalists could only talk to each other when their busy routine would let them. If they published at all, their publications ran to popular accounts rather than papers in learned journals: anyway, the journals were published far away in London where they would have to compete with those to whom they had some allegiance as collectors. Science, then, in the corporate and institutional sense, could not be practised in New Zealand in those days. Its practitioners were too few, too isolated, too preoccupied and lacking the confidence that experience and an institutional base can provide. In spite of these severe handicaps they made valuable contributions to the study of natural history, but before the 1860s very few (and Colenso was one), would emerge with scientific distinction enough to tilt at the hegemony of an English natural history world that rested on institutional, aristocratic, and commercial bases. But if there were strong objections to their being the vassals of imperial science, then these were not loudly voiced — scientific independence would come with time.

It is around 1840-1841 that the first real node of scientific activity can be observed — a mixture of individuals, expeditions and events, neither planned nor coordinated and with scarcely a common thread of motivation between them beyond an interest in natural history.[1] D'Urville, Wilkes and Ross and their naturalists came and went, the last with Hooker particularly stirring the enthusiasm of the missionaries at the Bay of Islands. Dieffenbach and Sinclair stayed and collected there. Walter Mantell arrived at Wellington followed by William Swainson, a once-prominent but now fugitive figure from British natural history circles. Lady Franklin, wife of Tasmania's Governor and naturalist Sir John Franklin, and partner in the patronage of science, visited Wellington and the Bay of Islands soliciting New Zealand interest in the Tasmanian Society.[2] Also in Wellington Dr Frederick Knox was beginning a lengthy period of scientific publishing through the medium of the local newspaper. At Akaroa, the French nourished a small community, with scientific interests represented by St Croix de Belligny and the botanist Etienne Raoul from Lavaud's *L'Aube*, on station there.[3] With collectors such as Captain Sir Everard Home, and Percy Earl not far away, the collective scientific effort in New Zealand was not insubstantial. However it was thinly spread; the naturalists were geographically scattered and many of the visits and activities, while swelling the current of scientific endeavour, did not always exactly coincide.

Other positive signs of growth were the attempts to form societies and the first examples of indigenous scientific publication. Australian societies, which could provide a model if not a start to similar New Zealand ventures, had shaky beginnings. The short-lived

1. Apart from the voyagers who were to some extent commonly motivated to explore the high southern latitudes.

2. Hoare, 1969, p.205.

3. Simpson, 1984, pp.66-67.

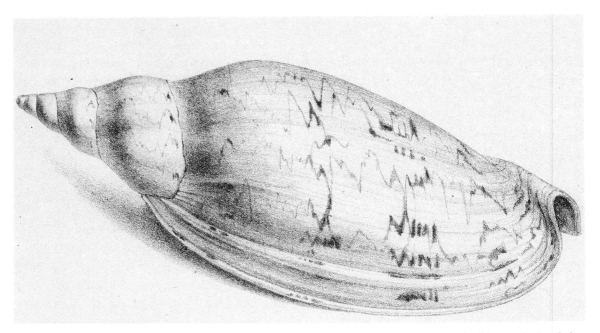

The Arabic Volute (*Alcithoe arabica*), a lithograph proof prepared by William Swainson for his *Exotic Conchology* . . . . v. 21 (pls. 20, 21). An able illustrator, almost all of Swainson's work on New Zealand animals was carried out prior to his arrival in that country. Published by permission of the Alexander Turnbull Library (E 217/55). *(see p.165)*

4. Hoare, 1969, p.199.

5. Brereton, 1948, pp.11-12; Reed, 1979, p.62.

6. Parkinson, 1984, p.54.

7. The objects of the Mechanics Institute of Richmond (near Nelson) were to promote literature and science with elementary education. No discussion of religion and politics was allowed; Brereton, 1948, p.27.

8. Historically, botanical interests and societies tended to precede zoological ones (see Allen, 1976, p.6).

Philosophical Society of Australasia (1821) did not have much influence in New Zealand but the successor to the Van Diemen's Land Scientific Society, the Tasmanian Society — which spanned the years 1837-1849 — embraced several colonial outposts of which New Zealand was one.[4] And yet in spite of the Australian presence, similar societies in New Zealand were slow in coming and when they did were intitially spontaneous and localized. In spite of its favoured location and visits from overseas naturalists, it was surprisingly not the Bay of Islands that provided the first New Zealand setting for a scientific society. Societies then needed an infrastructure of interested lay people, and the busy missionaries, the Maoris, the traders, transients and habitués of gin-shops were not a promising audience for a scientific meeting. Instead the societies were intitially located much further to the south, in areas of planned settlement.

In 1841, the Literary and Scientific Institute of Nelson was formed by immigrants on the *Whitby* as she approached her destination and in 1842 they invited their fellow Nelsonians to join them.[5] Across Cook Strait, the Wellington Horticultural and Botanical Society held its first meeting in November, 1841, with William Swainson as a committee member.[6] The following years saw the inauguration of Mechanics Institutes and Libraries in several provincial centres including Wellington, New Plymouth and Auckland, as well as another horticultural society in New Plymouth. The Mechanics Institutes were an idea imported with the immigrants, a 19th century form of worker education in the physical sciences that incorporated a library of useful reference works and a place to read them.[7] They were a symbol of the democratisation of science, and their libraries in New Zealand held natural history works. The botanical and horticultural societies might have answered in part the need for natural history societies, but they also served the more practical and decorative demands of creating a garden in alien soil.[8]

These institutes and societies waxed and waned with members' enthusiasms and with the capacity for local support — in the troubled early 1840s, the Wellington Mechanics Institute went into virtual recess and it may have been these circumstances that prevented the Institute from taking advantage of Swainson's offer of his collections to form the basis of a museum.[9] That such societies and institutions should emerge at all from the the raw cultural landscape of the new colony might be seen as something of a miracle, but in the settlers' few leisure hours there were limited opportunities for entertainment. Some of them brought with them that enthusiasm for science — and natural history in particular — that had already pervaded Victorian England, a godsend particularly to those enduring the tedium of winters in isolated high country stations. But how difficult it must have been, at the end of a weary day breaking ground or planting crops, to come in on horseback from Lower Hutt, or to walk the muddy quayside to hear the evening lecture or exchange notes on their rambles in the bush or among the rocks of the harbour.

In spite of this early round of activity, the coming-of-age of New Zealand science was still some distance away. The support and enthusiasm was derived from a scanty but proportionately well-educated population resident in a Crown colony in which planning for the future could now begin.[10] But the years to follow would be made difficult by the diverting demands of the raw state of colonial life, by disputes with the Maoris and with each other, and by an economic depression that affected the Australasian colonies in the early 1840s.[11] Venues for publication, a more practical and local alternative to the *Tasmanian Journal of Natural Science...*, were needed; for a time only newspapers such as the *New Zealand Gazette* and *Wellington Spectator* offered any sort of forum for science publication. The resources for scientific study — books and equipment — were scarce, and news of discoveries and emergent scientific philosophies in Europe were slow in arriving, as were the answers to questions or descriptions of specimens sent to colleagues abroad. Also significant for the biological and geological sciences was the absence of a repository for specimens or a collection of reference. But for a while at least it would be the natural sciences, with their amateur support and practical underpinning from surveying, mining and horticulture, that would be the mainstay of science in New Zealand. It was mainly the novelty and excitement of the giant fossil birds and amateur interest in ornithology, insect and shell collecting that kept zoology at the forefront; a practical base in agriculture, crop pests and conservation would emerge later.

As the New Zealand Company vessels and others discharged their immigrant passengers on shore in various parts of New Zealand, the increasing human resource might at least help sustain and develop the initiative so recently begun. But the years of FitzRoy's governorship and the chaotic beginnings of the New Zealand Company's settlements were not fertile ground for a flourishing of science. Witness the protestations of the immigrant naturalist William Swainson who, although no stranger to personal and financial difficulties, was thoroughly disillusioned with the New Zealand Company, Wellington, and the prospects for science in the country.[12] Thwarted ambitions at home had left Swainson a bitter and not unbiased observer in New Zealand, but he demonstrates that a particular vigour, enthusiasm and spirit of enquiry were required to motivate the scientist of the day.

In better times, FitzRoy — whom we remember as the Captain of the *Beagle* — might have proved to be a patron of the sciences in the same mould as Sir John Franklin of Tasmania.[13] But science was left to his Colonial Secretary, Andrew Sinclair; it was FitzRoy's successor George Grey who was to come closer to filling that role. The unsettled decade had seen the arrival of several whose collective interests in natural history were to later become of greater significance, but in the meantime the plan for colonization kept them isolated in provincial settlements. New Zealand was perceived by some colonial planners such as Wakefield as a return to the pastoral England of yesteryear — a place in which large cities (and by implication their intellectual trappings) had little part.[14]

9. Hudson, 1851, p.221. *The New Zealand Colonist*, 19.5.1843, p.3; Parkinson, 1984, p.54.

10. Many early immigrants were well educated and professionally trained doctors, lawyers, clerics, etc. some of whom rose to the top of New Zealand commercial and administrative life. Several demonstrated an active interest in natural history and science in general.

11. Hoare, 1977a, p.20. There was also a depression in Britain at this time.

12. Parkinson, 1984, pp.55, 56.

13. It was no accident that these ex-explorer colonial Governors could be found to have qualities that included a strong interest in science.

14. J. Gruber *pers. comm.*

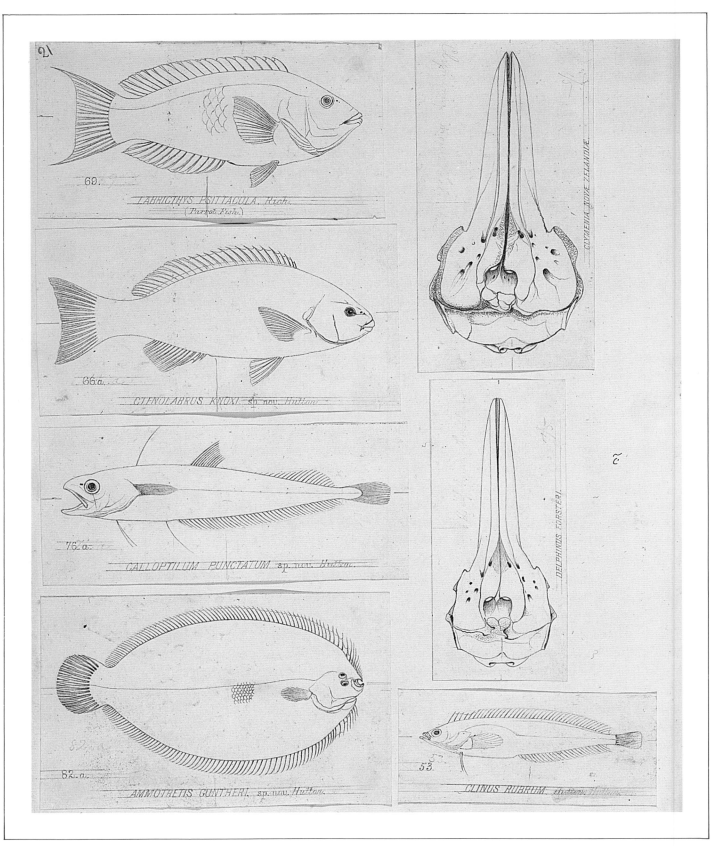

A group of original pencil and ink drawings of fish and marine mammal skulls by John Buchanan, for many years principal illustrator and lithographer for the New Zealand *Transactions*. Clean line drawings of this type were to become characteristic of zoological illustration. Published by permission of the Alexander Turnbull Library (E 208). *(see p.169)*

Wellington had Knox, Mantell and Swainson; Nelson, the pastoralist Dr David Munro and solicitor W. T. L. Travers; Auckland had Andrew Sinclair, with Governor Grey and Lieutenant-Colonel David Bolton:[15] the zoological acquisition records of the British Museum testify to the collecting activities of Knox, Sinclair, Grey and Bolton particularly. John Buchanan had landed in Otago; not a professional man like the others, he would make his mark in botany and scientific illustration. And finally Thomas Henry Potts, a businessman in early retirement who turned to natural history, had now settled in Christchurch. Other collectors included the Reverend J. F. Churton from Auckland, and Dr. D. Greenwood who sent shells to the British Museum.[16] Visiting collectors such

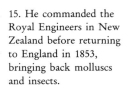

15. He commanded the Royal Engineers in New Zealand before returning to England in 1853, bringing back molluscs and insects.

16. Suter, 1913, p.ix. T. H. Potts (1882) would author the first locally produced book on natural history.

The nest of a Saddleback (*Philesturnus carnuculatus*) from T. H. Pott's *Out in the Open* . . . . . 1882 (p. 201). *(see above)*

as Earl, Strange, Thomson the military surgeon, artist and collector George French Angas and others, together with the missionaries and the *Acheron* and her scientists, moved through this loose network of amateur naturalists.

The cloth was still not woven tightly enough when in 1851 Governor Grey, Mantell and others attempted to form an association of those interested in science and the New Zealand Society was born. It was Wellington based, its membership including many prominent Wellington citizens who met at the Athenaeum, which had evolved from the Mechanics Institute.[17] But in spite of its broad objectives, including 'the development of the physical character of the New Zealand group: its natural history, resources and capabilities', it was not able to attract support from the other provincial centres, and with the departure of Walter Mantell for the South Island and Governor Grey from New Zealand, some of the driving force behind the Society was lost.[18] Thereafter it was fated to evolve into the Wellington Philosophical Society in 1868. Amongst its members and early officers there was a young man with a precocious interest in science, particularly ornithology; his name was Walter Lawry Buller and his present significance is that he was one of the first of the New Zealand-born naturalists.

It was possibly the arrival in Auckland of Haast, the *Novara* and Hochstetter that sparked off the next round of scientific activity. Although they were practical economic considerations that dragged Hochstetter into New Zealand, he and his friend Haast cut a swathe for science through New Zealand and when he left, Haast remained in Canterbury to carry on his scientific mission. Others to arrive at about this time were

17. Fleming, 1968, p.100.

18. Dick, 1951, p.139; 1957, p.12. Dick's (1951) excellent essay says much that is equally true today.

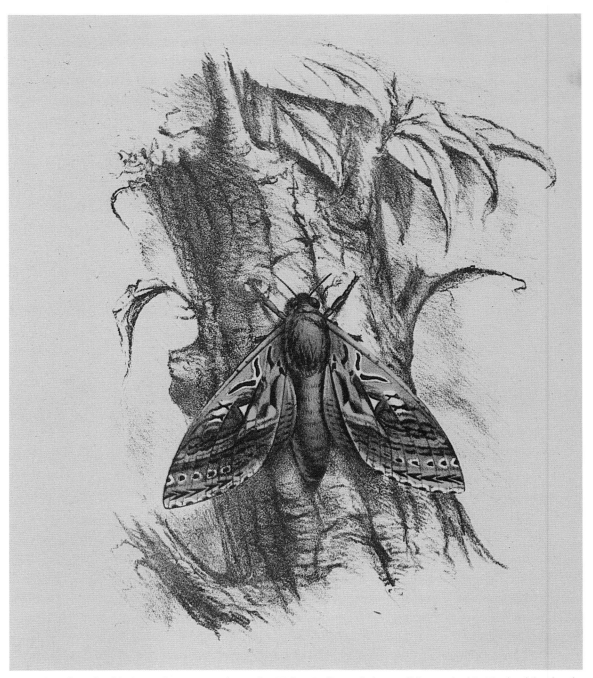

A species of moth of indeterminate genus drawn by Walter Buller and chromolithographed in England (under the supervision of the lepidopterist A. G. Butler); from Buller's *Notice of a new species of Moth . . .* in *Transactions and Proceedings of the New Zealand Institute*, 1873 (p. 279). Subsequent chromolithography for the *Transactions* was done in New Zealand. The moth has not been recorded since, and the ship carrying the type (and only) specimen to England sank. *(see p.169)*

19. An interest he shared with his position as Registrar of the University of New Zealand.

20. Scottish medical training was behind several who took up natural history in the colonies; J. Gruber, *pers. comm.*

John Enys who became a Canterbury runholder as well as an authority on New Zealand Lepidoptera, and William Maskell, another pastoralist who later developed an interest in agricultural entomology.[19] These two, later with G.V. Hudson and others, would form an entomological counterweight to ornithology. Then in 1862 — the same year that Haast founded the Philosophical Institute of Canterbury — James Hector arrived; if the traditions of a Scottish medical training had stood New Zealand in good stead with Knox, Munro, Sinclair and A. S. Thomson, then so it would with this newcomer.[20]

Given the scattered nature of scientific effort and endeavour, provincial rivalries and

Butterflies including: (1, 2) The Black Mountain Ringlet (*Percnodaimon pluto*); (3-5) Tussock Ringlet (*Agyrophenga antipodum*); (6, 7) Southern Blue (*Zizina otis oxleyi*); and (8, 9) Common Blue (*Zizina otis labradus*); reproduced by chromolithography from A. G. Butler's *Catalogue of the Butterflies of New Zealand* (with a preface by J. D. Enys), 1880. (pl. 1) Published by permission of the Alexander Turnbull Library. *(see p.170)*

One of several steps in the process of chromolithography is demonstrated in this lithograph proof plate of the above illustration from *Catalogue of the Butterflies of New Zealand* with pencil annotations by J. D. Enys. Published by permission of the Alexander Turnbull Library. (A9/E254) *(see p.170)*

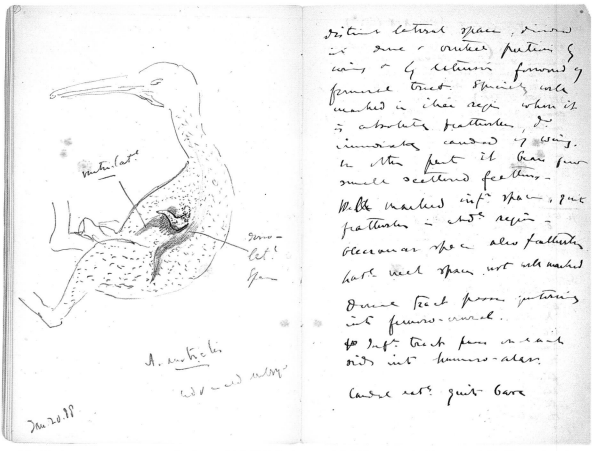

The kiwi was amongst many animals to which T. J. Parker extended his anatomical interests. This illustration shows two pages from Parker's working notes. From the collection of the Zoology Department, Victoria University of Wellington. *(see p.173)*

21. Hutton, originally a geologist, would prepare several catalogues of the New Zealand fauna and become one of this country's best known naturalists.

22. Fleming, 1968, p.100.

limited communications, some sort of centralization was needed to bind the parts together. New Zealand science demanded a leader. The faltering indigenous scientific society was peopled with those whose commitment was rather less than full time. As New Zealand moved towards administrative and social stability and recognition of what science could achieve for the country, full time leadership was recognized as an essential to further and coordinated development. When Parliament moved to Wellington, and the Colonial Museum and Laboratory were established in 1865 with Hector as its Director and staffed by Buchanan, McKay and others, the first steps had already been taken.

This was also the year of the Dunedin Industrial Exhibition — a brief but important focus for science — and with further new arrivals including Captain Frederick Woolaston Hutton the stage was set for another attempt at forming a national scientific society.[21] This time it was successful. Parliament and the Upper House had sympathetic members such as W.T.L. Travers and Walter Mantell and with their support the New Zealand Institute was established in 1867. James Hector was appointed as its 'Manager' and Editor of the *Transactions and Proceedings of the New Zealand Institute*, the country's first journal of science. The momentum was sufficient and federation was successful. Existing provincial societies in Nelson, Wellington and Canterbury were joined by like groups from other provincial centres and collectively supported the national body.[22] Developing, they added to a growing professionalism, begun earlier by provincial surveyors and geologists and reinforced by natural and other scientists. There followed the addition of a Museum at Canterbury and then in 1869 the founding of the University of Otago, followed by

Cuvier's Whale (*Ziphius cavirostris*) from a paper by J. H. Scott and T. J. Parker *On a specimen of Ziphius* . . . . in the *Transactions of the Zoological Society of London*, 1889, (pl.48); in which both authors assisted in the preparation of the plate. *(see below)*

the initiation of the University of New Zealand.

Although New Zealand was becoming more self-sufficient as the 19th century wore on, links with British scientists and their establishments continued to be maintained. From 1860 through to the end of the century the volume of donated specimens to the British Museum decreased significantly but never dried up, and private collections continued to be supplied as well.[23] Hector, followed by Sinclair, Haast, Parker, Buller and Benham, visited England and in some cases Europe and America in their capacity as scientists, and many of the scientists domiciled in New Zealand still sent some of their work to be published in the *Proceedings* or *Transactions* of the Zoological Society of London.[24] The court of British science did not predominate in quite the same way as it had in mid century, but it was still necessary to visit it. Trans-Tasman links were strengthened with visits by scientists such as Haast; the Australasian Association for the Advancement of Science started in 1886 and was formally established in 1888.[25]

The pioneering group of scientists represented by Mantell, Potts, Travers and others retreated into elder statesmanship during the 1860s while a new coterie emerged composed of Hector, Hutton, Haast, Buller, Parker and others, many of whom worked almost full time for science. Although enthusiastic amateurs still provided the backbone for research, publications and attendance at meetings of the Philosophical Institutes, science was well on the professional road with a speed and liberality of endowment and Government support that must be regarded as remarkable.[26] Compare progress in the mother country: it was only in the second half of the 19th century that professional science in Britain became established beyond its position in the relatively few museums and poorly paid chairs at the universities. Even later in the century the research worker remained an isolated figure and civil servants were reluctant to find financial support for something as unaccountable as scientific research.[27]

Unfortunately this pace of development in New Zealand did not last. From 1879 to

23. e.g. Buller supplied S. W. Silver and Canon Tristram.

24. This state of affairs continued well into the 20th century, and even quite recently there have been occasions where New Zealand specimens were described from material sent to the British Museum.

25. Its successor was ANZAAS.

26. Dick, 1951, p.140 plots this progress and the amount of money involved.

27. Allen, 1976, pp.83-84; Poole and Andrews, 1972, pp.8, 9.

Scale insects: (1) The Felted Pine Scale (*Eriococcus auraucariae*) and (2) *Noteococcus hoheriae* drawn by W. M. Maskell for his *An account of the Insects noxious to agriculture* . . . . 1887 (pl. 14). The chromolithography was done in Wanganui, New Zealand, to a high standard by the firm of A. D. Willis. *(see p.175)*

28. The effects were felt everywhere – in Nelson the president of the Literary and Scientific Institute noted sadly the decline in membership (which could probably be attributed to the cost of belonging); Brereton, 1948, p.40.

29. Dick, 1951, p.140.

1896 there was a long economic depression and changes in the personalities and constitution of the Government which slowed the progress of science.[28] Informed pressure groups and an understanding legislature are essential to its development and it was by now noticeable that the well-educated, amateur scientist-legislators of the earlier years were a decreasing presence in Parliament.[29] The sometimes spectacular ignorance of the representatives was embodied in the member for Motueka, Mr Kerr, who on more than one occasion was astray on the subject of animal acclimatisation. Responding to a Nelson Borough Council request to import half a dozen Venetian gondolas for a lake in the public gardens, Kerr expostulated: 'Why not import a pair and let Nature take its course?'[30]

Happily by now there was sufficient impetus amongst scientists and their organizations to survive the indifference or ignorance of their political masters. A similar state of affairs had existed in Britain where a depression and a lack of consensus over the future direction of science (including the natural sciences) slowed progress.[31] In New Zealand however the signposts were becoming visible, and once again it was a nation's economy that pointed the direction that science should take. But this time it was the indigenous economy — one that was clearly going to be heavily reliant on agriculture and forestry — that was dictating the trend. Voices including that of the botanist A. Lauder Lindsay called for the promotion of natural history as a necessary adjunct of colonization. Apart from being an end in itself to the 'higher classes', to 'other classes of minds' it could be shown to have practical uses: geology and botany and other allied sciences 'will pay in a young colony'. Even museums and universities could be oriented towards practical and local problems.[32]

But as the scientific demands of the economy grew in the last two decades of the 19th century (along with a rising tide of biological problems), so it was revealed that the legislature was 'biologically impotent' and the scientific civil service, such as it was, inadequate to the task.[33] Men such as the ones who discovered Moa bones, and the writers of handsome bird books — and others like them — were suddenly less needed than humble entomologists, applying themselves laboriously to a once disregarded fauna. And yet the entomologists in question, G. V. Hudson and William Maskell, were by day respectively a clerk in the Post Office and Registrar of the University of New Zealand, entomologists only at night and in the weekends or when any other spare time presented itself. Something more had to be done.

The mood was translated into a paper by Maskell, read to a meeting of the Wellington Philosophical Society in July 1891, in which he proposed the establishment of an 'Expert Agricultural Department'.[34] In an onslaught against the methods of politicians in dealing with animal pests, Maskell's motion was seconded by a Mr R. C. Harding who rose to say that: '. . . he could not speak with practical knowledge, like the gentlemen who had already given their views, but, even without such knowledge the necessity of such a department would be recognized. Many present would remember the cumbrous and oppressive Thistle Acts of the old provincial days, when every province had a different ordinance. These Acts were now acknowledged to have been ill-advised, while in regard to the Thistles they were about as effective as a Papal Bull against a comet. The present fashion seemed to be to launch a separate and voluminous Act against each individual nuisance. The public suffered from ill-considered legislation, and the pests flourished apace. The Act just passed relating to small birds was a case in point. It was crude and unwieldy to the last degree, and would prove an intolerable nuisance, and would probably produce all manner of effects other than those intended. Parliament was not qualified to deal with such matters; they lay altogether outside of the scope of its duties. If only in the interests of the economy, Mr Maskell's proposition deserved all support'.[34]

The logic of the arguments of the applied biologists Maskell, Wight, G. M. Thomson and their supporters was obvious, even to Parliament, and in 1892 T. W. Kirk, long an assistant in the Colonial Museum, was appointed as 'Clerk and Acting Biologist acting as Entomologist' in the Agricultural Branch of the Lands Department.[35] Later, in 1894, Thomas Broun — the industrious describer of hundreds of beetles — was appointed Government Entomologist. Marine interests tended to follow agriculture and G. M. Thomson had to wait longer for his marine biological station.[36] These gestures of official

30. Thomson, 1922, p.61.

31. McLeod, in Mathias, 1972, pp.127-131.

32. Lindsay, 1862, pp.5-7.

33. Dick, 1951, p.141.

34. Maskell, 1892, pp.625-691; and Harding *in* Maskell, 1892, p.691. Similar feelings were expressed later in relation to other, similar Acts (e.g. the Codlin Moth Act of 1884).

35. Miller, 1953, p.4.

36. Hoare, 1984, p.26.

support might have lacked the open-handedness of earlier years, but they were a beginning. For the biological sciences, it was a further step away from the dominance of collecting and classifying, a swing from natural history to applied biology. For the biologists themselves, no longer would this field be the province of a small band of enthusiastic part-timers, university or museum staff or those like Buller, able enough to make a quick fortune to allow early retirement and a lifetime's pursuit of a hobby.

There was a lingering official suspicion of scientific research that trailed into the 20th century, with the result that the emphasis was placed on extension staff rather than those who would carry out more fundamental or even applied research.[37] The conflict between the Premier Richard Seddon's Government and the scientists was one that was universal and timeless; it was the conflict between the roles of basic research and the scientist as a provider of quick answers to practical problems. With a few exceptions, science has never been able to provide the instantaneous and spectacular solutions expected of it, leading to impatient and mistrustful legislators and public.

To put this apparently feeble progress in New Zealand in some perspective, it is worth noting that applied science had similar difficulties in getting under way in Britain and was not well set on the path until the late 19th century.[38] Economic recession was partly responsible there too. For the biological sciences there was some growth generated as a result of Britain's development of her colonies. Tropical diseases were a blight on local and expatriate populations and a hindrance to economic development. Investigation of the biological causes and means of control of these diseases led to the establishment of schools of tropical medicine and hygiene, with an emphasis on research. In New Zealand, codlin moth, small birds and pastoral problems, although important did not have quite the same punch or urgency; traditionalist views and lack of funding impeded scientific research in the universities, and the fledgling scientific civil service was facing the difficulties already described.[39] Attempts at reform, including the launching of the *New Zealand Journal of Science* in 1882, were off to a slow start.

In spite of the difficulties, the New Zealand sciences were now set on the course that they would follow into the next century. Their links with the country's economy — with agriculture, fisheries, forestry, mining — were now established. The Board of Science and Art and the Department of Scientific and Industrial Research were around the corner; science would grow in the universities and museums and specialist research institutes. The day of the individual was almost over, more and more they would work together and individual effort in natural history would be replaced by team-work in basic and applied biology.

But what of zoology, as the individuals started to melt into their institutes and departments? Apart from one or two highlights, the years from 1865 until the century's close will be remembered for solid productivity and consolidation, rather than for spectacular discoveries or splendid feats of zoological detective work.

It might have been expected that the institutional activities of 1865 and the occasion of the Dunedin Exhibition that same year would have generated some like activity in the zoological sciences — and indeed they did. Walter Buller put himself and ornithology in the limelight with his *Essay on the Ornithology of New Zealand* prepared especially for the Exhibition, while proceeding with plans for a much larger and more ambitious work.[40] Hector, considered an authority in many branches of science including zoology, led by example and published in a variety of fields. The availability of the New Zealand Institute's *Transactions* rapidly attracted contributors: predominantly ornithological ones at first, but a rapid build-up of papers on fish, marine mammals and then insects followed, and

37. Dick, 1951, p.141.

38. Poole and Andrews, 1972, pp.10,11.

39. Hoare, 1977a, p.27 puts some of the blame on scientists themselves for their lack of innovation, although many of the more productive scientists were working on their own, able to influence only their local provincial Philosophical Institute, or their Parliamentary representative. They achieved a great deal (Hoare, p.28) but only a well-organized and central lobby could be really effective, and it seems that Hector and the New Zealand Institute were not up to it. G. M. Thomson's suggestion for improvements were not acted on for some time (see also Hoare, 1977b, pp.4-10). Hoare, 1977a, p.20, believed New Zealand had the 'scientific-technological manpower capabilities to face her problems' — but it is evident that some of these were biological problems whose foundations were generally poorly understood and the resources to cope with them rather inadequate.

40. Buller, 1869a, pp.213-231. He was also one of the *Transactions* most prolific authors.

by the 1870s the broad spectrum of zoology was being represented in its issues.[41] On the face of it, the period from 1870 to 1900 was not a promising one for the continuing development of any branch of science. But if numbers of published papers in the *Transactions* are any guide, zoology flourished, if somewhat unevenly with respect to its branches.

Ornithology dominated until 1880, no doubt enlivened by the appearance of Buller's first edition of the handsomely illustrated *A History of the Birds of New Zealand* in 1873. Interestingly, a similar flush of bird papers accompanied the appearance of the second edition in 1888. Otherwise ornithological publication, while continuing to provide the mainstay for papers in vertebrate zoology, levelled off for the remainder of the century. Marine mammals, an important but diminishing resource during New Zealand's early economic history, occupied a prominent place in the early issues, and papers on fish appeared steadily throughout the period. But New Zealand's small but interesting reptile fauna, a victim of past indifference, was still the subject of a mere trickle of papers.

From about 1880 onwards there was a gradual swing towards invertebrate animals, and here insects took a leading position. Entomology had been strengthened by the arrival of J. D. Enys and then R. W. Fereday in 1862, who early on saw the significance of economic entomology and published in 1871 on the importance of crop pests. By 1876 Fereday's published contributions were numerous. In the leaner years of the early 80s, entomology was given a lift by the presence of a visiting lepidopterist from England, E. Meyrick, who arrived in 1879 to stay for seven years before returning to England where he maintained an important link with local naturalists. Although he published little, the practical achievements of Alan Wight of Paeroa in economic entomology place him among the pioneers of this field in New Zealand. Wight was in the forefront of biological control — the use of insects or other biological agents to control plant or animal pests. He was instrumental in collecting Cardinal Ladybird Beetles, the natural enemy of the citrus tree pest called Cottony Cushion Scale, a nuisance that was active in South Africa and California.[42]

In 1881 a precocious entomologist by the name of George Vernon Hudson arrived in Wellington. He was only 14, but would soon begin his untiring labours in descriptive entomology that would last through this century and well into the next. Although in some respects he was a completely different personality, Hudson was to insects as Buller was to birds, both publishing numerous papers and substantial illustrated volumes in their specialist fields. Also in the van of entomological workers were Thomas Broun, an ex-Army man whose beetle descriptions in the order of 3500 species created something of a taxonomic jungle for his successors, and William Maskell, whose experience with the economically important scale insects led him to propose the formation of an agricultural department. Insects also received attention from those less specialized zoologists whose interests and publications covered many groups: Colenso followed by Hutton and T. W. Kirk were examples.

In the rush towards professionalism (and in the case of some of the dedicated part-timers there was very little that separated them from the professional) the role of the true amateur and hobbyist was recognized: encouraging others and supporting local microscopic clubs and natural history societies. Hochstetter remarked: 'Nearly in every house, in every family, I became acquainted with, there was somebody making collections. Here it was the husband, who had a collection of insects; there the wife, who pressed mosses and ferns neatly between the paper; or the sons and daughters, that gathered shells and seaweeds . . .'[43] There were many individuals, unpublished and unsung, except

41. We note at this point the beginning of the age of the zoological specialist. It was possible then, and for some years later, for one person to embrace more than one science or, if interested in zoology alone, more than one animal group. But as the numbers of species and information about them increased, so did the need to specialize, most notably in the insect and other invertebrate groups. This made it more difficult for generalists like Hector to keep abreast of developments in a variety of fields. Many of the *Transactions* papers were illustrated by John Buchanan, whose work has been described by Adams, N. M., 1983, MS. *The Life and Work of John Buchanan, F.L.S.* History of Science in New Zealand Conference, unpublished paper.

42. Miller, 1953, p.2.

43. *Ibid.*, p.3; Von Hochstetter, 1867, p.270.

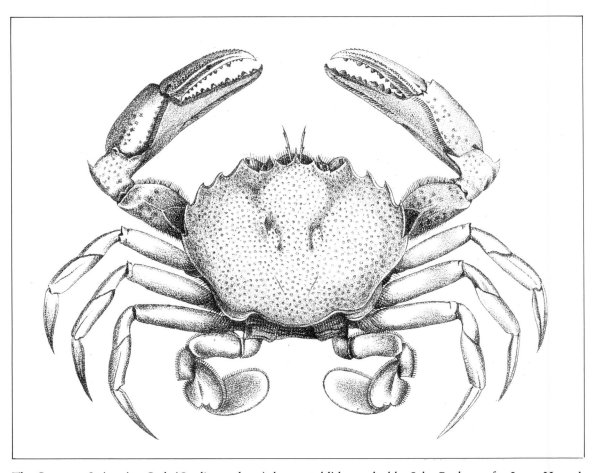

The Common Swimming Crab (*Ovalipes catharus*) drawn and lithographed by John Buchanan for James Hector's *Notes on New Zealand Crustacea* in *Transactions and Proceedings of the New Zealand Institute*, 1877 (pl. 27). *(see p.177)*

in the memoirs of more prominent associates who remembered their contributions. Albert Norris was an example; a bootmaker who 'lacked literary proclivities', he was an outstanding field naturalist whose observations on the New Zealand Glow-worm were highly regarded.[44] His collection was deposited with the Colonial Museum after he died at the early age of 28.

Apart from its applied aspects, entomology in New Zealand was destined for some years to produce the dogged rather than the spectacular. By its very nature it could do little more. It had given the lie to the sweeping generalizations of the early explorer-naturalists, but it did lack variety, and in the main the species were not eye-catching or highly coloured, the stuff of a collector's cabinet or iconography. But there were some gems: the Forest Ringlet, one of the most attractive although less well known of the New Zealand butterflies, was discovered in the forests of the West Coast by R. Helms in 1881, and its description formed one of the early zoological papers in the new *New Zealand Journal of Science*.[45]

The insect pests that gave New Zealand entomology its early importance were inevitable arrivals on imported fruit and vegetables, animal feed and stored products; they stayed to attack home-grown orchards, crops and garden plants. The Woolly Aphis attacked apple trees until grafting on to resistant stock solved the problem; *Phylloxera* attacked grape vines, and the Cabbage Aphis a variety of leaf crops. Particularly damaging was the Codlin Moth, supposedly imported in some Tasmanian apples[46] after which it spread rapidly in orchards throughout the country. Citrus and garden trees were susceptible

44. Hudson to McLachlan (undated) correspondence, Hope Entomological Library, Oxford University.

45. Butler, 1884, pp.159-160. A description was published a year earlier by Fereday, but for nomenclatural reasons was unacceptable (Fereday, 1883, pp.193-195; Gibbs, 1980, pp.77, 78).

46. Thomson, 1922, p.305.

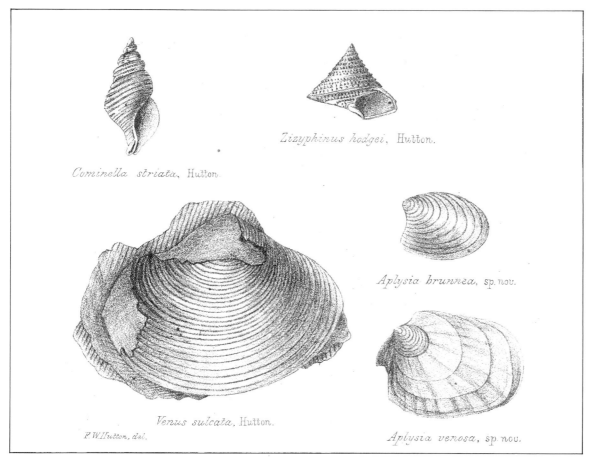

Cominella striata, Hutton.

Zizyphinus hodgei, Hutton.

Aplysia brunnea, sp. nov.

Venus sulcata, Hutton.

F. W. Hutton, del.

Aplysia venosa, sp. nov.

Although much of the *Transactions* illustrating was done by John Buchanan, occasionally the authors themselves would contribute illustrations, such as this one by F. W. Hutton for his *Description of three new Tertiary Shells . . .* in *Transactions and Proceedings of the New Zealand Institute*, 1876 (pl. 21). *(see below)*

to scale insects. Vigorous attempts at control were made to rid the country of these scourges: legislation such as the American Blight Protection Ordinances was passed by the Nelson and Otago Provincial Councils and the New Zealand Parliament passed the Codlin Moth Act 1884.[47] The contempt in which some of these Acts and Ordinances were held has already been mentioned in connection with Maskell's recommendation to form an agriculture department. Needless to say, biological control measures using the natural enemies of the pest animals were often more effective. Some attempts at control appeared a little bizarre, as the *New Zealand Herald* of 7 June, 1897, reported that the Department of Agriculture was contemplating the importation of 'toads, swallows and English bats' into the colony for the purposes of checking the spread of insect pests. Plants were not the only object of insect attention as animals including humans were attacked by native and introduced insect species including sandflies, mosquitoes, fleas, botflies, lice and others. But control of these was not widely practised until the following century.

While the majority of publications in invertebrate zoology dealt with insects, other groups — crustaceans, molluscs, echinoderms, annelids and others — were not entirely neglected. Crustaceans benefited from the researches of G. M. Thomson and Charles Chilton, and there were several papers between 1880 and 1883. Molluscs, which were such a feature of early New Zealand zoology, were by comparison neglected in the early issues of the *Transactions*, although F. W. Hutton published a catalogue of these animals

47. *Ibid.*, pp.330, 549.

48. Suter, 1913.

in 1873 and contributed several papers on them in 1881. But during the last decade of the century, the Swiss immigrant Henry Suter began publishing papers on molluscs, which led towards publication of his classic *Manual of New Zealand Mollusca* 22 years later.[48] Interest in other, less popular invertebrate groups also picked up in the last years of the century. These may have held little significance to collectors or have had little economic value, but to the science of ecology which at that time was sensed and felt rather than studied or discussed, knowledge of these less obvious animals was to become of greater importance.

Throughout the second half of the 19th century, and on occasions even before that, the elements of what we know today as ecology manifested themselves here and there as snatches of thought expressed as phrases in books and journals: Sparrman's concern about the introduction of rats by Cook's vessels, Dieffenbach's conception of the food chain, the growing recognition by a number of authors of the importance of conserving animals and plants, and the dangers of introducing foreign species. Although such thinking was not yet knit up into a science, there were signs here that the concepts of ecology in New Zealand were understood some time before their formal articulation in 1869.[49]

49. The term 'ecology' was coined by Haekel in 1869 (see Allee, Emmerson, Park, Park, and Schmidt, 1949, pp.1-72 for an introductory history of this multifaceted subject).

50. Simpson, P.1983, MS., *A History of Ecological Thinking in New Zealand*, History of Science in New Zealand Conference, unpublished paper. Richardson, 1843, p.12.

Of even greater antiquity was the cultural embrace of ecology by the Maori — an understanding born of centuries of close contact with the fauna and flora and their survival dependent on an understanding of the nature of their resources.[50] Human relationships to nature are inherently culturally based; since before the beginning of the 19th century the sometimes passive, sometimes exploitive culture of the Maori had to sit with the aggressively exploitive culture of the European until, out of the collective European subconscious, an appreciation of the aesthetics and balance of nature was brought to the surface and the dangers of what they were doing were recognized. And it was not just exploitation that was a cause for concern, but also the introduction of foreign species of animals and plants.

Unsurprisingly, it was the Maoris who first began to associate the disappearance of several bird species, including kiwis, with the presence of pigs, dogs, cats, rats and mice and '. . . the voracity and numbers of these foreign pests'. Colenso, in whose paper this comment appears, enjoyed a span of years from which to view the changes in the New Zealand fauna and flora and unlike some of his contemporaries had strong conservationist instincts. The problem was also recognised from England, by the astute prediction of John Richardson of a 'change . . . in the distribution of animals. Some will become rare or perhaps entirely disapper, while others, casually or intentionally introduced, and finding appropriate food and protection, will increase, and people the land.'[51]

51. Colenso, 1846, pp.81-107.

Richardson, 1843, p.12.

Visitors such as Hochstetter sounded warnings on the depletion of the fauna, and early issues of the *Transactions* carried the rumblings of local naturalists against the follies of plundering animal life and haphazardly introducing foreign species. Potts regretted that '. . . many species will have become extinct ere their habits can be sufficiently studied by the naturalist for their use, economy, and position in our Fauna to be correctly ascertained. To the future student of the natural history of our country vague, unreliable traditions, a conflicting nomenclature, and the contorted productions of the taxidermist mounted in acrobatic and weird-like attitudes will perhaps alone remain to fill up the hiatus.'[52] In a lecture at the Colonial Museum in August 1869 W. T. L. Travers, by then a convinced Darwinist, recognized that the 'struggle for existence' that had gone on for millions of years had been carried on in New Zealand 'under peculiarities little likely to be observed in other places, and the results already caused by the introduction of new and rival organisms satisfies me that the indigenous flora and fauna even on their

52. Potts, 1870, p.42.

own ground, are unable to cope with the intruders. I cannot but think that the former had reached a point at which, like a house built of incoherent materials, a blow struck anywhere shakes and damages the whole fabric.'[53] How perceptive this was, this recognizably ecological concept applied to New Zealand's native and introduced flora and fauna; even at this late stage the warnings might be considered timely but they went largely unheeded even by other naturalists.

Introductions of foreign species to New Zealand passed through several phases. The Maoris and early Europeans concerned themselves with a few plants and animals that conventionally provided food or clothing, but it was inevitable that these first and subsequent human arrivals would accidentally bring with them pests or parasitic animals that would spread as suitable habitat became available for them. With the establishment of European settlements came a demand for animals for food and companionship, although in the case of some, like the cat, more practical motives may have been in mind. Continued settlement saw a demand for introduced animals for aesthetic and hunting needs, by which time further animals were required to control some of those that were already introduced and had got out of hand.[54] The destructive agency of the earliest of introduced animals — the human — has already been mentioned. As an exploiter of the existing fauna and flora and an agent of other animal and plant introductions, human beings had a detrimental influence that was predominant.

There was very little official control over the importation of animals or plants, or the wanton destruction of native species. The Protection of Certain Animals Act 1865 and its successor, 'an Act to provide for the Protection of Certain Animals and for the Encouragement of Acclimatisation Societies in New Zealand 1867', prohibited certain animals, such as birds of prey and venomous snakes, from introduction as well as allowing for the whipping of small boys who infringed its provisions. It protected some native species that came under the heading of 'game' and exercised a limited control over the emerging acclimatisation societies. But in spite of the presence of the Act there was virtually an open door on plant and animal importations, with the result that bizarre assortments of animals were launched in the direction of New Zealand, with mixed results. Perhaps fortunately many died en route, or failed to establish, otherwise we might today be gazing at herds of Alpaca, Bharal, Zebra and Gnu, at who knows what expense to the native biota.[55]

It was partly to bring order to the chaos of introductions that the acclimatisation societies were initiated. The idea of such societies was not new; like many of the animals and plants themselves, it was an importation from Europe. They had their beginnings in France and then were later institutionalized in Britain in 1860, encouraged by a fashion for importing and eating exotic animals.[56] Frank Buckland established the movement with the support of leading naturalists including Richard Owen. But enthusiasm for roast Armadillo, Ostrich, consommé of Japanese sea-slug and other such delicacies was to wane, while the societies remained and continued their importations for reasons other than gastronomy.

In New Zealand, a meeting to form an acclimatisation society in Auckland was held as early as 1861; one was founded in Nelson in 1863, and in Canterbury and Otago the following year.[57] Other provincial centres soon followed suit. Unfortunately the societies did not stem the flow; indeed their mission was to ensure the success and maintenance of introductions, many of which would have benefited from a cooler appraisal of the consequences. Latter-day arks bulging with European birds were arriving in New Zealand to the cheers of acclimatisation society members while the animal traders such as Richard

53. Travers, 1870, p.312; it would be 1892 before the Huia was protected, and 1896 the kiwi. Wynn, 1977 discusses conservation in New Zealand in the late 19th century, but is concerned largely with forests.

54. Thomson, 1922, pp.6-23; describes the history of these introductions.

55. De Vos, Manville and Van Gelder, 1956, pp.181-183; list the successful as well as the failed introductions; Wodzicki, 1950, focuses on the more successful introductions;

56. Barber, 1980, p.147.

57. Sowman, 1981, pp.8-10; it was eventually formed in 1867. Druett, 1983, p.97.

58. Druett, 1983, pp.110, 112.

59. *Ibid.*, p.94.

60. Bathgate, 1898, p.266; although Bathgate was misguided in some respects, e.g. the rabbit.

61. Thomson, 1922, p.22.

62. Walsh, 1893, p.435.

63. Fereday, 1877, p.462.

64. Potts, 1870, p.43

65. King, 1981, p.142. Probably the largest single collection of animals taken from this country. King portrays Reischek as an accomplished but unfortunate figure, left stranded by time and circumstance.

66. *Ibid*, pp.110-111.

Bills who were responsible for these shipments took out native species in exchange.[58] And it was no help at all that prominent scientists like Julius Haast threw their weight behind the movement to stock the countryside with game and the streams and rivers with European fish.[59] As their mistakes began to come to account, so then were voices raised in criticism: 'It is a matter for regret that the zeal of the early acclimatisers was greater than their knowledge, and that mistakes were made by them fraught with evil results of a far-reaching and permanent nature', was one view put to the Otago Institute in 1897[60] and many years later G. M. Thomson was to remark '. . . I know the enthusiasm, unalloyed by scientific considerations, which animates the members'.[61] Even given the difficulties of scientific prediction of the outcome of an introduction, the one certain means of avoiding the problem, abstinence, would scarcely have occurred to them.

As the century neared its end, animal introductions continued, storing up problems for future generations and putting up a wake of legislative attempts to correct the mistakes. The first of the Rabbit Nuisance Acts was passed in 1876; the Small Birds Nuisance Act 1882 and the amendments to the Animal Protection Act 1895 were among the signs of continuing concern at the Parliamentary level. However, in their sum, the Acts were a long way short of being the teeth of a faunal protection policy or an absolute control on animal importations. By the 1890s some of the animals imported to control others had themselves become pests and in 1892 the Reverend P. Walsh bemoaned the action of those who were responsible for the introduction of rabbits, stoats and small birds and feared the worst from the introduction of deer. He had carefully observed the succession of events that took place in the forest as a result of the depredations of these animals and domestic stock, and: 'correctly understood the significance of introducing herbivores into a plant environment that had no experience of such animals'.[62] Walsh suggested fencing as a means of excluding deer from selected areas of the forest — an idea that was not taken up on grounds of cost.

In the meantime while the societies had their hands full with introduced animals, the destruction of habitat by pastoralists and town developers in New Zealand continued unabated. The lawyer and entomologist, R. W. Fereday, complained that the former bush, which afforded excellent entomological collecting ground, '. . . has, alas! disappeared, and cultivated fields now occupy the place where but a few years ago stood a dense forest'.[63] Some of the naturalists and collectors were of no help to the conservationist cause and Potts complained '. . . even the learned *savant* will sometimes be tempted to destroy both old and young, especially if they are rarer birds, a favourable opportunity of procuring choice and desirable specimens being too great for resistance; scientific zeal thus overcomes good policy, and consideration for the future'.[64] In justifying themselves some of the naturalists and collectors barely escape the charge of hypocrisy. Prominent among those with this difficulty was Andreas Reischek, an Austrian taxidermist recommended to Haast by Hochstetter.

Reischek spent 12 years in New Zealand from 1877 before returning to Austria with an enormous collection of artifacts and specimens taken on several major expeditions in the North, South and offshore islands.[65] He contributed to the scientific literature with several largely ornithological papers in the *Transactions*, but he was instinctively a collector in an age which was fast losing its enthusiasm for this sort of activity — at least on the scale that Reischek worked. His most notorious effort was the shooting of 150 Stitchbirds on Little Barrier Island, which act was followed by an orgy of self-reproachment that he extended to humankind in general.[66] It is perhaps as well for the safety of the *Notornis* that his expeditions in search of it were unsuccessful. More positively he provided local museums with specimens and shared his knowledge with local ornithologists such as Buller, and to some degree he redeemed himself for the Little Barrier episode by promoting the idea of a sanctuary there, although his proposal left the way open for limited collecting.[67]

In fairness to Reischek it must be remembered that he lived in times when assaults on the flora and fauna were commonplace. Even Walter Buller, while extolling the virtues of conservation, would have numbers of birds collected for public and private collections abroad.[68] It was not only representatives of the living that were exported for private or public reward — Haast, one of the czars of Moa bone collecting, used these as a medium of exchange. One such deal concerned a Professor Marsh of Yale, with whom Haast exchanged correspondence and specimens. Professor MacMillan Brown from Canterbury University encountered Marsh on a visit to America and found that the latter had 'a bone to pick' with Haast. 'A Moa bone I suppose?' responded MacMillan Brown. 'Yes,' said Marsh, '. . . he wrote to me and said . . . that if I remitted £50, he would send me a complete skeleton when the swamp was drained. I sent him the money, and instead of sending the skeleton he sent me only a heterogeneous lot of bones.' 'Oh,' chuckled MacMillan Brown, 'so you were the Marsh he was draining.'[69]

The loss of specimens, extinct or extant, to overseas collections cannot be regarded as some brutalizing effect of colonial life. It was not until the mid 1860s that a conservationist movement in Britain was able to make headway against prominent naturalists and enthusiastic amateurs whose idea of preserving a bird for posterity was to shoot it out of the sky.[70] Thus the attitudes and philosophies of Europe were still having a significant effect on the ways in which the fauna and flora were managed in New Zealand. The exploitation of a native biota and the development of acclimatisation societies might have had ends and means modified to suit the colonists, but the origins of these movements lay elsewhere.

The scene, therefore, at the end of the century was rather tumultuous. With the centralized authority of Hector declining, an unsympathetic Government, hard times, and only the beginnings of a scientific civil service, biological problems were heaping up with neither the institutions nor the resources ready to deal with them. Heading the list were: rampant animal and plant introductions, accidentally introduced crop and other farm pests, and a threatened fauna and flora. And if their institutions were not ready, neither with a few exceptions were their minds. The stock-in-trade of zoology, description and classification, needed the addition of a deeper understanding of principles and processes; classical biology needed to make way for the new disciplines of ecology and behaviour. Not until these were grasped would the principles underlying animal management and control be understood. Understanding would come with learning, but by remaining tied to classical biology for some years the universities were unable to speed up progress. The words of G. M. Thomson describing the contemporary scene in 1921 are relevant here: 'There is still no general principle underlying the work, and not sufficient knowledge of the possibilities of each problem.'[71] Zoological research would not be allowed to go its own way in the 20th century and in due course it would have to react to the problems of the 19th.[72]

While biological questions were accumulating in New Zealand, there were developments in scientific thinking abroad that had sparked a wide public debate of an entirely different kind that sooner or later might be expected to spread to New Zealand. A joint paper from Charles Darwin and Alfred Russell Wallace on natural selection was read to the Linnean Society on 1 July, 1858 and was followed by publication of Darwin's *On The*

67. An idea successfully taken up later by Walter Buller. Angehr, 1984, p.308 goes some way towards rescuing Reischek's reputation and suggests that his activities did not cause lasting damage to the Stitchbird population.

68. See Note 23 above.

69. Von Haast, 1948, p.655.

70. Allen, 1976, p.197.

71. Thomson, 1922, p.23

72. It does still.

Origin of Species. . . in 1859. Hard on the heels of this came the almost inevitable outcry from conservative scientists and churchmen, resulting in a diversity if not a polarisation of strongly held viewpoints that would last some 20 years until the protagonists had worn themselves out. The controversial ideas of Darwin and his colleagues inevitably found their way to New Zealand. How did they settle on this periphery of the learned world? The answer is, surprisingly, with apparent ease and lack of fight. It might have been that New Zealand lacked a Thomas Huxley and fellow radicals on one side and a Sir Richard Owen with an alliance of religious and scientific dogmatists on the other. Certainly there were things about New Zealand scientists and churchmen that differentiated them from their British counterparts. There were fewer established reputations to be lost, and the democratising tendencies of the Darwinian debate[73] were less noticeable as colonisation had selected a generally more egalitarian group, willing to reach out to '. . . all classes of colonial society in carrying out the really noble objects of the [New Zealand] institute',[74] and ready-made for the acceptance of new philosophies. Here at the frontier, with their sophisticated interests in natural history, some of them were more aware of the dynamics and frailties of nature. There was also the practical question of the time and place at which they learned of Darwin's work and their response to it. The public responses of both conservatives and moderates were individual rather than collective and spread out over some 10 or more years. The respondents were further separated by their location in provincial centres, with the result that debates tended to be localized in the relevant institutes rather than conducted nationally.

Jointly, then, these factors encouraged a compatability of viewpoint between potentially warring factions.[75] Although the arguments and the rhetoric were familiar, there was a softening of the edges of debate which resulted in those with more extreme views being chided by their colleagues or dismissed as cranks. In looking for an example on which to focus more closely, it has been seen that Otago was the place where the issue stood in somewhat starker relief than elsewhere.[76]

Hutton, author of faunal catalogues and Otago's provincial geologist, was a leading proponent of Darwin's theories. He had published a favourable review of the *Origin* before his arrival in New Zealand and in September 1872 read a paper to the Wellington Philosophical Society in which he established himself as New Zealand's leading theorist in matters of zoogeography and evolution.[77] While fundamentally Darwinian, the paper was not a fervent nor even a strongly worded plea for Darwin's theories. Based on the geological evidence and Hutton's unsurpassed knowledge of the breadth of New Zealand animal life, it proposed a series of geological elevations and subsidences resulting in the isolation of an ancient continental fauna, with subsequent faunal recruitment from the surrounding regions. But even his thoughts on the speciation of Moas, on isolated islands over geological time, were presented as a cautious appraisal of the possibilities rather than as a nail in the coffin of religious dogmatists. It was thus that he came to the Chair of Natural Science at Otago University in 1877, a Darwinian, but in moderation and with a continuing adherence to his religious beliefs. His successor at Otago University was Thomas Parker, who clarified his support for Darwinism at his inaugural lecture in 1880.

It might be expected that in Otago, imbued with strong Presbyterian and other church influences, there would be some rebuttal of such views however mildly expressed — possibly even outrage. But in the event, the fire in the argument over Darwinian theory died reasonably quickly. Following the opening rounds, the ultimate temper of the debate is demonstrated by the following events. Robert Gillies was President of the Otago

73. Desmond, 1982, p.17.

74. Firth, 1876, p.420.

75. Stenhouse, 1984, p.162.

76. Parsonson, G. S., MS. *The Darwinian Debate in Otago 1876*, 1983, History of Science in New Zealand Conference, unpublished paper.

77. Hutton, 1861, pp.1-11; 1873, pp.227-256.

Institute during 1876-1877 and gave a talk on 22 August, 1876, at one of their so-called 'popular' meetings, which were usually given over to more speculative subjects. The title was 'The Pedigree of Man'. The subject came from a recent reading of an important book called the *History of Creation* by Haekel, who was hailed by some as 'the German Darwin', and his book as 'the chief source of the world's knowledge of Darwinism'. These claims were probably exaggerated but the work had major significance in the development of the concept of 'phylogeny' or 'pedigree' of animals.[78] Gillies' analysis of this work was bound to promote discussion. He said later: '. . . the aspect of the combatants has been very different from the fierce, uncompromising attitude with which the sword first leapt from its scabbard, and has become more that of trying to see how far an agreement can be arrived at without an absolute surrender and acknowledgement of defeat.'[79] During the debate he was indebted to Hutton and Bishop Neville in particular, who stood at opposite sides of the matter but had allowed clarity of thought and fairness in discussion to defuse the issue somewhat. The Reverend William Salmond, Professor of Divinity, took a rather firmer stand against Darwinism, although his shell of orthodoxy cracked sufficiently to allow the theory in part, limiting it to 'the lower forms of animal and vegetable life'.[80] Even more extreme, although not taken seriously, was J. G. S. Grant, whose pamphlet *Evolution the Blackest Form of Materialism* was published in 1881 — its rhetoric, abusive of evolution, was similar in style to the more rabid opponents of evolution in Britain.[81]

Analagous positions were taken and similar scenes enacted in other institutes throughout New Zealand, the protagonists perhaps lacking the authority of a Hutton or the spleen of a Grant. From his station Mesopotamia, Samuel Butler sent out his first articles on Darwinism to the *Press* in Christchurch and began to make notes that would later in England lead to *Erewhon* and then a succession of books critical of Darwin's theories.[82] But whatever the basic elements of the debate or the quality of its participants, the outcome was in each case much the same — Darwinism, or some modification of it, was accepted by the scientific community. The speed of its absorption varied. For some it was a question of relishing a good philosophical argument, for others it was a matter of putting the discussion aside and getting on with the job. The latter camp did not see that their position on this issue was an urgent matter of public declaration or essential to the progress of their work on the identification of species and elucidations of life histories. The issues of Darwinism were complex and time-consuming; philosophy was no substitute for hard work or the fulfilment of ambitions. Two who embraced Darwin later rather than sooner are the subjects of our final chapter.

78. Desmond 1982, p.151, 152.

79. Gillies, 1877, p.657.

80. Parsonson, *in* Stenhouse, 1984, p.153

81. Stenhouse, 1984, p.143.

82. Festing Jones, 1919, p.39. The *Press* 13.6.1863; 29.7.1865. Stenhouse, 1985, pp.158-159, describes Butler's progress from being an admirer of Darwin to an opponent on his return to England and the writing of *Erewhon* then *Life and Habit, Evolution Old and New, Unconscious Memory* and *Luck or Cunning*. Butler's conflicts with evolution and his religion are described in detail in Willey, 1960.

Chapter Twelve

AMONG THE LAST OF THE 'GENTLEMEN AMATEURS' – BULLER AND HUDSON

WALTER BULLER STARTED HIS LIFE in circumstances which – from a naturalist's point of view – were vastly different from those of his younger contemporary, G. V. Hudson. Not for him the proximity of the Bethnal Green Museum or the British Museum of Natural History, with their public collections and experts, or a choice of learned texts and journals. Instead, he was to spend his early years on a turbulent New Zealand frontier where one pursued natural history in isolation from all but a scanty literature and the enthusiasm of friends or family. At the time of Buller's birth Richard Taylor was still on his way to New Zealand; the *Tory* and Dieffenbach had not yet left England; collectively the United States Exploring Expedition, d'Urville on his last voyage, and the *Erebus* and *Terror* were still on the horizon. Compared with Hudson then, in terms of natural history resources Walter Buller might be said to have come into this world a pauper.

He was born on 9 October, 1838, at the Wesleyan mission station at Pakanae on the Hokianga Harbour, where his father, the Reverend James Buller, was a missionary. The family then moved up the Wairoa River to Tangiteroria, where James Buller remained in charge of the mission station for sixteen years. They were some 40 or so miles (64 km) distant from the Bay of Islands, cradle of so much early New Zealand natural history, but it is unlikely that this would have meant a great deal to Walter, as the influence of this locality was a declining one during his formative years. In spite of these circumstances, by the time he had completed his education at the Wesleyan College in Auckland, he was already regarded as having a developed ability in the collection and observation of natural history.[1] He began working in a bank, but shortly afterwards in 1855 moved with his family and insect and bird collections to Wellington.

On the face of it, a move to Wellington could only benefit Walter Buller's scientific interests. It was after all the home of the New Zealand Society, Walter Mantell, Knox, and (close by) Swainson. But in the mid 1850s the New Zealand Society was faltering, and Swainson – who could have been of enormous help to Buller – was not long returned from Australia, crusty and querulous, and decrying the possibility of the existence of a scientific community in Wellington.[2] Swainson would live only until 7 December, 1855, and contact between the two must have been rather limited. Once again, Buller would have to have been largely self-reliant; but in due course for those interested in science, things would improve in Wellington.

Following a visit to the Chatham Islands, Buller became an official Maori interpreter in the Magistrates Court in Wellington. His facility in Maori, acquired during his youth in the far north, and his new association with the administrative and legal process, were the foundation of a successful and lucrative, if at times controversial, career.[3] Unlike G. V. Hudson, whose ambitions were limited to his science, Buller had already started to show an unashamed duality of purpose.

A reinvigoration of the New Zealand Society found him as its Secretary in 1859,[4]

1. Galbreath, 1984, p.7.

2. Parkinson, 1984, pp.56-57.

3. Bagnall, 1966, pp.272-274. Both Bagnall and Galbreath, (see Note 1 above) give useful summaries of Buller's life as does Turbott, 1967, pp.xi-xviii, in his introduction to the reprint of the second edition of Buller's *History of the Birds of New Zealand*. Buller's obituary, suitably laudatory, was anonymously published in the *Transactions*, 1907, pp.iii-v.

4. Bagnall, 1966, p.272.

although it would fade again. The advent of the Colonial Museum and the New Zealand Institute were imminent, and for one who wished to move quickly in the scientific and administrative world, Wellington, with all its contacts was now the right place to be. In 1862 Buller married Charlotte Mair, and was also on the way to preparing a fine illustrated work on the New Zealand birds — an ambitious undertaking for a young man who had yet to prove himself. He did have a smaller project — a *New Zealand Ornithological Manual* — in preparation, but that did not immediately come to anything. There would be no problems of financial backing, as there would be for G. V. Hudson's insects — he was moving in an age which still counted amongst its influential politicians and administrators those who were followers and practitioners of science. Sir George Grey, that earlier sponsor of science and co-founder of the New Zealand Society, had offered to raise at least £1000 from English friends to support the project.[5] But before any books saw the light of day an essay was written by Buller for the Dunedin Exhibition of 1865.

Regarded as a 'preliminary canter', it was called *An Essay on the Ornithology of New Zealand* and was later published in the *Transactions and Proceedings of the New Zealand Institute* of 1869. For this work and an accompanying collection of birds, the Exhibition's Commissioners awarded Buller a silver medal.[6] Partly to secure personal recognition, and more so than his contemporaries, he sent his work to Europe where it came to the attention of Otto Finsch of Bremen, who had a particular interest in the ornithology of Buller's part of the world. The essay, which contained 'some new and original matter on the birds of New Zealand and their habits', was the first of his works to be circulated and attracted a detailed criticism from Finsch. It was, Finsch thought, altogether too derivative of G. R. Gray's earlier works, and one would have expected rather more from an ornithologist established in New Zealand.[7] He went on to raise a number of rather more specific points. Compared with some of the weightier offerings written by his contemporaries for the Exhibition, Buller's *Essay* indeed seemed rather slight, and even within the bounds set by the Commissioners, it could have added more substantially to that knowledge already acquired by the British Museum ornithologists.

In a style that was to become characteristic, Buller published a point-by-point rebuttal of the detail of Finsch's criticism and on many of these points he stands correct. On the larger question of whether he should have added more on the habits of the birds listed, he reserved his position by stating that such matters would be dealt with in the not-too-distant future 'in the form of a general work on the birds of New Zealand, illustrated by numerous coloured drawings by an eminent zoological artist'.[8] In the end, the Cambridge ornithologist Alfred Newton felt that the combatants had both reached the same conclusion and that the debate had been conducted in better spirit than most.[9] Buller had it seemed at last emerged from his novitiate. The *Essay* was later the basis for the award (*in absentia*) of a Doctorate of Science from the University of Tübingen which Finsch helped arrange for his New Zealand colleague, but it would not be the last time that he would hit out at Buller's work.

The second occasion followed Buller's description of several new bird species in the British ornithological journal *The Ibis* in 1869.[10] Finsch replied in the same year, demolishing Buller's species and eliciting a response from the latter in a subsequent issue.[11] Buller again had the best of the encounter, condemning Finsch's reliance on specimens identified and sent by Haast, and disparagingly referred to 'cabinet' zoologists. After this rather frank and pointed exchange the two protagonists appeared to have adopted a degree of mutual respect, and Finsch later visited New Zealand to see its birds at first hand.

5. Galbreath, 1984, p.8.

6. Bagnall, 1966, p.273; he was not however awarded the New Zealand Cross that year as popularly supposed (R. Galbreath, *pers. comm.*). In 1865 he was also gazetted a Judge of the Native Land Court. Then, a legal qualification (yet to come for Buller) was not required for officers of administrative courts.

7. Finsch, 1869a, p.58.

8. Buller, 1869b, p.50.

9. Buller, 1890, p.9

10. Buller, 1869c, p.37; also Buller, 1870a, pp.390-392.

11. Finsch, 1869b, pp.378-381; Buller, 1870b, p.455.

Three of the birds described by Buller in *The Ibis* still have subspecies status today: the Grey Teal, Chatham Islands Snipe, and the Chatham Islands Fernbird. The last two were discovered by Charles Traill, described as 'a gentleman devoted to conchology', while on a visit to the Chathams, the Fernbird being taken by the rather unorthodox means of being hit with a stone.[12] Buller had himself disposed of the Grey Teal, a male and a female, with one shot.

While jousting with foreign adversaries, there was also a certain amount of jostling for position at home. When Buller entered the field of ornithology in New Zealand he was by no means alone: Potts, Travers and Hutton were active at about the same time. Hutton, the great cataloguer of New Zealand species, who was only a little older than Buller, crossed swords with him more than once.[13] The ornithological works of these three authors dominated the *Transactions* initially, but in due course it was Buller who came to the fore.

In 1865 he had transferred to Wanganui, where he was Resident Magistrate, but work towards his opus was unaffected. He collected either through his own enthusiastic efforts or by those of the numerous professional collectors and taxidermists he had at his call. His standards here were always high and he insisted that the skins be 'first class'.[14] Progress continued, and by 1868 he had arranged a London publisher, Van Voorst, and a note appeared in *The Ibis* of that year advertising the work and calling for subscribers.[15] Buller's preparation had now reached a point where he needed to go to London himself to see to the illustrations and publication.

Publishing in London was not a piece of colonial snobbery. Although some competent chromolithography was being done in New Zealand, the resources for large-scale hand-colour or chromolithographic work were not sufficient. Even G. V. Hudson was compelled to have his insect works published in London some years later.[16] The increasing costs of the undertaking compelled Buller to apply to the Government for assistance, and finally, in 1871, he was granted 18 months leave on half pay and his return fare. Approval had been neither swift in coming nor universal, and Hector wrote to say that there had been a row in the House about his 'birds'.[17] A £300 grant towards the cost of publication was conditional on Buller's handing over his bird collection to the Colonial Museum, and as well there was a recommendation by Hector that the Government receive 25 copies of the finished work.[18] While in London, Buller was to act as an agent for the Flax Commissioners and as secretary to the Agent-General, Dr I. E. Featherston, for whom he had previously worked in Wellington.[19] Thus constrained, he left for England where he not only attended to publication of his book but also drummed up subscribers, ably assisted Featherston, and qualified himself for the practice of law. Any one of these assignments would have tried the resources of most, but the seemingly boundless energy and ambition of Buller ensured that he ultimately carried all to successful completion.

While the manuscript for the book was Buller's responsibility, a significant, if not the major part of the work was its illustration, and he had already signalled that he was looking for an artist of considerable reputation and talent. London had long been a mecca for bird artists: Swainson, a pioneer in bird lithography practised there, and for Audubon's plates for the *Birds of America* only Scottish or English engravers would do. Edward Lear, John Gould, Joseph Wolf, Henry Richter and William Hart — all featured in the magnificent publications that crowned an era that can be referred to as the 'golden age of illustrated works'; an age that was already showing signs of fading. Relatively few of these established artists were now engaged on large projects; the outstanding Joseph

12. Buller, 1870a, p.392.

13. e.g. Hutton to Buller, 15.4.1875, Buller Correspondence, MS. 2213/1, Alexander Turnbull Library.

14. Buller to Hawkins, 8.10.1892, Buller Letterbook, National Museum Archives.

15. Anon, 1868, p.504.

16. A very little chromolithography was done for the *Transactions* overseas, e.g. Buller, 1873a, p.280, but work of a high standard was carried out in New Zealand, e.g. Maskell, 1896, Pl.25-35.

17. Hector to Buller, 29.10.1871, Buller Correspondence, MS. 48/22, Alexander Turnbull Library.

18. Bagnall, 1966, p.273.

19. Appendix to the Journals of the House of Representatives, 1871, G4, p.101.

Wolf might have been a possibility, but if approached he must have declined the commission. And yet a relative newcomer had set up in London and was already making a name for himself. His name was Johannes Gerardus Keulemans, later anglicised to John Gerrard Keulemans. As he was instrumental to the success of Buller's work, it is necessary to describe him in more detail.

When Keulemans arrived in London from Holland in 1869 he was already a well-published naturalist and a highly regarded artist with particular flair for portraying birds. It was this skill that encouraged R. Bowdler Sharpe, then librarian of the Zoological Society of London and himself an ornithologist, to persuade Keulemans to leave Holland for England. When approached by Buller, Keulemans was already at work on Dresser's *Birds of Europe*.[20]

The technique employed by Keulemans and his contemporaries was lithography, long since demonstrated as ideal for bird studies by Swainson and given further prominence by the works of Lear and then John and Elizabeth Gould with their fellow artists and lithographers. Audubon's *Birds of America* proved that dramatic effect could be achieved in the usually more stilted medium of copper engraving, but the crayon and the stone were easier to handle and cheaper as well.

Keulemans would first make a pencil sketch, usually of a bird that was stuffed and mounted — he used live material infrequently — and the sketch was then sent off to the author with a series of questions on colour, attitude and form. The author then came back with corrections on plumage, the colour of feet or beak, or a recommended change of posture. An exchange between Keulemans and Buller began thus: 'Sunday Island Petrel: (Fig. 1) Colour of feet pale yellow? (Fig. 2) Irides brown?', Buller replying: '(Fig. 1) Yes (Fig. 2) Yes and dark brown. Approved.'[21] The artist then made a drawing finished in watercolours or pastel, and from this he prepared the lithograph with a greasy crayon on Bavarian stone. The proof copy he would colour as a guide to the colourists.[22] For all this he was paid £2 per plate (there were 35 in the first edition) and the colourists £2 per 100 copies — these charges changed little throughout Keulemans' career.[23]

Colouring was skilled work, but as each plate of each copy of the book had to be coloured it usually required more than one person to do the job and consistency became a problem.[24] It was finally the delays in colouring and his unfinished law studies that postponed Buller's return to New Zealand. A request to the New Zealand Government for an extension of his leave and changes in his financial support were not enthusiastically received by Parliament, but he was subsequently allowed his stay, although he had to resign his position as magistrate.[25] He cannot have been too dismayed about the loss of this post as by now he must have seen that the way to fame and fortune were through his ornithology and ultimate practice in law.

As was commonly the case with subscriber works, *A History of the Birds of New Zealand* came out in parts, the first appearing in 1872, the last a year later. The reviews were very favourable, and most satisfying must have been that from Otto Finsch who found it: 'an undertaking that I desire to recommend most warmly to all friends of and experts in Ornithology . . .'; he places the *History. . .* in the ranks of the great ornithological works.[26] The British Museum's Günther saw it as a means for future generations to compare New Zealand's ornithology with that of times past;[27] several found Keulemans' work to their liking with comments ranging from: 'the Plates are good' to 'the most life-like and beautiful . . . we have seen'[28] and more than one reviewer found that it served the needs of science as well as being an 'elegant drawing-room companion'.[29] It also put the colony well out in front in the 'race for scientific honours'.[30] Apart from some detailed

20. The *Birds of South Africa* would follow the New Zealand work. An illustrated biography of Keulemans has been privately published in subscriber edition by Keulemans and Coldewey, 1982; the former is the artist's great-grandson. Keulemans illustrated other New Zealand work: e.g. Forbes, 1893, a work on the birds of the Chatham Islands.

21. Keulemans and Coldewey, 1982, p.36.

22. *Ibid.*, pp.36, 37.

23. Galbreath, 1984, p.8.

24. Keulemans and Coldewey, 1982, p.37 indicate that the daughters of Bowdler-Sharpe hand-coloured the 1st edition plates, whereas it was more likely to be those of the *Supplement*, (1905) (R. Galbreath *pers. comm.*); Bagnall, 1966, p.273.

25. Galbreath, 1984, p.8; Bagnall, 1966, p.273.

26. Buller, 1890, pp.21, 22.

27. *Ibid.*, p.14.

28. *Ibid.*, p.17; *The Zoological Record* and *The Field*.

29. *Ibid.*, p.20; *The Home News*.

30. *Ibid.*, pp.19, 20; *The Zoological Record*.

criticism from Hutton, to which Buller made his typically robust replies, it was not slow to find favour in New Zealand.[31] The reviewer of the *Nelson Examiner* reported: 'It will be a beautiful ornament to a drawing-room table, as well as a means of education, and its price (£3.3s.), large though it may seem for a book, ought not to deter that large part of the people of New Zealand who buy pianos, and pay from year to year four, eight, twelve, even sixteen guineas for the instruction of each daughter in music'.[32]

The significance of all this acclaim cannot have been lost on the author, as it allowed him to return to New Zealand in 1874 in a virtually unassailable position — what Government could now take him to task for his previous actions? His newly acquired qualification in law would provide him a means of earning a living that was largely independent of official whim. He did however oblige his former employers with the 25 copies that they had asked for, in part-payment of the favours granted him. He was now thoroughly well set up.

Buller was elected President of the Wellington Philosophical Society in 1875 and in 1879 was made a Fellow of the Royal Society, the first New Zealander to receive this distinction. Amongst his proposers were Darwin, Günther and Gould. John Gould was by then the grand old man of British ornithology, and if one were seeking an analogy to Buller, Gould would come the closest. He too had humble beginnings but quickly put his talents in drawing and natural history to good use, rising rapidly in the ranks of British zoologists. Gould was also hard-nosed and ambitious, with an instinct for commercial and scientific success. His talent for organization led to enormous productivity, far greater than Buller's, but he too attracted criticism and accusations that he had thrived on the work of others. However, examination of original Gould illustrations leaves no doubt that his was the guiding genius in a succession of works published under his name.[33]

To his credit Buller did not rest on his laurels after he returned but launched himself again into a new round of studies and the publishing of the results in the *Transactions*: these would be the basis of a new edition of the *History*.... His dispute with Hutton continued, this time over the anticipation of Buller's work by Hutton's *Catalogue of the Birds of New Zealand*, published in 1871. Finally Buller wrote to Hutton saying that 'Unless we all manage to pull more together, I fear we shall lose the good name we have at Home for co-operation'.[34]

Buller's profitable but controversial practice of law at the Maori Land Court was sufficient to enable him to retire ten years after his return. Shortly after retirement he embarked for London to see through the stages of illustration and publication the second and larger edition of the *History*.... This time his official responsibility was as one of the Commissioners for the Colonial and Indian Exhibition of 1886, in whch he assisted with the setting up of the New Zealand court. For his work here he was awarded the KCMG, the first New Zealander to be thus honoured.[35]

The new edition was again to be illustrated by Keulemans, but this time the work was to be printed on a larger paper size and there were to be 13 additional illustrations.[36] As before, the edition was to be limited to 1000 copies, and another round of chasing up subscribers took place, a chase that was no doubt eased by the success of the first edition. This procedure was peculiar neither to Buller nor his works. Even the eminent John Gould found it necessary to importune the wealthy and titled in order to find buyers for his works which were degrees larger and more sumptuous, and consequently a great deal more expensive than Buller's, and Lear and Audubon had done the same before him. Later, in New Zealand, G. V. Hudson was faced with the same problem,

31. Hutton, 1874, pp.126-138.

32. Buller, 1890, p.27.

33. The character of the illustrations in the Gould works is more clearly exemplified by the original drawings of John Gould than by those of his wife Elizabeth (seen at Knowsley Hall). See McEvey, 1973, for a discussion of the attribution of Gould paintings.

34. Buller to Hutton, 19.4.1875, Buller Correspondence, MS. 2213, Alexander Turnbull Library. R. Galbreath *(pers. comm.)* suggests that the post-1874 flush of papers helped cement his claim to a F.R.S.

35. Galbreath, 1984, p.8.

36. Imperial instead of Royal Quarto.

Walter Buller himself appears to have been a competent illustrator as indicated by these drawings for *On the Ornithology of New Zealand* in *Transactions and Proceedings of the New Zealand Institute*, 1875 (pl. 8). *(see p.189)*

one that was probably heightened by his more retiring nature and less aggressive salesmanship, as well as by the limitations of remaining in New Zealand. It would not have helped that his books were on insects rather than birds.

The high cost of the labour-intensive processes that went into the production of fine books restricted their market to a point where even the comparatively well off sometimes jibbed at the prices asked. It was fortunate perhaps for Buller and those who sought

The Kea (*Nestor notabilis*) an original painting by J. C. Keulemans used as the basis of a chromolithograph in Walter Buller's *A History of the Birds of New Zealand*, 1888, (v. 1, pl. 18). The paintings and the final lithographs differ mainly in the clarity of colour. Published by permission of the Trustees of the British Museum. *(see p.194)*

37. Although the English market was important to all such publications. There was however an increase in subscriber numbers, allowing a half-guinea discount on the original price; Turbott, 1967, p.xv.

his publications that the New Zealand bird fauna was so limited in its number of species, otherwise a multivolume publication such as Gould's *Birds of Australia* might have found few subscribers in a country like New Zealand with a small population.[37] Even the increasing use of chromolithography which was enjoying a vogue at the time of the second edition of the *History. . .* was not able to reduce the cost. The process was complex involving up to 12 and sometimes more stones, each for a separate colour, for each plate. The results, although providing an excellent finish, were sometimes found wanting where the exact colouration of a bird or insect was required, especially as live material was seldom to hand for exact comparisons, and there was sometimes a slight cast or dullness to the print.

These problems were present in the second edition of the *History. . .* only to a limited extent, and the plates on the whole were outstanding. Very few bird books of distinction were printed by the process and the second edition stands out as a fine example of the printer's art.[38] Keulemans' style was different this time; his treatment of background vegetation was more detailed and formal, and although never in conflict, it came more strongly into balance with the bird which was the object of the plate. This tendency to formalization was heightened by the setting of the plate well within the page, surrounding it with a broad white margin on which the name of the bird was printed at the bottom. This effectively 'framed' the plate and the whole gave the impression of being more finished, formal and stylized than the plates of the first edition. These, 'unframed', had a lighter, more dynamic look about them, although somewhat overlarge for the page, and they featured plants that were frequently less identifiable as New Zealand species. As well there was greater variation in quality between the plates. Keulemans had also taken the opportunity to improve the posture of some of the birds in the second edition, the New Zealand Pigeon being a notable example; rather stooped in the first edition it was revived considerably in the second.

38. They became the standard New Zealand bird portraits, unchallenged until a recent proliferation of bird artists.

There was also substantial development of the text and Buller provided a lengthy introduction which included details of avian classification and relevant morphology. The accounts of each bird included an updated description of the status and habitat of each species, and in some cases a lively account of the means of capture. Even for the times, Buller dwelt with too much relish on this part of his activities, and on the last page of the *History. . .* he recalls: 'the bright dewey morning, now five-and-thirty years ago, when I shot my first Koheperoa in the old Mission-garden at Tangiteroria, and found my beautiful prize lying on the sward with its banded wings and tail stretched out to their full extent. I have remembered the delight with which, almost as long ago, I shot in the Tangihua mountains my first Piopio, a bird so rare at the far north, even at that time, that it was entirely unknown to the natives of the district.'[39] Since that time the Piopio or North Island Thrush has become extinct.

39. Buller, 1888, v.2. p.340.

Such considerations did not dampen the enthusiasm of the reviewers of the second edition when its first part appeared in 1887 and it was completed in 1888. Buller had become, in the words of one paper, 'the Audubon of New Zealand'.[40] The reviews that appeared throughout this period were even more enthusiastic than those of the first edition, with Keulemans' plates coming in for particular praise. *The Ibis* in 1888 testified to Buller's thoroughness as well as to Keulemans' skill, and later reported that the issue had been sold out.[41] Newspaper reviews in the *Times* and *Australian Times* were equally favourable, the latter containing a statement from Sir Richard Owen that one of the plates was the best he had ever seen, although the *Anglo-New Zealander* preferred the plates of the first edition.[42]

40. Buller, 1890, p.37.

41. *Ibid.*, p.35

42. *Ibid.*, pp.36-38.

The North Island Fantail (right) and South Island Fantail (left) (*Rhipidura fuliginosa placabilis* and *R. fuliginosa fuliginosa*, respectively), an original painting by J. G. Keulemans used as the basis for a chromolithograph in Walter Buller's *A History of the Birds of New Zealand*, 1888 (v. 1, pl. 8). Published by permission of the Trustees of the British Museum. *(see p.194)*

Once again Buller was the recipient of awards and honours, amongst which was Knight Commander of the Crown of Italy. But his unfortunate illustrator received only his fee for the plates (usually around £2 per each) which was insubstantial considering the work involved, and when he died on 29 March, 1912, it was in impoverished circumstances.

Meanwhile, continuing to reap the benefits of the success of his new work, legal problems brought Buller back to New Zealand in 1890. He quickly re-established contact with his network of bird collectors and before long his studies, as well as the traffic of birds and other animals from New Zealand to Europe, began anew. But one of his first tasks was to chase up defaulting subscribers, and in a letter to an agent he wrote: 'Will you kindly go over the enclosed list of subscribers to the "Birds of New Zealand" marking off 1. those who have paid. 2. those who are bankrupt. 3. those who in your opinion are *good* for the amount, if pressed; . . . I have a large sum of money outstanding.' In the same letter he referred to some volumes lost when the ship carrying them sank.[43]

Buller's collection and sale of birds to private collectors and museums was a contentious issue that refused to leave him so long as he remained in New Zealand. The collections of Silver, Rothschild, Tristram, Sclater and Cambridge University, and of museums in Austria, Hungary, Germany, together with the Carnegie Museum of Pittsburgh, all received specimens from Buller. They could be very persuasive: 'I know you will always do me a good turn and you will agree that a collection which contains a female stitchbird and a Scolopax pusilla ought not to be without an Apteryx haasti'.[44] For years Buller had been plagued by charges that he was a dealer in birds, something that he strenuously denied: 'My object after completing my own private collection is to get the fullest possible record procurable of all our native species before they have finally passed away (with many of these only a question of a few years now!) and to aid our public museums in procuring specimens . . .'[45] It is unquestioned that he actually sold birds overseas, but in doing so he was merely recovering the charges made by his collectors and the cost of shipping, and if he pocketed anything at all it was probably small change. He did however use some of these transactions to personal advantage, in that they kept his name in front of those who could secure him recognition and honours as well as other assistance in the British and European world of ornithology.

There was more than a touch of self-interest in Buller's ambivalent attitude towards the conservation of rare species, in that the promotion of conservation measures proceeded hand in hand with a determination to collect the few remaining examples of species that he thought were doomed to extinction anyway. He was very active in promoting Little Barrier Island, ransacked by Reischek and others, as a sanctuary for the Stitchbird. In a letter to the Premier, John Ballance, on 2 June, 1892, he pleaded for the island to be saved from further devastation,[46] but a week later he himself wrote to one of his collectors to cancel instructions for the collection of Stitchbirds.[47] Three years after this he found himself having to deny that he had collected Stitchbirds from the island.[48] Although he committed no transgressions here, Buller's actions suggest that he had some difficulty in separating his personal interests from his more publicly expressed philosophies and behaviour.

His wife died in November 1891 and the following year he once again took up the Presidency of the Wellington Philosophical Society. Towards the end of his term a short trip to England in 1893[49] postponed his address as retiring President of the Society but the return journey on the boat gave him the leisure to consider his topic. When he finally spoke to them on 27 June, 1894, his approach to the subject – Illustrations of Darwinism

43. Buller to Burke, 8.6.1890, Buller Letterbook, National Museum Archives.

44. Tristram to Buller, 26.8.1893, MS. 48/27, Buller Correspondence, Alexander Turnbull Library.

45. Buller to Dall, 6.7.1892, Buller Letterbook, National Museum Archives.

46. Buller to Ballance, 2.6.1892, Buller Letterbook, Alexander Turnbull Library.

47. Buller to Spencer, 7.6.1892, Buller Letterbook, National Museum Archives.

48. Buller to McKenzie, 17.10.1895, Buller Letterbook, Alexander Turnbull Library.

49. From March 1893-March 1894.

— was tempered by the inaugural address of the incumbent President, Major-General Schaw, and a paper by Coleman Phillips published in the *Transactions* the previous year.[50]

Buller sailed into both these gentlemen with characteristic vigour. Schaw's views on this subject were described as unorthodox and pernicious, with passages of his address worthy of the 'Dark Ages of Science'. Coleman Phillips, whose partial reading of Darwin had filled Buller with 'pain', was worse. This unscientific paper should not have graced the *Transactions* — Buller had already lightly chastised Coleman Phillips in a letter written a few days earlier.[51] Later declaring himself to be a 'thorough disciple of Darwin in the higher sense of that term',[52] Buller went on to give example after example from his knowledge of the New Zealand fauna to support the theory of natural selection. Kiwis, kokakos, thrushes, fantails, lizards and others — all had speciated; who could doubt their common parentage?[53] In short, he could find no alternative to Darwin's theories, but like many others before him found no conflict with a belief in the spiritual nature of man.

This confidently written and presented paper was shortly to be followed by submissions of a different kind. Following accusations by John McKenzie, Minister of Lands, a Royal Commission was set up to investigate an allegation of Buller's fraudulent involvement with land at Horowhenua. Buller was incensed and in 1896 petitioned Parliament and the Legislative Council, praying for the clearance of his name, and (with ultimate success) he invited his accusers to take the matter up outside the House. The newspapers of the day were on his side, but he was never completely exonerated and the controversy remains, yet to be settled by a detailed investigation. Following the issuing of several pamphlets to put his case, Buller left for England in 1898, this time for good. He had already accomplished in one lifetime what most would do in two and with continuing conflicts at home, a retreat to the country that had flattered him most would provide balm for his battered self-esteem.

He was not to be disappointed as Cambridge University awarded him an honorary Doctorate in Science in 1900.[54] Further, he had plans for another work. Even in his declining years he found time to publish a *Supplement to the Birds of New Zealand* in 1905, containing twelve hand-coloured lithographs by Keulemans of species encountered since the publication of the earlier volumes. Included also were some photographs taken by his daughter. Sometimes more decorative than relevant, together with the lithographs they presented an interesting juxtaposition of illustrative style. Keulemans's plates, differing from his previous work for Buller, were not his best work and the colours of the backgrounds put a gloomy cast on a number of them.

The introduction included Buller's thoughts on Darwinism published earlier in the *Transactions*. Here he also expanded his views on conservation, where he accused the Government of 'supine neglect' which the introduction of the Wild Birds Protection Act of 1864 only partially remedied. He had a great deal to say about the shortsightedness of introducing ferrets, stoats and weasels to the Colony for the purpose of controlling rabbits.[55] He quoted Alfred Newton, the eminent British ornithologist who had noted: 'Regret . . . against that sentiment which prompts our colonial fellow-subjects indiscriminately to stock their fields and forests not only with the species of their Mother-country, but with all the fowls of heaven whencesoever they can be procured'.[56]

Buller's last communication published in the *Transactions* was a letter read to the Wellington Philosophical Society in 1902 by Sir James Hector. It described a drake sent by Buller to the Colonial Museum. If he bore his country or countrymen any ill will he showed no sign of it. He died in 1906.

With Buller's passing New Zealand science lost one of its most flamboyant and best-

---

50. Buller, 1895, pp.76, 77; Philips, 1894, pp.604-609.

51. Buller to Philips, 23.7.1894, MS. 48/22, Buller Correspondence, Alexander Turnbull Library.

52. Buller, 1895, p.103.

53. *Ibid.*, p.90; In fact most are now subspecies.

54. Venn, 1940, Pt.2, p.441.

55. Buller, 1905, pp.xxviii, xxix; In condemning importations, Buller was quick to defend those in which he had played some part, e.g. p.xxiii.

56. *Ibid.*, p.xxix.

A page of G. V. Hudson's diary with drawings and notes on the caterpillar of the Convolvulus Hawk Moth (*Agrius convolvuli*); Diary, 1890, p. 177. Published by permission of the Gibbs Family. *(see p.199)*

known figures. Buller dominated New Zealand ornithology for more than thirty years and he left a valuable legacy of publications, books and scientific papers, some of which had an influence on ornithology and zoology well into the next century and even today constitute a valuable record of species whose populations were being dramatically affected by European colonization. He added seven new subspecies to the literature, including apart from those mentioned earlier, the North Island Laughing Owl, Auckland Island Shag and the Pitt Island Shag.

His relationship with his scientific colleagues, although sometimes tense, was generally typical of the manners of the period and less abrasive than some. That he was prepared to lend a helping hand to those on their way up – G. V. Hudson for example – is evidence that he was not entirely self-centred. Hudson himself later demonstrated that it was possible to publish well-illustrated books without the need for great personal wealth or government support, although his personal input into the work through doing his own drawings was far greater than Buller's. The fact remains, however, that whatever his motives and methods, Buller's succession of works on the birds was a task unlikely to have been picked up in the same way by any other contemporary New Zealand ornithologist, and the results were widely acclaimed and a source of pride to the colony and its scientific establishment.[57]

Buller reigned in a period the latter part of which appreciated scientific virtuosi less, especially when so nakedly ambitious. His probity in commercial affairs and his apparent hypocrisy in the wholesale collection of specimens was and always will be a subject of controversy. It seems that over the years he did nothing to remove himself to a position where he would be entirely above suspicion, preferring instead to issue strenuous denials. Regrettably, whatever the merits of his case, he did this with sufficient frequency to foment the notion that he must have been guilty of wrongdoing. Finally, he was a model for future locally-born professional people who were not content to remain in New Zealand, but believed it necessary to prove their worth in an England which was then, and for many years later, regarded as 'home'. It is thus that we see an interesting reversal of roles in the next subject of our attention, G. V. Hudson – a man who was born and bred in England and came to reside permanently in New Zealand.

57. At the point Buller came into prominence, New Zealand was setting an example to the other British colonies (e.g. Australia and Canada) in terms of scientific achievements and institutions (particularly in natural history). Buller was very much part of this.

The notice of retirement of George Vernon Hudson bore a subheading which was significant. Carried in the *Dominion* of 1 February, 1919, it read 'Scientist and Civil Servant'. But it was the latter position he was giving up after 35 years of good and faithful service to the post office, finally as Principal Clerk of the Postal Division. This part of his life would pass into obscurity and with a minimum of regret on his part, but he would never retire from his unpaid position as scientist, nor for his work would he ever be forgotten. When G. V. Hudson walked out of the General Post Office for the last time on 23 December, 1918, he continued his study of insects with the same dedication with which he began it over 40 years previously in England.

George Hudson was born at 31 Penton Place, London, on 20 April, 1867, Easter Sunday. His mother died in his early childhood leaving him in the care of older brothers and sisters and his father Charles, a skilled decorator and painter of glass as well as an occasional collector of insects, a man who was to provide continuing guidance and support for his son in the forthcoming years.[58] G. V. Hudson's memories of early childhood were fragmented: a glimpse of the pattern of a dress, Uncle Hopkin's drapers shop, the tyranny of the nurse, fear of falling off a rocking-horse. Then there was the first school, Miss Hall's in Claremont Square, where the principal, with the connivance of his older sister Jeannie, 'used fear of Hell as a deterrent from wrong doing',[59] — a fear that was to torment him for several years.

His first recollection of insects was around 1874 when he was fascinated by some tortoiseshell butterflies; it was at about this time he was exposed to some of the influences that would lead him to entomology. On a visit to Ledbury in Herefordshire he captured a dragonfly, which fired him with the determination to make an insect collection of his own. This was the summer of 1878; Hudson was eleven years old and he had begun a collection that would last a lifetime and span two hemispheres. 1878, incidentally, was also notable for the loss of his fear of hell, probably as a result of his taking up astronomy and generally widening his experience of life. Whatever the cause, this made him a happier person.

At about the same time his father's decorative work was at its peak and impressed George who himself possessed natural skills in drawing and painting, already well developed. Within two years his sketches of insects would start to show a delicacy and accuracy that were to be features of his later work. His brother Will was also collecting insects and George soon began to recognize some of the common species.

A change of schools in October 1878 took him from one that was harsh to one that ridiculed his interest in natural history. Once again the support of his father won through and school became almost enjoyable. The move to Busby Place allowed many hours of collecting on the nearby Hampstead Heath, with every bit of spare time spent on his hobby: evenings, Wednesday and Saturday afternoons and holidays. Ledbury during vacations was another source of specimens.

On 13 May, 1879, George Hudson started a diary. Its first entry read: 'Went to school morning and afternoon found 3 larvas supposed the[m] to be the larva of the ground beetle or else the cocktail found an earwig. The Weather was fine and warm. Heavy rain fell about nine o'clock'.[60] Over the many years that were to follow the terse, unpunctuated entries would remain the same. A brief reference to the day's routine ('office' was eventually substituted for 'school') would be followed by a summary of his entomological activity and a brief report on the day's weather. Occasionally an astronomical observation, an outing, gardening or some activity might take the place of his beloved insects, but they were never far away. If something prevented him from

58. A summary of G. V. Hudson's life is contained in his autograph *Early Life and Summary of Diaries* (unpaginated). Two further outlines of Hudson's life are by Salmon, 1946, pp.264-266 and Sharell, 1982, pp.208-217. Most of the Hudson MS. material (Diaries, original illustrations) is held in the Gibbs Family collection and correspondence in the National Museum Archives and the Hope Entomological Library, Oxford University. Charles junior was referred to in the diaries as Charlie to distinguish him from the father, Charles senior.

59. Hudson, 1873, 1874, 1875, in MS. *Early Life and Summary of Diaries.*

60. Hudson, 1879, MS. Diary, P.1.

looking to them on one day you could be sure they would be attended to the next. Passages written in red were important events or observations. Often the flow of the diary would be interrupted by a full description of an insect or of a life history, frequently accompanied by a detailed watercolour illustration. Later in life, when time was at a premium, his wife wrote up the diary for him.

The entries were descriptive and lack introspection. Hudson scarcely reflected on the virtues and failings of his colleagues, and if there were any scientific scandals of the day, they did not surface in the diaries. His rare bursts of anger or pique he saved for his correspondence when he attacked a problem fearlessly and head on.

He must have enjoyed writing and illustrating; in fact the whole business of assembling a book. An early unpublished attempt, *A Book on Insects. . .* , at the age of twelve years was followed in 1880 at the age of thirteen by a small hard-covered notebook which became *The Common Aquatic Insects of England*; neatly written with 88 pages including a glossary and superbly detailed colour illustrations on many of its pages. It was also the year of his first published paper in the *Entomologist*.[61] His manuscript *Common British Diptera* appeared in 1881, a somewhat longer effort with five coloured plates which were immaculate considering his age.

By now he was affecting a maturity of style and sense of purpose that was to remain almost unchanged during his lifetime. It was as if, like one of his insects, he suddenly metamorphosed to adulthood enabling him to lengthen a productive working life by many more years than the normal span.

In April 1881, Hudson learned that a great change was about to take place in his life; he and his family were to leave for New Zealand, driven by the depressed economic circumstances in Britain that had already forced them to reduce the standard of their accommodation. Three of the older members of the family, Jim, Charlie and Jeannie, had left already, while Hudson with his father, Will and Mary, were to follow. It was a decision and a journey that Charles Hudson senior had been driven to twice before, once with his young family, but circumstances prevented them from establishing in the colony and they returned to England.

Amidst the hectic business of packing the young George Hudson made his personal preparation for this event by searching the bookshops for literature on New Zealand insects to find that such information was available only in the *Transactions and Proceedings of the New Zealand Institute* — then rare in Britain.[62] He also visited the British Museum and the museum at Bethnal Green, examining and measuring the New Zealand specimens in the public collections, and a notebook of New Zealand insects was begun. He spent over 15 shillings in purchasing pins, net, cork and boxes to take with him, as well as stocking up on watercolours and brushes at Windsor and Newtons, and Rowneys.[63] Thus fortified, and with a substantial library of reference works and lighter reading, he embarked on the *Glenlora* with his brother Will on 16 June, 1881, and they moved downriver to Gravesend where his father and sister Mary came aboard. They left England on 16 June, 1881, with George engrossed in reading Darwin's journal of the voyage of the *Beagle*.

Separated from his main sources of inspiration, Hudson found the long sea voyage soon lost its novelty; he took his amusement from his books, finishing the Darwin and moving onto Burmeister's *A manual of Entomology* whose ponderous detail he attempted to absorb in the heat of the sun off Morocco. On the lighter side there was *Vanity Fair*, sketching, the occasional natural history observation, and of course his diary.

The voyage finally ended at Wellington on 23 September, where they were reunited

61. Hudson, 1880, p.217.

62. He visited Wheldon's now Wheldon and Wesley, still in business as booksellers specializing in natural history; Hudson, 1950, p.160.

63. Hudson, 1881, MS., Diary, v.1, pp.89-90.

with Charlie who greeted George with a small collection of New Zealand insects. Some of these were the worse for wear, but a huhu beetle and a cicada were in excellent condition. He lost no time in getting down to business and the next day, Saturday, he visited the Colonial Museum, where he was disappointed to find that the collections displayed very few New Zealand insects although he later found a weta.[64] He illustrated his first New Zealand specimen, a caterpillar, in his diary the following day, although: 'Papa says it may be a little too thin and possibly too small',[65] a sign that his father was still keeping an eye on his son's progress in drawing.

Shortly thereafter he left for Nelson to join his brother Jim and sister Jeannie[66] and here, in Collingwood Street, he took his first New Zealand butterfly, a Red Admiral. Now at the age of fourteen, he had to go to work: a position on a farm was arranged by Jeannie. Fortunately his employers took kindly to his interest in natural history, but as routine gradually took over, his diary entries — overwhelmingly entomological to begin with — were replaced in part by a record of cleaning stables, grubbing thistles and feeding stock. His entomology, a stable rock in an uncertain world around him, he carried on in the spare time available. Further encouragement came from a Mr A. S. Collins who gave him a copy of the illustrated *Catalogue of New Zealand Butterflies* which had been published by J. D. Enys in 1880 and Captain Broun's *Manual of New Zealand Coleoptera*. From Nelson came his first paper to be published in the *Transactions*... on a parasitic fly, *Nemoroea nyctemerianus*. Hudson worked out the life history of the Magpie Moth and the Red Admiral, which he sent to London to be published in the *Entomologist* of 1883, and also made a very early sighting of the Forest Ringlet.[67]

In spite of the warmth and hospitality of those around him, it was decided that George should go back to Wellington, where in 1883 he entered the post office as a cadet. The first year was not a particularly happy one. The transfer to Palmerston North, then in the midst of a 'primeval forest', separated him from his family and his hours of work left him little time for his insects. But he was soon recalled to Wellington where conditions of work improved although his joy at the new arrangements was tragically interrupted by the death of his father. New Zealand entomology owes a great deal to this man whose quiet guidance and support were instrumental in his son's rapid development as an entomologist and illustrator.

In spite of his loss, George Hudson managed to work out a number of life histories that year and was now able to re-establish his connection with the Colonial Museum, where although the collections were poor, he was helped by T. W. Kirk (later affectionately referred to as 'Old Kirk'), then an assistant there. He gradually established contact with other entomologists in New Zealand: Fereday in Christchurch, Broun in Auckland, and locally, Maskell. He was visited by entomologists from England, the most significant of whom was E. Meyrick who visited Wellington in 1886 and was described as '. . . a very entertaining man & undoubtedly the best systematic zoologist south of the line' — which was about as much in the way of personal remarks that he ever made about anybody in his diary.[68] He also maintained English contacts amongst whom was R. McLachlan, an authority on Neuroptera and editor of the *Entomologist's Monthly Magazine*. Over the years the two exchanged many letters and specimens.

The year 1885 marked further progress. Improved hours at the post office gave him more daylight in which to collect; he had joined the Wellington Philosophical Society and was hard at work on the *Elementary Manual of New Zealand Entomology*. The manuscript was finished the following year and sent to the Premier, Sir Robert Stout, in the hope that the Government would subsidise its publication.[69] The wages of a 19-year-old postal

64. There must have been more somewhere in the Museum.

65. Hudson, 1881, MS. Diary, v.2, p.3.

66. A Dr J. Hudson was for some years a member of the Nelson Literary and Scientific Institute and occasional contributor to the *Transactions*.

67. *Dodonidia helmsii*.

68. Hudson, 1886, MS. Diary, p.183.

69. Hudson, 1886, MS, *Early Life and Summary of Diaries*.

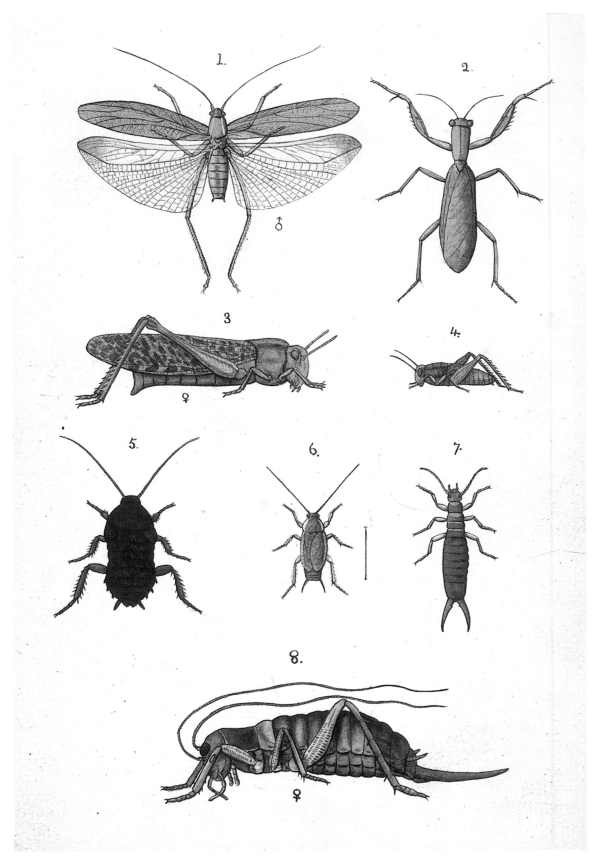

Original watercolour illustrations for Plate 17 of G. V. Hudson's *Elementary Manual of New Zealand Entomology*, 1892, including: (1) The Katydid (*Caedica simplex*); (2) Praying Mantis (*Orthodera ministralis*) and (8) The Wellington Weta (*Hemideina crassidens*). Published by permission of the Gibbs Family. *(see p.201)*

clerk left very little over for publishing ventures, and when Stout turned him down he had nowhere else to go. Happily his case was later taken up by Walter Buller and the President of the Wellington Philosophical Society, Charles Hulke. Buller in 1890 was brim-full of success, having just returned to New Zealand following his second major publishing venture; with Hulke he persuaded the Government to buy one thousand copies.[70] Broun had been pessimistic about sales of the book: 'People, as a rule, do not seem to care very much for books or insects'. He himself did not have the money for its purchase — or so he said.[71] Nevertheless the manuscript together with Hudson's drawings went at last to London and West Newman and Hudson sought also the help of R. McLachlan in obtaining subscribers in Britain.[72] Meanwhile his research continued with sixteen papers published by 1890; the insect collection in the Colonial Museum benefited from his sorting and curation, and he would occasionally examine crops for insect pests at Hector's request. He was only 23 years old.

Wellington in the decade of the 80s was still a collector's paradise. Many of the hills were covered in forest, and the fallen logs and gullies were found rich with insect life if the collector was as thorough and enthusiastic as Hudson. But sadly, day after day, the summer's sun would set on a smoke-laden sky as bush fires, set by the colonists, cleared the Karori, Makara, and Terawhiti hills for the development of pasture. The few conservationists who raised their voices in protest were declared 'cranks' and the burning went on. As the bush disappeared so did the insect and other life it supported.

G. V. Hudson spent much of his time tramping these hills in search of specimens. Sunday, 16 December, 1888 reads typically: 'Went up the MacKenzies spur onto the bushcovered, made height 1100 owing to slightly falling barometer, from there along the spur till the 'peaks' and then along the road until it crosses on to the SE side of the range then down the big gulley into the Wadestown gullys over the Tinakori Range home went with Harry P. Took abt 6 specns of D. metallifera in the gully. On the Range 1 specn of Gym Co-ardatella & a fine var of productata . . . & a lot of Pyronota festiva which was forming on the manuka on the top of the hill in the hot sunshine . . . very few blossoms to be seen & I only beat one bush! Fine day Killed insects & wrote diary evening.'[73]

There were excursions to Plimmerton and Pukerua Bay to the north, and across Cook Strait to Nelson, where he spent much time in the region of the Mount Arthur tableland; even his marriage provided an excuse for a field trip to the south of the South Island. But not all his collecting demanded an exhausting expedition. He wrote about *Vanessa cardui*: 'This insect is phenomenally abundant 2 seen in Botanicals [Botanical Gardens] on 9 th & one at the steps. 1 seen outside the Athenaeum on Lambton Quay & 1 brought as having been caught in the Guards van of the Palmerston train'.[74] Even more quotidian was: 'Got Supplejack for beating carpets. Did a little beating this produced a small additional weevil in considerable numbers & 2 specimens of a Beetle which I think is allied to the Caribidae but have not settled on its affinities'.[75]

In 1891, after a succession of temporary houses, Hudson finally moved into one of his own to become known as 'Hillview', in Karori, and his domestic situation further improved with his engagement to Florence Gillon. The previous year he had become disenchanted with the post office and contemplated the study of dentistry, but he received a favourable transfer and stayed on. With marriage in December 1893 and his domestic and working life secure, he settled down to an even more ambitious entomological work, the *New Zealand Moths and Butterflies*.

As a respected authority, with 'rather many visitors at times',[76] he went on through

70. *Ibid.*, 1890.

71. Brown to Hudson, 10.5.1890, 29.7.1890; Hudson Correspondence, National Museum Archives.

72. Hudson to McLachlan, 21.3.1890; McLachlan Correspondence, Hope Entomological Library, Oxford University.

73. Hudson, 1888, MS. Diary, p.184.

74. *Vanessa cardui* = Painted Lady Hudson, 1889, MS. Diary, p.276.

75. Hudson, 1889, MS. Diary, p.126.

76. Hudson 1894, MS. *Early Life and Summary of Diaries*.

77. Brown to Hudson, 6.7.1895; Hudson Correspondence, National Museum Archives.

78. Hudson to Meyrick, 8.3.1885; Hudson Correspondence, National Museum Archives.

79. Meyrick to Hudson, 12.5.1895; Hudson Correspondence, National Museum Archives.

80. Hudson, 1897, MS. Diary, p.45.

81. It was published in 1904.

82. Hudson, 1899, MS. Diary, p.213.

83. Hudson, 1901, pp.383-395.

84. Although professional natural history collectors such as Docherty and Reischek suffered enormous hardship. Docherty (who collected for Haast) relates: 'Fancy me all this winter sleeping on the wet ground as I had to shift my camp every two days, always wet feet, suffering from toothache so that I drew two of my own teeth, and came to Hokitika to get a third one drawn. But all this brings philosophy' (Von Haast, 1948, p.548).

85. Hudson, 1901, p.384.

86. *Dominion*, 11.5.11.

87. Hudson, 1920, MS. *Early Life and Summary of Diaries.*

the 90s, corresponding with fellow entomologists in New Zealand and overseas, preparing and exchanging specimens, and attending to office and domestic chores. His correspondence was always formal and very occasionally testy, such as when Broun thought Hudson had trespassed on his work[77] and more seriously when Hudson erupted in a letter to Meyrick over delays in the identification of specimens. He was 'much disappointed' and put to 'a great deal of inconvenience'. Meyrick's actions, he said, would lead to no end of trouble and confusion. He concluded by threatening to go elsewhere for assistance unless Meyrick improved his performance.[78] Meyrick's temperate reply defused the issue but he thought Hudson's preoccupation with New Zealand entomology had blinkered his view of the science as a whole.[79]

Year after year Hudson added to *New Zealand Moths and Butterflies* and once again there was a need to find subscribers, but this time the job seemed easier, no doubt because of his increasing reputation. In August 1897 he was finally able to send off the prospectuses and the manuscript to West Newman.[80] A daughter, Stella, was born a month earlier, and at this stage Hudson would have been excused his entomological labours for a well-earned rest, but instead he began a book on the Neuroptera which would occupy him for the next five years.[81]

New Zealand Moths and Butterflies was published in England in 1898 and on 18 July of the following year the *Rakaia* arrived in New Zealand with the copies of the book amongst its cargo. The next evening Hudson and his wife packed them for dispatch to subscribers until late in the evening, when: 'String and ourselves' were 'used up'.[82] The book was grander than his previous effort and the plates numerous and of high quality, although the chromolithography did not always do justice to his originals. And so he reached the end of the century, an occasion appropriately celebrated with a substantial promotion at the post office and election to the Presidency of the Wellington Philosphical Society.

For his presidential address he chose a subject with which he was closely identified, and one by which he set a great deal of store: entomological fieldwork.[83] Of all the collectors from this century, only Colenso in pursuit of his plants comes quickly to mind as a match for Hudson in his relentless hunt for insects.[84] In the company of others, Hudson emphasized the importance of a good basic collection in a country where the fauna was not well known, and particularly where changes 'through the agencies of civilization' were inflicting changes on the fauna. The destruction of the Wellington forest was a case in point, and in his lifetime he had seen once common species dwindle to the point of rarity. Insect collecting was not for the sluggish or faint-hearted: 'many interesting species . . . frequent the overhanging banks of deep forest ravines, and can only be dislodged by a vigorous probing with a stick into all kinds of nooks and crannies'.[85] Only persistence would reveal some of the beautiful forest-dwelling lepidoptera, their delicate colouring a camouflage against the forest moss. Careful field observations would enable accurate interpretation of structures 'acquired' by natural selection; was it not the case that Darwin's philosophies came out of collection and meticulous observation? Had not fieldwork triumphed over book learning in the revelation of far-reaching principles?

In this vein he later attacked the education system for its emphasis on the commercial value of scientific studies rather than supplying the mind 'for those great generalizations of science . . .'.[86] This was Hudson's own philosophy, carved from experience, unshakeable and often repeated. He even put it into practice by encouragement of the young, amateur naturalists: Albert Norris, Frank Hawthorne, the hapless Ted Clarke — 'a failure as an understudy'[87] — and many others. 'Hillview', his home in Karori, became a mecca for

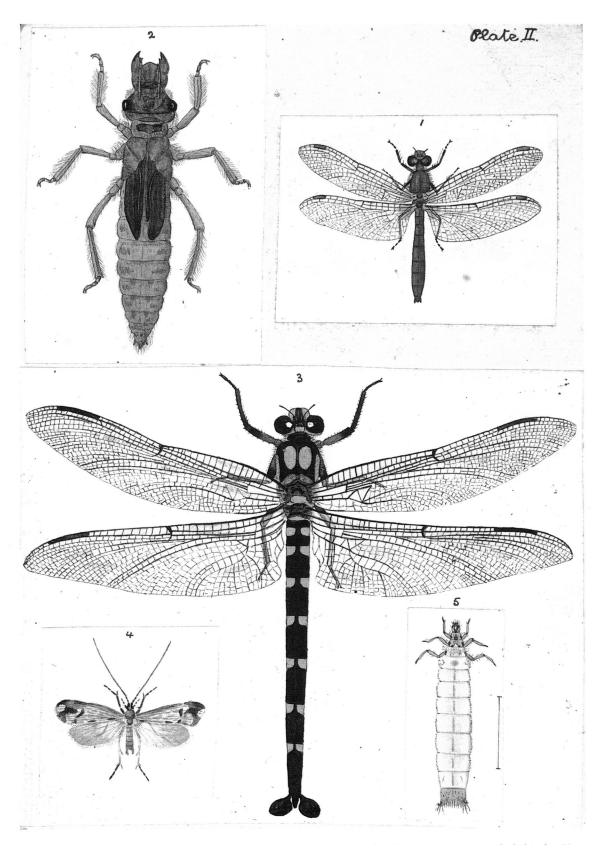

Original watercolour illustrations for Plate 2 of G. V. Hudson's *New Zealand Neuroptera*, 1904; included is the Giant Dragon-fly (*Uropetala carovei*) and its nymph stage (2, 3). Published by permission of the Gibbs Family. *(see p.204)*

apprentices in entomology.

Subsequent years would see five more books including the classical revised enlarged edition of the *Moths and Butterflies* and many more journal publications. Following his death in 1946, his last book was published in 1950 through the agency of his daughter Stella.[88]

Hudson had carried on the amateur tradition well into the twentieth century, but he was no dilettante, nor did he consider his major works as iconographies for the drawing-room table or the shelves of wealthy collectors. A firm believer in the accurate, coloured representation of the entire insect,[89] his butterflies and beetles lay on the page in severely practical rows and columns, rather than resting jewel-like on some fitting background of flower or leaf. He could have easily produced a work of art and doubled the number of subscribers and increased his reputation, but it would have been a corruption of his objectives and would have taken up valuable time, better spent in fieldwork. It would also have put the price of his works well beyond those who might make use of them.

As the years passed he became a critic of modern entomological technique, including its emphasis on a highly specialized, diagrammatic approach to description and the development of a jargon and detailed system of classification that excluded the amateur. In certain respects he was justified, but in the name of progress his criticisms were put to one side, only to be resurrected when the time-honoured argument over applied versus basic research arose for discussion. There is today still room in entomology for the amateur in spite of the developments Hudson so bitterly opposed; along with ornithology and shell collecting, it has managed to retain something of this sort of following. If Hudson's ideas were not immediately acceptable or were ridiculed he would simply bide his time until his critics came around to his point of view. His ideas for daylight saving were an example.

If the productivity of Hudson's twentieth century was almost to eclipse that of the nineteenth, he could nevertheless look back on those early years with pride and a sense of fulfilment. There would be joy in recollecting days spent on the Mount Arthur tableland, in the Botanical Gardens studying the glow-worm with Albert Norris, and observing the insects and habitat that would never be seen in the same way again as under smoke-laden skies the bush-covered hills and mossy gullies of his youth disappeared first into pasture land and gorse and then under the blanket of twentieth century urban development.

Thus passed a life unselfishly dedicated to a science, one in which the work was its own reward and the medals, fellowships and honorary degrees that might have come his way were not worth the time and effort of pursuit. It was not the work that needed the recognition, it was the man, and, as is so often the case, the greatest appreciation of him would come after he had gone.[90]

88. Hudson, 1950, 188 pp.

89. *Ibid.*, p.168.

90. The Royal Society of New Zealand (Wellington Branch) now has the Hudson Lecture in his honour; see also Miller, 1953, p.80.

CENTURY'S CLOSE

NOW AT THE END of the century we can stand and look back at the lengthening shadows cast by Banks, the Forsters, Lesson, Quoy and Gaimard, Owen, Gray and others over the several ages of New Zealand zoology. There had been an age of vertebrates, overtaken by the invertebrates; an age of fossil birds, to be replaced by that of ornithology and entomology; an age of animal introduction and exploitation. We can talk of an age of explorers, of missionaries, of imperial zoologists, of amateurs and professionals. We can recall the human qualities of the scientists — their stamina and energy, and their skirmishes; the persistence of the British, the flair of the French and the restless and solitary scholarship of the Germans and Austrians. From all of this emerged the individuals and the institutions of an indigenous science. The fauna itself was now revealed as distinctive: The Short-tailed Bat; the extinct Moas and other flightless birds; the Tuatara and native frogs; the freshwater fish and the Peripatus. With its archaic and southern faunal elements, together with contributions from neighbouring zoogeographic regions, New Zealand would provide a source of debate for biogeographers for many years to come.

But how difficult it is to make some concluding statement about 130 years of scientific endeavour, even in a field as confined as New Zealand's natural history! We know, as we leave the protagonists, that at one point the sum of their endeavours had given zoology in this country a flying start. But the natural sciences, which were a large part of New Zealand's science, retreated from a position as part of a developing world view to one of increasing introspection and utilitarianism. Ignoring the warnings, the country was placed at the mercy of introductions and exploitation; the insult to the fauna, flora and landscape has still not been forgiven. At this critical time the certainties of taxonomy needed to be exchanged for the imponderables of ecology and conservation. Underlying principles had to be established and applied, and a few, a very few, knew almost instinctively some of the answers. But no one was prepared to listen. The agenda for the sciences in the twentieth century was clearly being written in the nineteenth.

The tug of provincialism, the demands of agriculture, the cries for technology and research to be applied to trade and manufacturing problems — all helped to contribute to the decline of centralist science, suggesting that this aspect of a nation's cultural identity cannot be sought from a parochial standpoint or from rafts of individual enterprise. To a great extent the choice lay between pragmatic and abstract or philosophical science. Politicians and the state of the economy would not favour both. The choice for pragmatism was also a choice for an *ad hoc* approach, with science reacting rather than science leading. And if one believes in the sacralisation of science in the nineteenth century,[1] then in New Zealand the ecclesiastical trappings were not allowed to flourish and grow — latterly there were no spiritual leaders or temples of science, industry, and art such as could be found in the northern continents and to which several of New Zealand's scientists and others were drawn. Without this stable and central core, providing sanctuaries of basic and fundamental research and a reference point for all, the integration of science into the culture — and indeed of intellectual pursuits in general — would be that much more difficult in the times that lay ahead.

1. Russell, 1983, p.259.

REFERENCES

Allee, W. C., Emmerson, A. E., Park, O., Park, T., and Schmidt, K. P.	1949	*Principles of Animal Ecology.* W. B. Saunders Co.; Philadelphia.
Allen, D. E.	1976	*The Naturalist in Britain : A Social History.* Allen Lane, London.
Andrews, J. R. H.	1982	*A Miscellany of Ornithologists and Illustrators* in *The Summer Book*, B. Williams and R. Parsons (eds.). Port Nicholson Press, Wellington.
Angas, G. F.	1847	*Savage life and scenes in Australia and New Zealand: Being an artist's impressions of countries and people at the antipodes*, 2 v. Smith, Elder and Co., London.
	1847	*The New Zealanders Illustrated.* Thomas M'Lean, London.
Angehr, G. R.	1984	A bird in the hand: Andreas Reischek and the Stitchbird. *Notornis*, 31: 300-311.
Anon. 1775	1775	*A Catalogue of ... Natural Curiosities ... of Shells, Petrefactions, Insects, Animals in Spirits etc. ... The Property of Sir John Dalston ... which will be sold by Auction, by Mr Walsh ... Wednesday May 3, 1775.* 24 pp.
Anon.	1868	[note on forthcoming issue of W. Buller's 'Birds of New Zealand'] *The Ibis*, 4 : 504.
Anon.	1907	In Memoriam [obituary notice, Sir Walter Lawry Buller]. *Transactions and Proceedings of the New Zealand Institute*, 39 : iii-v.
Anon.	1910	*Gray, John Edward* in *Encyclopaedia Britannica* 11 ed., v.12. Cambridge University Press, London.
Anon.	1911	*Maximilian* in *Encyclopaedia Britannica* 11 ed., v. 17. Cambridge University Press, London.
Anon.	1958	*Earl, George Samuel Windsor* in *The Australian Encyclopaedia*, v.3. Angus and Robertson, Sydney.
Audubon, J. J. L.	1827—1838	*The Birds of America from original drawings*, 4 v. Published for the author, London.
Ayling, T., & Cox, G. J.	1982	*Collins Guide to the Sea Fishes of New Zealand.* Collins, Auckland.
Bagnall, A. G.	1966	*Buller, Walter* in *An Encyclopaedia of New Zealand*. A. H. McLintock, ed., Government Printer, Wellington.
Bagnall, A. G. (ed.)	1980	*New Zealand National Bibliography to the year 1960*, 4 v. Government Printer, Wellington.
Bagnall, A. G., & Peterson, G. C.	1948	*William Colenso – Printer, Missionary, Botanist, Explorer, Politician, his life and journeys.* A. H. & A. W. Reed, Wellington.
Banks, J.	1962	*The Endeavour Journal of Joseph Banks 1768 – 1771*, 2 v. J. C. Beaglehole, ed., Trustees of the Public Library of New South Wales in Association with Angus and Robertson, Sydney.
Barber, L.	1980	*The Heyday of Natural History 1820 – 1870.* Jonathan Cape, London.
Barclay, A.	1854	*Life of Captain Andrew Barclay* (as dictated to Thomas Scott). Privately published, Edinburgh.
Barratt, G.	1979	*Bellingshausen, a visit to New Zealand: 1820.* The Dunmore Press, Palmerston North.
Bartlett, A. D.	1852	On the genus Apteryx. *Proceedings of the Zoological Society of London*, 18: 274 — 276.

	1868	On the Incubation of the Apteryx. *Proceedings of the Zoological Society of London* [1868] : 329
Bathgate, A.	1898	Notes on Acclimatisation in New Zealand. *Transactions and Proceedings of the New Zealand Institute*, 30: 266 – 278.
Beaglehole, J. C.	1974	*The Life of Captain James Cook.* Adam and Charles Black, London.
Begg, A. C., & Begg, N. C.	1966	*Dusky Bay.* Whitcombe and Tombs Ltd., Christchurch.
Bell, G. E.	1976	*Ernest Dieffenbach.* The Dunmore Press, Palmerston North.
Benham, W. B.	1902	Note on an Entire Egg of a Moa, now in the Museum of the University of Otago. *Transactions and Proceedings of the New Zealand Institute*, 34: 149 – 151.
Bennett, E. T.	1835	[letter to E. T. Bennett from A. MacLeay]. *Proceedings of the Zoological Society of London*, 3: 61.
Bennett, E. W.	1964	The Marine Fauna of New Zealand: Crustacea Brachyura. *New Zealand Department of Scientific and Industrial Research Bulletin*, 153: 120 pp.
Bettany, G. T.	1908	(see Stephen and Lee).
Bladen, F. M.	1901	*Historical Records of New South Wales*, v. vii *Bligh and Macquarie 1809, 1810, 1811.* Government Printer, Sydney.
Blair, K. G.	1913	*Tribolium castaneum*, Herbst = *ferrugineum* Auct (nec. Fab.) *Entomologists Monthly Magazine*, 49: 222-224.
Bloch, M. E.	1801	*M. E. Blochii, Doctoris Medicinae Berolensis, et societabilis literariis multis adscripti, Systema Ichthyologiae*, etc. Schneider, Berlin.
Boisduval J. B. A. D. de		(see Dumont d'Urville).
Bonnaterre, J. P.	1788	*Tableau encyclopédique et méthodique des trois régnes de la nature ... Ichthyologie.* Panckoucke, Paris. (This is part of the Zoological portion of the *Encyclopédie Méthodique* 1782-1832).
Boog Watson, R.	1883	New Zealand Mollusca of the 'Challenger' expedition. *New Zealand Journal of Science*, 1:319-321.
Boulenger, G. A.	1910	*Ichthyology* in *Encyclopaedia Brittanica* 11 ed., v. 14. Cambridge University Press, London.
Brockway, L. H.	1979	*Science and Colonial Expansion : The Role of the British Royal Botanic Gardens.* Academic Press, London.
Brady G. S.	1882	New Zealand Copepods of the 'Challenger' Expedition. *New Zealand Journal of Science*, 1: 35-39.
Brereton, C. B.	1948	*History of the Nelson Institute.* A. H. & A. W. Reed, Wellington.
Brisson, M. J.	1763	*Ornithologia sive synopsis methodica sistens Avium divisionem in ordines etc. ... Ornithologie ... etc.* 6 v. T.Haak, Paris.
Brosse, J.	1983	*Les tours du monde des explorateurs, les grands voyages maritimes, 1764-1843.* Bordas, Paris.
Broun, T.	1880–1893	Manual of the New Zealand Coleoptera. Colonial Museum, Wellington.
Broussonet, P. M. A.	1780	Mémoire sur les différentes espèces de chiens de mer. *Mémoires de l'Académie royale des Sciences Paris*: 641-680.
Brown, P.	1776	*New Illustrations of Zoology containing fifty coloured plates of new, curious, and nondescript birds, with a few quadrupeds, reptiles and insects together with a short and scientific description of the same.* B. White, London.
Bruguière, J. G.	1782–1832	*Histoire naturelle des vers. Dictionnaire Encyclopédie Méthodique*, 1789. Panckoucke, Paris.
Buffon, G. L. L. de	1749–1804	*Histoire naturelle ... avec la description du Cabinet du Roi*, 44 v. De l'Imprimerie royale, Paris.

Buick, T. L.	1928	*The French at Akaroa, an adventure in colonization.* N.Z. Book Depot, Wellington.
	1936	*The Discovery of Dinornis, The story of a man, a bone, and a bird.* Thomas Avery and Sons Ltd., New Plymouth.
	1937	*The Moa-hunters of New Zealand.* Thomas Avery and Sons Ltd., New Plymouth.
Buller, W. L.	1865	*On the Ornithology of New Zealand.* Commissioners of the New Zealand Exhibition, 1865. Dunedin. (Published again in the *Transactions and Proceedings of the New Zealand Institute*, 1869.)
	1869a	Essay on the Ornithology of New Zealand. *Transactions and Proceedings of the New Zealand Institute*, 1: 213-231.
	1869b	Notes on Herr Finsch's Review of Mr Walter Buller's Essay on New Zealand Ornithology. *Transactions and Proceedings of the New Zealand Institute*, 1: 49-57.
	1869c	On Some New Species of New Zealand Birds. *The Ibis*, 5: 37-42.
	1870a	On Some New Species of New Zealand Birds. *Transactions and Proceedings of the New Zealand Institute*, 2 : 390-392.
	1870b	Remarks on some disputed species of New Zealand Birds. *The Ibis*, 6 : 455-460.
	1871	On the Katipo, or venomous spider of New Zealand. *Transactions and Proceedings of the New Zealand Institute*, 3 : 29-34.
	1873a	Notice of a New Species of Moth in New Zealand. *Transactions and Proceedings of the New Zealand Institute*, 5: 279-280
	1873b	*A History of the Birds of New Zealand.* John van Voorst, London.
	1876	Remarks on Dr Finsch's paper on New Zealand Ornithology. *Transactions and Proceedings of the New Zealand Institute*, 8: 194-204.
	1888	*A History of the Birds of New Zealand*, 2 v. 2nd ed. Published by the author, London.
	1890	*Reviews and other notices of Sir Walter Buller's 'Birds of New Zealand'* . . . Taylor & Francis, London.
	1895	Illustrations of Darwinism; or, The Avifauna of New Zealand considered in Relation to the Fundamental Law of Descent with Modification. *Transactions and Proceedings of the New Zealand Institute*, 27: 75-104.
	1905	*Supplement to the 'Birds of New Zealand'*, 2 v. Published by the author, London.
Burdon, R. M.	1941	*New Zealand notables*, (Series 1). Caxton Press, Christchurch.
	1950	*New Zealand notables*, (Series 3). Caxton Press, Christchurch.
Burmeister, C.H.C.	1836	*A manual of Entomology.* W. E. Shuckard (trl.), London.
Busk, G.	1852	*An Account of the Polyzoa, and Sertularian Zoophytes, collected in the Voyage of the Rattlesnake, on the Coasts of Australia and the Louisade Archipelago, etc. in v.1. Narrative of the Voyage of H.M.S. Rattlesnake, commanded by the late Captain Owen Stanley, R.N., F.R.S. etc during the years 1846-1850 . . . etc.* 2 v. T. & W. Boone, London.
Butler, A. G.	1874	*Catalogue of the Lepidoptera of New Zealand.* E. W. Janson Ltd., London.
	1880	*Catalogue of the Butterflies of New Zealand . . . with a short preface by John D. Enys, F.G.S.* G. Tombs & Co., Christchurch.
	1884	On a new genus of butterfly from New Zealand. *New Zealand Journal of Science*, 2: 159-160 (also published in *Annals and Magazine of Natural History*, 1884, p.171).
Callaghan, F. R. (ed.)	1957	*Science in New Zealand.* A. H. & A. W. Reed, Wellington.

Cameron, H. C. — 1952 — Sir Joseph Banks, K.B., P.R.S., The autocrat of the philosophers. The Batchworth Press, London.

Carolin, R. C. — 1971 — *The Natural History of the Endeavour Expedition* in *'Employ'd as a Discoverer'*, *Papers presented at the Captain Cook Bicentenary Historical Symposium, Sydney*. J. V. S. Megaw, ed. A. H. & A. W. Reed, Sydney.

Carr, D. J. (ed.) — 1983 — *Sydney Parkinson, Artist of Cook's Endeavour Voyage.* Nova Pacifica with the British Museum (Natural History) and Australian National University Press, Wellington.

Cernohorsky, W. — 1974 — Type specimens of Mollusca in the University Zoological Museum, Copenhagen. *Records of the Auckland Institute and Museum*, 11: 143-192.

Chemnitz, J. H. — 1783 — Von den Südländischen Conchylien welch sich in der Sammlung des Herrn Pastors Chemnitz in Kopenhagen befinden und bei den Cookischen Seereisen gesammelt werden. *Der Naturforscher*, 19: 177-208.

Cole, F. J. — 1975 — *A History of Comparative Anatomy, from Aristotle to the Eighteenth Century.* Dover, New York.

Colenso, W. — 1846 — An account of some enormous Fossil bones of an unknown Species of the Class Aves, lately discovered in New Zealand. *Tasmanian Journal of Natural Science, Agriculture, Statistics, etc.*, 2: 81-107. (Also, 1844, with minor changes in the *Annals and Magazine of Natural History*, 14: 81-96).

— 1880 — On the Moa, *Transactions and Proceedings of the New Zealand Institute*, 12: 63-108.

— 1892 — *Status quo:* a Retrospect — A few more words by way of Explanation and Correction concerning the First Finding of the Bones of the Moa in New Zealand; . . . etc. *Transactions and Proceedings of the New Zealand Institute*, 24: 468-478.

Cook, J. — 1777 — *A Voyage towards the South Pole, and round the world. Performed in His Majesty's Ships the Resolution and Adventure, In the years 1772, 1773, 1774 and 1775.* W. Strahan and T. Cadell, London.

— 1955 — *The Journals of Captain James Cook on His Voyages of Discovery: The Voyage of the Endeavour 1768-1771.* v.1. J. C. Beaglehole, ed. Cambridge University Press for the Hakluyt Society, Cambridge.

— 1961 — *The Journals of Captain James Cook on His Voyages of Discovery: The Voyage of the Resolution and Adventure, 1772-1775.* v.2 J. C. Beaglehole, ed. Cambridge University Press for the Hakluyt Society, Cambridge.

— 1967 — *The Journals of Captain James Cook on His Voyages of Discovery: The Voyage of the Resolution and Discovery, 1776-1780.* v.3. (2 Pts.) J. C. Beaglehole, ed. Cambridge University Press for the Hakluyt Society, Cambridge.

Craw, R. C. — 1978 — Two Biogeographical Frameworks: Implications for the Biogeography of New Zealand. A Review. *Tuatara*, 23 (2): 81-114.

Cruise, R. A. — 1823 — *Journal of a ten months' residence in New Zealand.* Longman, Hurst, Rees, Orme and Brown, London.

Cunningham, A. — 1839 — Rough notes . . . on the habits of the *Apteryx Australis.* . . . *Proceedings of the Zoological Society of London*, 7:63-64.

Cust, L. — 1909 — *Frederick P. Nodder* in *Dictionary of National Biography*, v. 14, S. Lee, ed. Smith, Elder and Co., London.

Cuvier, G., & Valenciennes, A. — 1828–1849 — *Histoire naturelle des poissons* 22 v. Levrault, Paris.

Dall, W. H. — 1905 — Thomas Martyn and the Universal Conchologist. *Proceedings of the U.S. National Museum*, 29 : 415-432.

— 1908 — Supplementary Notes on Martyn's Universal Conchologist. *Proceedings of the U.S. National Museum*, 33: 185-192.

	1915	An Index to the Museum Boltenianum. *Smithsonian Institution Publication* 2360 : 1-61.
	1921	Molluscan species named in the Portland Catalogue, 1786. Part II, Foreign Species. *Nautilus*, 34 : 124-132.
Dance, S. P.	1966	*Shell Collecting, An Illustrated History.* Faber and Faber, London.
	1971	The Cook Voyages and Conchology. *Journal of Conchology*, 26 : 354-379.
	1972	The Cook Voyages and Conchology: a Supplementary Note. *Journal of Conchology*, 27: 357-358.
	1978	*The Art of Natural History – Animal Illustrators and their Work.* Overlook Press, New York.
Darwin, C. R.	1839	*Journal of Researches into the Geology and Natural History of the various countries visited by H.M.S. Beagle under the command of Captain FitzRoy, R.N. from 1832 to 1836.* Henry Colburn, London. (2nd ed., 1845.)
	1839–1843	*The Zoology of The Voyage of H.M.S. Beagle during the years 1832–36.* Smith, Elder & Co, London. (with contributions from Owen, Gould, Jenyns and Bell)
	1859	*On the Origin of species by natural selection, or The preservation of favoured races in the struggle for life.* Murray, London.
Dawson, G. E., & Dawson, E. W.	1958	Birds of the Cook Strait Islands, collected by Professor Hugo Schauinsland in 1896 and 1897. *Notornis*, 8 (2): 39-49.
Dawson, W.	1958	*The Banks letters: a calendar of the manuscript correspondence of Sir Joseph Banks.* Trustees, British Museum, London.
Debenham, F.	1945	*The Voyage of Captain Bellingshausen to the Antarctic Seas 1819 – 1821*, 2v. Cambridge University Press for the Hakluyt Society, London.
Desmond, A.	1982	*Archetypes and Ancestors – Palaeontology in Victorian London, 1850 – 1875.* Blond and Briggs, London.
De Surville, J.	1981	*The expedition of the St Jean-Baptiste to the Pacific, 1769 – 1770: from the journals of Jean de Surville and Guillaume Labé.* J. Dunmore, ed. Cambridge University Press for the Hakluyt Society, London.
De Vos, A., Manville, R. H., & Van Gelder, R. G.	1956	Introduced mammals and their influence on native biota. *Zoologica, N.Y.* 41 (4): 163-194.
Dezallier d'Argenville, A. J.	1780	*La conchyliologie, ou Histoire naturelle des coquilles de mer . . .* 3rd ed. by M. M. Favanne de Montcervelle père et fils, 3 v. G. de Bure, Paris.
Dick, I. D.	1951	The History of Scientific Endeavour in New Zealand. *New Zealand Science Review*, 9 (9) pp.139-143.
	1957	*Historical Introduction to New Zealand Science* in *Science in New Zealand*, F. R. Callaghan, ed. A. H. & A. W. Reed, Wellington.
Dieffenbach, E.	1841	An Account of the Chatham Islands. *Journal of the Royal Geographical Society*, 11: 195-215.
	1843	*Travels in New Zealand; with contributions to the Geography, Geology, Botany, and Natural History of that country.* 2 v. John Murray, London.
Die Historische Commission bei der Königliche Akademie der Wissenschaften	1875–1912	*Allgemeine Deutsche Biographie*, 56 v. Duncker and Humblot, Leipzig.
Dodge, E. S.	1973	*The Polar Rosses, John and James Clark Ross and their Explorations.* Faber and Faber, London.
Donovan, E.	1805	*An epitome of the Natural History of the Insects of New Holland, New Zealand, New Guinea, Otaheite, and other islands in the Indian, Southern, and Pacific Oceans, etc.* The author and F. C. and J. Rivington, London.

| | 1823–1834 | *The Naturalists Repository, or miscellany of exotic Natural History, exhibiting . . . specimens of foreign Birds, Insects, Shells, etc. . . .* 5 v. The author and Simpkin and Marshall, London. |

Dresser, H. E. | 1871–1896 | *A History of the Birds of Europe, including all the species inhabiting the Western palaearctic region,* 9 v. Published for the author, London.

Druett, J. | 1983 | *Exotic Intruders, The Introduction of plants and animals into New Zealand.* Heinemann, Auckland.

Dugdale, J., & Fleming, C. | 1969 | Two New Zealand cicadas collected on Cook's Endeavour voyage, with description of a new genus. *New Zealand Journal of Science,* 12: 929-957.

Dumont d'Urville, J. S. C. | 1830–1833 | *Voyage de la corvette L'Astrolabe Exécuté par Ordre du Roi, pendant les années 1826 – 1827 –1828 –1829, etc. . . . Histoire du voyage,* 5 v. J. Tastu, Paris.

| 1830 | *Voyage de découvertes de L'Astrolabe . . . etc. . . . Zoologie* par MM. Quoy et Gaimard, 4 v. J. Tastu, Paris.

| 1832–1835 | *Voyage de découvertes de L'Astrolabe . . . etc. . . . Faune Entomologique de l'Ocean Pacifique, avec l'illustration des insectes nouveaux recueillis pendant le voyage.* Par le docteur Boisduval . . . etc. 2 pts. J. Tastu, Paris.

| 1834 | *Voyage de la corvette L'Astrolabe exécuté pendant les années 1826 – 1827 – 1828 – 1829, etc. . . . Atlas, Zoologie.* J. Tastu, Paris.

| 1841–1846 | *Voyage au Pôle Sud et dans l'Océanie sur les corvettes L'Astrolabe et la Zélée, exécuté par Ordre du Roi pendant les années 1837 – 1838 – 1839 – 1840, etc. . . . Histoire du Voyage,* 10 v. Gide, Paris.

| 1846–1854 | *. . . Zoologie* par MM. Hombron et Jacquinot 5 v., Gide et Cie, Paris. [authors include Honoré Jacquinot, Pacheran, Quichenot, Lucas, Blanchard and Rousseau.]

| 1853 | *. . . Atlas . . . Zoologie* Gide et Cie, Paris.

Duncan, F. M. | 1937 | On the Dates of Publication of the Society's 'Proceedings', 1859 – 1926. With an Appendix containing the dates of Publication of 'Proceedings', 1830 – 1858, compiled by the late F. H. Waterhouse, and of the 'Transactions', 1833 – 1869, by the late Henry Peavot, originally published in the P.Z.S. 1893, 1913. *Proceedings of the Zoological Society of London* Ser. A: 71-84.

Dunmore, J. | 1965 | *French Explorers in the Pacific,* v.1. Oxford University Press, Oxford.

| 1969 | *French Explorers in the Pacific,* v.2. Oxford University Press, Oxford.

Duperrey, L. I. | 1826 | *Voyage autour du monde, Exécuté par Ordre du Roi, sur La Corvette de sa Majesté, La Coquille, pendant les années 1822, 1823, 1824 et 1825, etc. . . . Histoire naturelle Zoologie.* [Atlas]. Arthus Bertrand, Paris.

| 1828–1838 | *. . . Zoologie,* Par MM. Lesson et Garnot, 2 v. Arthus Bertrand, Paris. [authors include F. E. Guérin-Méneville].

Du Petit-Thouars, A. A. | 1840–1845 | *Voyage Autour du Monde sur la frégate La Vénus, pendant les années 1836 – 1839, etc. . . .* 4 v. Gide, Paris.

Earle, A. | 1832 | *A Narrative of a nine months' residence in New Zealand, in 1827; Together with a Journal of a residence in Tristan D'Acunha, an island situated between South America and the Cape of Good Hope.* Longman, Rees, Orme, Brown, Green, and Longman, London. (1966 edition, E. H. McCormick, ed. Oxford University Press).

Edwards, G. | 1743–1751 | *A Natural History of Birds . . .* Published by the author, London.

| 1758–1764 | *Gleanings of Natural History, exhibiting figures of Quadrupeds, Birds, Insects, Plants, etc . . .* Published by the author, London.

Enys, J. D. | 1880 | see Butler, A. G. 1880.

Fabricius, J. C.   1775   *Systema entomologiae, sistens insectorum classes ordines, genera, species, adiectis synonymis, locis descriptionibus, observationibus.* Korte, Flensburgi et Lipsiae.

     1781   *Species insectorum exhibentes eorum differentias specificas, synonyma auctorum, loca natalia, metamorphosin adiectis observationibus, descriptionibus,* 2 v. C. E. Bohn, Hamburgi et Kilonii.

     1787   *Mantissa insectorum sistens eorum species nuper detectas adiectis characteribus, genericis differentiis specificus, amendationibus, observationibus,* 2 v. C. G. Proft, Hafniae.

     1792–1798   *Entomologia systematica emendata et aucta. Secundum classes ordines, genera, species, adiectis synonymis, locis, observationibus, descriptionibus,* 4 v., suppl. C. G. Proft, Hafniae.

     1801   *Systema eleutheratorum secundum ordines, genera, species, adiectis synonymis, locis, observationibus, descriptionibus,* 2 v. Bibliopolii Academici novi, Kiliae.

     1803   *Systema Rhyngotorum secundum ordines, genera, species, adiectis synonymis, locis, observationibus, descriptionibus.* Reichard, Brunsvigae.

     1805   *Systema Antilatorum secundum ordines, genera, species, adiectis synonymis, locis, observationibus, descriptionibus.* Reichard, Brunsvigae.

Farber, P. L.   1982   *The Emergence of Ornithology as a Scientific Discipline: 1760 – 1850.* D. Reidel, Dordecht.

Fell, H. B., Garrick, J. A. F., Sorensen, J. H., Stephenson, S. K., Stout, V. M., & Sullivan, G. E.   1953   *The First Century of New Zealand Zoology 1769 – 1868.* Department of Zoology, Victoria University College, Wellington.

Fereday, R. W.   1877   Brief observations on the genus *Chrysophanus,* as represented in New Zealand. *Transactions and Proceedings of the New Zealand Institute* 9: 460-462.

Ferro, D. N. (Convener)   1977   *Standard Names for Common Insects of New Zealand.* Entomological Society of New Zealand, Auckland.

     1883   Description of a Species of Butterfly new to New Zealand and probably to Science *Transactions and Proceedings of the New Zealand Institute,* 15: 193-195.

Festing Jones, H.   1919   *Samuel Butler, author of Erewhon (1835 – 1902). A memoir.* 2v. Macmillan and Co., London.

Fildes, H. E. M.   1984   *Selective Indexes to Certain Books Relating to Early New Zealand.* Victoria University of Wellington Library, Wellington.

Finsch, O.   1869a   Notes on Mr Walter Buller's 'Essay on the Ornithology of New Zealand'. *Transactions and Proceedings of the New Zealand Institute,* 1: 58-73.

     1869b   Remarks on Some species of Birds from New Zealand. *The Ibis,* 5: 378-381.

Firth, J C.   1876   [Presidential address to the Auckland Institute, 17.5. 1875] *Transactions and Proceedings of the New Zealand Institute,* 8: 420-425.

Fitzinger, L. J.   1861   Eine neue Batrachier-Gattung aus Neu-Seeland. *Verhandlungen der K. K. Zoologisch-botanischen Gesellschaft in Wien,* 11: 217-220.

Fleming, C. A.   1968   The Royal Society of New Zealand – A Century of Scientific Endeavour. *Transactions of the Royal Society of New Zealand.* 2 (6): 99-114.

     1982a   *George Edward Lodge, the Unpublished New Zealand Bird Paintings.* Nova Pacifica and the National Museum, Wellington.

     1982b   The Influence of Early German Naturalists and Explorers in New Zealand. *Alexander von Humbolt Stiftung, Mitteilungen,* 44 : 25-32.

Forbes, H. O.   1893   A List of the Birds Inhabiting The Chatham Islands. *The Ibis,* 5: 524-546.

Forster J. G. A.   1777   *A Voyage round the world, in His Britannic Majesty's Sloop, Resolution, commanded by Capt. James Cook, during the Years 1772, 3, 4, and 5.* 2 v. B. White, J. Robson, P. Elmsly, G. Robinson, London.

Forster, J. R. 1776 *Characteres generum plantarum, quas in itinere and insulas Maris Australis, Collegerunt, Descriptserunt, Delinerunt, annis MDCCLXXII – MDCCLXXV.* B. White, T. Cadell and P. Elmsly, London.

1778 *Observations made during a Voyage round the world, on physical geography, natural history, and ethic philosophy. Especially on 1. The Earth and its Strata, 2. Water and the Ocean, 3. The Atmosphere, 4. The Changes of The Globe, 5. Organic Bodies,* and *6. The Human Species.* G. Robinson, London.

1781 Historia Aptenodytae, generis avium orbi Australi Auctore Io Reinoldo Forster LLD. *Commentationes Societatis Regiae Scientiarum Gottingensis, III, Classis Physicae,* 3: 121-148.

1785 Mémoire sur les Albatros. *Mémoires de Mathématique et de Physique. Présentés à l'Academie Royale des Sciences, pars divers Savans, et lus dans ses assemblées,* 10: 563-572.

1788 *Enchiridion historiae naturali inserviens, quo termini et delineationes ad avium, piscium, insectorum et plantarum adumbrationes intelligendas et concinnandas, secundum methodum systematis Linnaeani continentur,* Hermmerde and Schwetschke, Halle.

1844 *Descriptiones animalium quae in itinere ad maris australis terras per annos 1772, 1773, et 1774 suscepto collegit observavit et delineavit Johannes Reinoldus Forster.* M. H. K. Lichtenstein ed. Dummler, Berlin.

1982 The Resolution *Journal of Johann Reinhold Forster.* 4v., M. E. Hoare, ed. Cambridge University Press, for the Hakluyt Society, London.

Fox, R. & Weisz, G. 1980 *The Organisation of Science and Technology in France 1808 – 1914.* Cambridge University Press, and Editions de la Maison des Sciences de l'Homme, Paris.

Freed, D., 1963 Bibliography of New Zealand Marine Zoology 1769 – 1899. *New Zealand Department of Scientific and Industrial Research Bulletin,* 148: 9-46.

Galbreath, R. 1984 Sir Walter Buller, The History of The Birds of New Zealand. *Forest and Bird* 15 (2): 7-8.

Gibbs, G., 1980 *New Zealand Butterflies, Identification and Natural History.* Collins, Auckland.

Gillies, R. 1877 [Presidential address, Annual General Meeting, Otago Institute, 16.2.1877] *Transactions and Proceedings of the New Zealand Institute,* 9: 656-664.

Gmelin, J. F. (Linnaeus, C.) 1788–1793 *Systema Naturae, sive regna tria naturae systematice proposita per classes, ordines, genera and species. Editio decima tertia, aucta, reformata, cura J. F. Gmelin,* 3 v. Lipsiae.

Godley, E. J. 1967 A Century of Botany in Canterbury. *Transactions of the Royal Society of New Zealand,* 1 (22): 243-266.

1970 Botany of the Southern Zone Exploration 1847-1891. *Tuatara,* 18 (2): 49-93.

Gould, J. 1837–1838 *The Birds of Australia, and the adjacent islands.* Published by the author, London.

1840–1869 *The Birds of Australia* 7 v. + supplement. Published by the author, London.

1847 On a new species of Apteryx. *Proceedings of the Zoological Society of London,* 15 : 93-94.

1848 On a new species of the genus Apteryx. *Transactions of the Zoological Society of London,* 3: 379-380

1852 Remarks on Notornis mantelli. *Proceedings of the Zoological Society of London,* 18: 212-214.

Gourlay, E. S. 1950 Auckland Island Coleoptera. *Transactions of the Royal Society of New Zealand,* 78 (2-3): 171-202.

Grant, J. G. S. 1876 *Philosophical Thoughts on evolution.* Published by the author, Dunedin.

| | ca. 1899 | *Evolution The blackest form of materialism*. Published for the author, Dunedin. |

Gray, G. R.   1844–1849   *Genera of Birds*, 3 v. Longman, London.

1847   Description of *Strigops habroptilus*. Proceedings of the Zoological Society of London, 15: 61-62.

1862   A list of the Birds of New Zealand and the adjacent Islands. *The Ibis*, 4: 214-252.

Gray, J. E.   1831   Notes on a peculiar structure in the head of an Agama. *Zoological Miscellany*, 1: 13-14.

1867   *The lizards of Australia and New Zealand in the Collection of the British Museum*. Bernard Quaritch, London.

1868   *Synopsis of the species of Whales and Dolphins in the Collection of the British Museum*. Bernard Quaritch, London.

Gregg, P.   1971   *A Social and Economic History of Britain, 1760 – 1970 6th ed.* George G. Harrap and Co. Ltd., London.

Gressitt, J. L. (ed.)   1971   Entomology of the Aucklands and Other Islands South of New Zealand. *Pacific Insects Monograph*, 27 : 1-339.

Günther, A. E.   1980   *The Founders of Science at the British Museum, 1753 – 1900.* The Halesworth Press, London.

Haast, J.   1862   Observations on the Birds of the Western Districts of the Province of Nelson, New Zealand. *The Ibis* 4: 100-106.

1869   On the measurements of Dinornis Bones, obtained from excavation in a swamp situated at Glenmark, on the property of Messrs. Kermode and Co., up to February 15, 1868. *Transactions and Proceedings of the New Zealand Institute*, 1: 80-89.

1872   Moas and Moa Hunters. Address to the Philosophical Institute of Canterbury; Additional Notes; Third Paper on Moas and Moa Hunters. *Transactions and Proceedings of the New Zealand Institute*, 4: 66-107.

1875a   Researches and Excavations carried on in and near the Moa-bone Point Cave, Sumner Road, in the year 1872. *Transactions and Proceedings of the New Zealand Institute*, 7: 54-85.

1875b   Notes on the Moa-hunter Encampment at Shag Point, Otago. *Transactions and Proceedings of the New Zealand Institute*, 7: 91-98.

Hall-Jones, J.,   1976   *Fiordland Explored*. A. H. & A. W. Reed, Wellington.

1979   *The South Explored*. A. H. & A. W. Reed, Wellington.

Hansen, T.,   1964   *Arabia Felix, The Danish Expedition of 1761 – 1767*. Collins, London.

Harding, R. C.,   1892   *In* Maskell, W.M. 1892. On the Establishment of an Expert Agricultural Department in New Zealand. *Transactions and Proceedings of the New Zealand Institute*, 24 : 691.

Hawkesworth, J.,   1774   *Relation des voyages entrepris par ordre de Sa Majesté britannique, et successivement exécutés par le Commodore Byron, le Capitaine Carteret, le Capitaine Wallis & le Capitaine Cook, dans les vaisseaux le Dauphin, le Swallow & l'Endeavour. 9 v. Traduite de l'Anglois* [by J. B. A. Suard] Saillant et Nyon, Paris. (*Endeavour*, v.3-5) (There were translations into Dutch in 1774, German 1775).

Heaphy, C.,   1880   Notes on Port Nicholson and the Natives in 1839. *Transactions and Proceedings of the New Zealand Institute*, 12 : 32-39.

Hector, J.   1872   On Recent Moa Remains in New Zealand. *Transactions and Proceedings of the New Zealand Institute*, 4: 110 – 120.

Herbst, J. F. W.   1782–1804   *Versuch einer Naturgeschichte der Krabben und Krebse nebst einer Systematischen Beischreibung ihrer verschieden Arten.* G. A. Lange, Berlin and Stralsund.

| | 1783–1806 | *Natursystem aller bekannten in-und ausländischen Insekten . . . fortgesetzt von J. F. W. Herbst . . . Der Käfer* 5 v. + atlas J. Pauli, Berlin. (with K. G. Jablonsky). |

Hermann, J.   1782   In Zweiter Brief über einige Conchylien, an den Herausgeber. *Der Naturforscher*, 17: 126-152.

Hoare, M. E.   1969   'All Things Are Queer and Opposite' Scientific Societies in Tasmania in the 1840's. *Isis*, 60 (2): 198-209.

   1975   *Three Men in a Boat – The Forsters and New Zealand Science.* The Hawthorn Press, Melbourne.

   1976   *The Tactless Philosopher. Johann Reinhold Forster (1729 – 1798).* The Hawthorn Press, Melbourne.

   1977a   *Beyond the 'Filial Piety' – Science History in New Zealand, A Critical Review of the Art.* The Hawthorn Press, Melbourne.

   1977b   *Reform in New Zealand Science 1880 – 1926.* The Hawthorn Press, Melbourne.

(ed.)   1979   *Enlightenment and New Zealand, 1773 – 1774: essays commemorating the visit of Johann Reinhold Forster and George Forster with James Cook to Queen Charlotte and Dusky Sounds.* National Art Gallery, Wellington.

   1984   The Board of Science and Art 1913 – 1930: A Precursor to the D.S.I.R. *Bulletin of the Royal Society of New Zealand*, 21: 25-48.

Hombron J.-B., and Jacquinot, H.   1841, 1846, 1853   see Dumont d'Urville

Home, E.   1815   On the structure of the organs of respiration in animals which appear to hold an intermediate place between those of the Class Pisces and Class Vermes, and in two genera of the last mentioned class. *Philosophical Transactions of the Royal Society*, 1815: 264.

Hudson, G. V.   1880   *Vanessa antiopa* at Dulwich. *The Entomologist*, 13: 217.

   1892   *An Elementary Manual of New Zealand Entomology, Being an Introduction to the Study of our Native Insects.* West, Newman and Co., London.

   1898   *New Zealand Moths and Butterflies (Macrolepidoptera).* West, Newman and Co., London.

   1901   Entomological Field Work in New Zealand. *Transactions and Proceedings of the New Zealand Institute*, 33: 383-395.

   1904   *New Zealand Neuroptera, a Popular Introduction to the Life-Histories and Habits of May-flies, Dragon-flies, Caddis-flies and Allied Insects inhabiting New Zealand – including notes on their relation to Angling.* West, Newman and Co., London.

   1928   *The Butterflies and Moths of New Zealand.* Fergusson and Osborn, Wellington.

   1950   *Fragments of New Zealand Entomology.* Fergusson and Osborn, Ltd., Wellington.

Hudson, J. W.   1851   *The History of Adult Education in which is comprised a Full and Complete History of the Mechanics' and Literary Institutions, Athenaeum, . . . etc.* Longman, Brown, Green & Longmans, London. (Reprinted 1969 by the Woburn Press, London.)

Hutton F. W.   1861   *Some Remarks on Mr Darwin's theory.* (Reprinted from the 'Geologist' Magazine, no. 40), London.

   1871   *Catalogue of the Birds of New Zealand, with Diagnoses of the species.* J. Hughes, Wellington.

   1872–1881   [Published numerous catalogues including Echinoderms, molluscs, insects and fish 1872 – 1881. See *New Zealand National Bibliography* v.1, pp.510 – 511.]

   1873   On the Geographical Relations of the New Zealand Fauna. *Transactions and Proceedings of the New Zealand Institute*, 5: 227-256.

	1874	Notes by Captain Hutton on Dr Buller's 'Birds of New Zealand' with the Author's Replies thereto. *Transactions and Proceedings of the New Zealand Institute*, 6: 126-138 (First published in *The Ibis*, 1874).
	1876a	Notes on the Maori cooking places at the Mouth of the Shag River. *Transactions and Proceedings of the New Zealand Institute*, 8: 103 – 108.
	1876b	On *Peripatus novae-zealandiae*. *The Annals and Magazine of Natural History*, series 4, 18: 361-365.
	1877	On the Structure of *Peripatus novae-zealandiae*. *The Annals and Magazine of Natural History*, series 4, 20: 81-83.
	1892	The Moas of New Zealand. *Transactions and Proceedings of The New Zealand Institute*, 24: 93-172.
	1904	*Index Faunae Novae Zealandiae*. Dulau and Co., London.
Huxley, T. H.	(1869)1901	*The Scientific Memoirs of Thomas Henry Huxley*. Foster, M., and Lankester, E. R., (eds.) 4v. Macmillan and Co., London.
Jenkinson, S. H.	1940	*New Zealanders and Science*. Department of Internal Affairs, Wellington.
Jones, W.	1785	*Papiliones Nymphales Gemmati et Phalerati delineati et picti*. (Bound but unpublished, held in Hope Entomological Museum Library, Oxford University).
Karrer, C.	1978	Marcus Elieser Bloch (1723-1799) Sein Leben und die Geschichte seiner Fischsammlung. *Sitzungsberichte der Gesellschaft Naturforschender Freunde zu Berlin (N.F.)*, 18: 129-149.
Keulemans, T. & Coldewey, C. J.	1982	*Feathers to Brush, the Victorian Bird Artist, John Gerrard Keulemans 1842 – 1912*. Privately published by the authors, Epse and Melbourne.
King C.	1984	*Immigrant Killers: Introduced predators and the conservation of birds in New Zealand*. Oxford University Press, Auckland.
King, M.	1981	*The Collector, A Biography of Andreas Reischek*. Hodder and Stoughton, Auckland.
King, P. P.	1939	*Narrative of the surveying voyages of His Majesty's Ships Adventure and Beagle between the years 1826 and 1836, describing their examination of the southern shores of South America, and the Beagle's circumnavigation of the globe*. 3 v. Henry Colburn, London.
Kinsky, F. C. (Convener)	1970	*Annotated Checklist of the Birds of New Zealand*. A. H. and A. W. Reed, Wellington.
Knight, D. M.	1972	*Natural Science Books in English 1600 – 1900*. B. T. Batsford Ltd., London.
	1977	*Zoological Illustration – An essay towards a history of printed zoological pictures*. Dawson – Archon Books, Kent and Hamden Connecticut.
Kuschel, G.	1970	New Zealand Curculionoidea from Captain Cook's Voyages (Coleoptera). *New Zealand Journal of Science*, 13 (2) : 191-205.
Labé, G.	1981	see De Surville, 1981.
Latham, J.	1771–1785	*A general Synopsis of Birds* 3 v. Benjamin White, London.
	1790–1801	*Index Ornithologicus, sive system ornithologiae*. Published by the author, London.
Lesson, R. P.	1828	*Manual d'Ornithologie ou Description des Genres et des Principales Espéces d'Oiseaux*. 2 v. Roret Libraire, Paris.
	1830	see Duperrey
	1838	*Voyage autour du monde entrepris par ordre du Gouvernement sur la corvette La Coquille* 2 v. P. Pourrat Fréres, Paris.
Lesson, R. P., & Garnot, P.	1828	see Duperrey.
Levaillant, F.	1796–1912	*Histoire naturelle des Oiseaux d'Afrique*, 6 v. J. J. Fuchs, Paris.

Lightfoot, J. 1786 *A catalogue of the Portland Museum, lately the property of the Duchess Dowager of Portland, deceased: which will be sold by auction, etc.* London.

Limoges, C. 1980 *The development of the Muséum d'Histoire Naturelle of Paris, c. 1800 – 1914* in Fox, R., and Weisz, G., *The organisation of science and technology in France 1808 – 1914*, Cambridge University Press and Editions de la Maison des Sciences de l'Homme, Paris.

Lindsay, W. L. 1862 *The place and power of Natural history in Colonization; with special reference to Otago: Being portions of a lecture prepared for, and at the request of, the 'Young Men's Christian Association' of Dunedin.* John Dick, Dunedin.

Linklater, E. 1972 *The Voyage of the Challenger.* John Murray, London.

Linnaeus, C. 1735 *Systema Naturae, sive regna tria naturae systematice proposita per classes, ordines, genera & species,* Lugduni Batavorum also 10th ed. 1758 – 1759 *Edito decima, reformata.* 2 v. Holmiae.

Lyall, D. 1852 On the Habits of *Strigops habroptilus* or Kakapo. *Proceedings of the Zoological Society of London,* 20: 31-33.

Lysaght, A. M. 1959 Some Eighteenth Century Bird Paintings in the Library of Sir Joseph Banks. *Bulletin of the British Museum (Natural History) Historical Series,* 1 (6): 253-371.

1971 *Joseph Banks in Newfoundland and Labrador, 1766: His Diary, Manuscripts and Collections.* Faber and Faber, London.

1979 *Banks's artists and his* Endeavour *collections* in *Captain Cook and the South Pacific.* British Museum Yearbook 3, British Museum Publications, London.

McClelland, C. E. 1980 *State, society and university in Germany 1700 – 1914.* Cambridge University Press, New York.

McCormick, R. 1884 *Voyages of discovery in the Arctic and Antarctic Seas, and Round the world: Being personal narratives of attempts to reach the North and South Poles; and of an open-boat expedition up the Wellington Channel in search of Sir John Franklin and Her Majesty's ships 'Erebus' and 'Terror', in Her Majesty's Boat 'Forlorn Hope', under the command of the author . . . etc.* 2v. Sampson Low, Marston, Searle and Rivington, London.

McDowall, R. M. 1978 *New Zealand Freshwater Fishes, a guide and natural history.* Heinemann, Auckland.

McEvey, A. 1973 *John Gould's contribution to British art: a note on its authenticity.* Sydney University Press (for the Australian Academy of Humanities), Sydney.

MacGillivray, J. 1852 *Narrative of the voyage of H.M.S. Rattlesnake commanded by the late Captain Owen Stanley, R.N., F.R.S., &c. during the years 1846 – 1850.* 2 v. T. and W. Boone, London.

McKay, A., 1875 On the Identity of the Moa-hunters with the present Maori Race. *Transactions and Proceedings of the New Zealand Institute,* 7: 98-105.

McKay, D. L. 1979 *A Presiding Genius of Exploration: Banks, Cook, and Empire, 1767 – 1805;* in *Captain James Cook and his times,* R. Fisher and H. Johnston, eds. Douglas and McIntyre, Vancouver; Croon Helm London.

McLeod, R. 1982 On Visiting the 'Moving Metropolis': Reflections on the Architecture of Imperial Science. *Historical Records of Australian Science,* 5 (3): 1 – 16.

McLintock, A. H. (ed.) 1966 *An Encyclopaedia of New Zealand,* 3v. Government Printer, Wellington, pp.179-90.

McMillan, N. F., & Cernohorsky, W. D. 1979 William Swainson, F.R.S., in New Zealand with notes on his drawings held in New Zealand. *Journal of the Society for the Bibliography of Natural History,* 9 (2): 161-169.

McNab, R. 1907 *Murihiku and the Southern Islands; a history of the West Coast sounds, Foveaux Strait, Stewart Island, the Snares, Bounty, Antipodes, Auckland, Campbell and Macquarie Islands, from 1770 to 1829.* W. Smith, Invercargill.

(ed.)	1908–1914	*Historical Records of New Zealand* 2 v. Government Printer, Wellington.
Malaspina, A.	1885	*Viaje politico – científico alrededor del mundo por las corbetas Descubierta y Atrevida al mando de los Capitanes de Navio D. Alejandro Malaspina y Don José de Bustamente y Guerra desde 1789 á 1794.* Viude é Itijos de Abienzo, Madrid.
Mançeron, C.	1983	*The French Revolution, IV: Toward the Brink, 1785 – 1787.* Alfred Knopf, New York.
Mantell, G. A.	1940	*The Journal of Gideon Mantell, surgeon and geologist, covering the years 1818 – 1852.* E. Cecil Curwen, ed. Oxford University Press, London.
	1852	Notice of the Discovery by Mr Walter Mantell in the Middle Island of New Zealand, of a Living Specimen of the *Notornis*, a Bird of the Rail Family, allied to Brachypteryx, and hitherto unknown to Naturalists except in a Fossil state. *Transactions of the Zoological Society of London*, 4: 69-74. (also an announcement in the *Proceedings*, pp.209-212)
Martini, F. H. W., and Chemnitz, J. H.	1769–1795	*Neues systematisches Conchylien – Cabinet*, 11 v. Gabriel Nicolaus Raspe, Nuremberg.
Martyn, T.	1784–1787	*The Universal Conchologist, exhibiting The Figure of every known Shell accurately drawn and painted after nature: with a New Systematic Arrangement by the Author . . .* Published by The Author, London.
Maskell, W. M.	1887	*An account of the Insects noxious to agriculture and plants in New Zealand. The scale-insects (coccididae).* Government Printer, Wellington.
	1892	[On the Establishment of an Expert Agricultural Department in New Zealand]. *Transactions and Proceedings of the New Zealand Institute*, 24: 691.
	1896	Further Coccid Notes: with Descriptions of New Species, and Discussions of Questions of Interest. *Transactions and Proceedings of the New Zealand Institute*, 28: 380-411.
Mathias, P. (ed.)	1972	*Science and Society, 1600 – 1900.* Cambridge University Press, Cambridge.
Mead, A. D.	1966	*Richard Taylor, Missionary Tramper.* A. H. & A. W. Reed, Wellington.
Meads, M. J.	1982	A proposed revision of the *Naultinus/Heteropholis* species complex *in* New Zealand Herpetology, D. G. Newman, ed. *New Zealand Wildlife Service, Occasional Publication No. 2.* : *321–325.*
Medway, D. G.	1976	Extant Types of New Zealand Birds from Cook's Voyages. Pts. 1 and 2, *Notornis*, 23: 44-60, 120-137.
	1979	Some ornithological results of Cook's third voyage. *Journal of the Society for the Bibliography of Natural History*, 9 (3): 315-351.
Mellersh, H. E. L.	1968	*FitzRoy of the Beagle.* Rupert Hart-Davis, London.
Miers, E. J.	1876	*Catalogue of the stalk– and sessile-eyed Crustacea of New Zealand.* E. W. Janson, London.
Miller, D.	1952	*Insect People of the Maori.* The Polynesian Society, Wellington.
	1953	Historical Review of New Zealand Entomology. *Report of the Seventh Science Congress, Royal Society of New Zealand, 1951*: 80-86.
	1956	Bibliography of New Zealand Entomology, 1775 – 1952, with annotations. *Department of Scientific and Industrial Research Bulletin*, 120 : 1–492.
	1971	*Common Insects in New Zealand.* A. H. & A. W. Reed, Wellington.
Moseley, H. N.	1877	Remarks on Observations by Captain Hutton, Director of the Otago Museum, on *Peripatus novae-zealandiae*, with Notes on the Structure of the Species. *The Annals and Magazine of Natural History*, Series 4, 19: 85 – 91.
	1879	*Notes by A naturalist on the 'Challenger', being An account of various observations made during the Voyage of H.M.S. 'Challenger' round the world, in the years 1872 – 1876, under the Commands of Capt. Sir G. S. Nares, R.N. K.C.B., F.R.S., and Capt. E. T. Thomson, R.N.,* Macmillan and Co., London.
Murison, W.	1872	Notes on Moa Remains. *Transactions and Proceedings of the New Zealand Institute*, 4: 120-124.

Murray-Oliver, A. A. St.C. 1969 *Captain Cook's artists in the Pacific, 1769 – 1779.* Avon Fine Prints, Christchurch.

Natusch, S. 1978 *The Cruise of the Acheron.* Whitcoulls, Christchurch.

Newman, D. G. (Ed.) 1982 New Zealand Herpetology (Appendix). *New Zealand Wildlife Service Occasional Publication,* 2: 441-443.

Newton, A. 1911 *Kiwi in Encyclopaedia Britannica* 11th ed., v. 15 Cambridge University Press, Cambridge.

Nicholas, J. L. 1817 *Narrative of a Voyage to New Zealand, Performed in the Years 1814 and 1815, in Company with the Rev. Samual Marsden, Principal Chaplain of New South Wales.* 2 v. James Black and Son, London.

Oliver, H. C. 1968 *Annotated Index to Some Early New Zealand Bird Literature* Wildlife Publication No. 106. Department of Internal Affairs, Wellington.

Oliver, W. R. B. 1949 The Moas of New Zealand and Australia. *Dominion Museum Bulletin No. 15:* 1-205.

Olivier, G. A. 1789–1800 *Entomologie, ou Histoire naturelle des Insectes . . . Coleoptères* 6 v. Baudoin, Paris.

Ollivier, I. 1983 French explorers in New Zealand, 1769 – 1840: a list of manuscript material; Pts. I, II. *Turnbull Library Record,* 36 : 4 – 19, 95 – 101.

Orchiston, D. W. 1969 The White Island Crater Valley in 1826. *Transactions of the Royal Society of New Zealand,* 2 (10): 139-142.

Owen, R. 1840a On the Anatomy of the Southern Apteryx *(Apteryx australis* Shaw). *Transactions of the Zoological Society of London,* 2: 257-302.

1840b Exhibition of a Bone of an Unknown Struthious Bird from New Zealand. *Proceedings of the Zoological Society of London,* 7: 169-171.

1842 Notice of a Fragment of the Femur of a Gigantic Bird of New Zealand. *Transactions of the Zoological Society of London,* 3: 29-32.

1843–1844 On *Dinornis Novae-Zealandiae. Proceedings of the Zoological Society of London,* 11: 8-10, 144-146.

1844 On Dinornis, an Extinct Genus of tridactyle Struthious Birds, with descriptions of portions of the skeleton of five species which formerly existed in New Zealand (Part 1). *Transactions of the Zoological Society of London,* 3: 235-276.

1848 On Dinornis (Part III): Containing a Description of the Skull and Beak of that genus. . . . and of two other genera of Birds *Notornis* and *Nestor;* forming part of an extensive series of Ornithic remains discovered by Mr Walter Mantell at Waingongoro, North Island of New Zealand. *Transactions of the Zoological Society of London,* 3: 345-378. (Also an announcement in the *Proceedings,* 16: 1-11, of the same year)

1879 *Memoirs on the Extinct Wingless Birds of New Zealand; with an Appendix on those of England, Australia, Newfoundland, Mauritius, and Rodriguez.* 2 v. John van Voorst, London.

1886 On Dinornis (Part XXV); containing a description of *Dinornis elephantopus. Transactions of the Zoological Society of London,* 12: 1-3.

Owen, R. (Rev.) 1894 *The Life of Richard Owen,* 2 v. John Murray, London. (By his grandson the Rev. Richard Owen.)

Pallas, P. S. 1766a *Elenchus Zoophytorum sistens generum adumbrationes generaliores et specierum cognitarum succinctus descriptiones . . .* etc. P. Van Cleef, The Hague.

1766b *Miscellanea Zoologie quibus novae imprimis atque obscurae animalium species describuntur et observationibus iconibusque illustrantur.* P. van Cleef, The Hague.

Parker, T. J. 1886 On the skeleton of *Reglecus argenteus. Transactions of Zoological Society of London,* 12 (1): 5-33.

Parkinson, P., 1984 William Swainson 1789 – 1855: Relics in The Antipodes. *Bulletin of the Royal Society of New Zealand*, 21: 49-63.

Parkinson, S., 1773 *A Journal of a Voyage to the South Seas, in His Majesty's Ship, The Endeavour. Faithfully transcribed from the Papers of the late Sydney Parkinson, Draughtsman to Joseph Banks, Esq., on his late Expedition with Dr Solander, round the World.* Stanfield Parkinson, ed. London.

Pennant, T. 1761–1766 *The British Zoology.* Cymmrodorion Society, London. also: 2nd ed. 1768 – 1770, 4 v. London and Chester.

Penniket, J. R. 1982 *Common Seashells.* A. H and A. W. Reed, Wellington.

Perry, G. 1811 *Conchology, or the natural history of Shells: containing a new arrangement of the genera and species . . . etc.* Miller, London.

Perry, T. M. 1966 *Cunningham, Allan* in *Australian Dictionary of Biography* v.1. A. G. L. Shaw and C. M. H. Clark eds. Melbourne University Press, Melbourne.

Phillips, C. 1894 On a Common Vital Force. *Transactions and Proceedings of the New Zealand Institute*, 26: 604-619.

Polack, J. S. 1838 *New Zealand: being a narrative of Travels and adventures during a residence in that country between the years 1831 and 1837.* Richard Bentley, London.

Poole, J. B. & Andrews, K. 1972 *The Government of Science in Britain.* Weidenfeld and Nicolson, London.

Potts, T. H. 1870 On the Birds of New Zealand. *Transactions and Proceedings of the New Zealand Institute*, 2: 40-80.

1872 Notes on a New Species of *Apteryx* (*A. Haastii*, Potts). *Transactions and Proceedings of the New Zealand Institute*, 4: 204-205.

1882 *Out in the Open: A budget of scraps of natural history, gathered in New Zealand.* Lyttelton Times, Christchurch.

Powell, L. 1871 On *Latrodectes* (Katipo) the Poisonous spider of New Zealand. *Transactions and Proceedings of the New Zealand Institute*, 3: 56-59.

Putnam, G. 1977 A Brief History of New Zealand Marine Biology. *Tuatara*, 22 (3): 189-212.

Quoy J. R. C. & Gaimard, J. P. 1830, 1833 see Dumont d'Urville.

Radford W. P. K. 1980 The Fabrician Types of the Australian and New Zealand Coleoptera in the Banks Collection at the British Museum (Natural History). *Records of the South Australian Museum*, 18 (8): 155-197.

Ramsay, G. W. 1979 Annotated bibliography and index to the New Zealand wetas (Orthoptera: Stenopelmatidae, Rhaphidophoridae.) *New Zealand Department of Scientific and Industrial Research Information Series* No. 144: 1-56.

Reed, A. W. 1979 *Two hundred years of New Zealand history.* The Reed Trust, Wellington.

Reingold, N. (ed.) 1966 *The Wilkes Expedition* in *Science in Nineteenth-century America.* Hill and Wang, New York.

Richardson, J. 1843 Report on the present state of the Ichthyology of New Zealand. *Report of the British Association for the Advancement of Science, 12th meeting:* 12-30.

1844–1875 *The Zoology of the Voyage of H.M.S. Erebus and Terror, under the Command of Captain Sir James Clark Ross, R.N., F.R.S., during the years 1839 to 1843.* J. Richardson and J. E. Gray, eds. 2 v. E. W. Janson, London.

1846 On Preserving specimens of fish &c. for exportation. *Tasmanian Journal of Natural Science, Agriculture, Statistics, etc.* 2: 72-73.

Röding, P. F., 1798 *Museum Boltenianum.* Privately published, Hamburg. (see also Dall, 1915).

Ross, J. C., 1847 *A Voyage of Discovery and research in the southern and Antarctic regions, during the years 1839-43.* 2 v. John Murray, London.

Rothschild, W., 1893 Notes on the Genus Apteryx. *The Ibis*, 5: 573-576.

Rowley, G. D. 1876 *Ornithological miscellany*, 3 v. Trübner and Co., Bernard Quaritch and R. H. Porter, London.

Russell, C. A. 1983 *Science and Social Change: 1700-1900* Macmillan, London.

Salmon, J. T. 1946 Obituary: George Vernon Hudson, F.R.S.N.Z. (1867-1946) *Transactions of the Royal Society of New Zealand* 76 (2): 264-266.

Sauer, G. C. 1982 *John Gould The Bird Man: A Chronology and Bibliography*. The University Press, Kansas.

Savage, J. 1807 *Some account of New Zealand; particularly The Bay of Islands, and surrounding country; with a Description of the religion and government, language, arts, manufactures, manners and customs of the natives, &c, &c.* J. Murray and A. Constable and Co., London and Edinburgh.

Sawyer, F. C. 1950 Some Natural History Drawings Made During Captain Cook's First Voyage Round the World. *Journal of the Society for the Bibliography of Natural History* 2(6): 190-193.

Scherzer, K. Ritter von 1861−1862 *Reise der Oesterreichischen Fregatte Novara um die Erde, in den Jahren 1857, 1858, 1859, unter den Befehlen des Commodore B. von Wüllerstorf-Urbair*. 3 v. Vienna. Also an English version 1861 − 1863: *Narrative of the Circumnavigation of the Globe by the Austrian frigate Novara, (Commodore B. von Wüllerstorf-Urbair) Undertaken by the Order of the Imperial Government, in the years 1857, 1858, 1859, under the immediate auspices of His I. and R. Highness the Archduke Ferdinand Maximilian, Commander-in-Chief of the Austrian Navy*. Saunders, Otley and Co., London. For detail of the multi-authored zoological and geological parts of the *Novara* voyage see Bagnall, A. G. (ed.), 1980, *New Zealand Bibliography* v. 1, p.925-929.

Schiner, J. R. 1868 *Diptera* in Scherzer, 1864−75, *Zoologischer Theil*, v. 2 - 3, of *Reise der. . . Fregatte Novara . . . etc.* Vienna.

Schönherr, C. J. 1806−1817 *Synonymia Insectorum*. Nordstrom, Stockholm, etc.
1833−1845

Sclater, P. L. 1858 On the General Geographical Distribution of the Members of the Class Aves. *Journal of the Proceedings of the Linnean Society of London (Zoology)*, 2, 1858: 130.

Schmarda, L. K. 1859 *Neue Turbellarian, Rotatorien und Anneliden beobachtet und gesammelt auf einer Reise um die Erde 1853 bis 1857*, 2 v. Wilhelm Engelman, Leipzig.

1861 *Ludwig K. Schmarda's Reise um die Erde in den Jahren 1853 − 1857*. George Westermann, Braunschweig.

Scholefield, G. H. 1940 *A dictionary of New Zealand biography*, 2 v. Department of Internal Affairs, Wellington.

Scott, J. H. & Parker T. J. 1889 On a specimen of *Ziphius* recently obtained near Dunedin. *Transactions of the Zoological Society of London*. 12(8): 241−248.

Sharrell, R. 1982 *New Zealand Insects and their Story* (2nd ed.) Collins, Auckland.

Sharpe, A. (ed.) 1971 *Duperrey's visit to New Zealand in 1824*. Alexander Turnbull Library, Wellington.

Shaw, G. 1792−1796 *Museum Leverianum, containing select specimens from the museum of the late Sir Ashton Lever, Kt. with descriptions in Latin and English*. James Parkinson, London.

Shaw, G. and Nodder, F. P. 1789−1813 *Vivarium naturae, sive rerum naturalium icones. The Naturalists Miscellany*, 24 v. Published by the authors, London. (Vols. 13 - 24 co-authored by E. Nodder, engravings attributed to R. P. Nodder).

Short, T. K. 1837 [letter from T. K. Short 10.8.1836] *Proceedings of the Zoological Society of London*, 5: 24.

Shortland, E.	1851	*The Southern Districts of New Zealand: a journal, with passing notices of the customs of the aborigines.* Longman, Brown, Green, & Longmans, London.
Sibson, R. B.	1983	Richard Laishley 1815–1897 – priest, painter, naturalist. *Notornis,* 30 (1): 29-33.
Simmons, D. R.	1968	Man, Moa and the Forest. *Transactions of the Royal Society of New Zealand.* 2 (7): 115-127.
Simpson, M. J. A.	1984	The contribution of the French to the Botany of Banks Peninsular. *Royal Society of New Zealand Bulletin No. 21:* 65-70.
Skinner, H. D.	1964	Crocodile and Lizard in New Zealand Myth and Material Culture. *Records of the Otago Museum: Anthropology,* 1: 1-43.
Smith, S. P.	1909	Captain Dumont D'Urville's visit to Tolaga Bay in 1827. *Transactions and Proceedings of the New Zealand Institute,* 41: 130-139.
	1913	The Lore of the Whare-wananga or Teachings of the Maori College on Religion, Cosmogony and History. Part 1, Te Kauwae-runga or 'Things Celestial'. *Polynesian Society Memoir No. 3:* 1-193.
Sowman, W. C. R.	1981	*Meadow, mountain, forest and stream: The provincial history of the Nelson Acclimatisation Society, 1863-1968.* Nelson Acclimatisation Society, Nelson.
Sparrman, A.	1786–1789	*Museum Carlsonianum, in quo novas et selectas Aves, coloribus ad vivum brevique descriptione illustras, suasu et sumtibus generosissimi possesoris, exhibet Andreas Sparrman . . . ex Typographia regia, Holmiae.*
	1953	*A Voyage round the World with Captain James Cook in H.M.S. Resolution.* Translated by H. Beamish and A. Mackenzie Grieve, Robert Hale, Ltd., London.
Spengler, L.	1776	Abhandlung von den Conchylien der Südsee überhaupt und einigen neuen Arten derselben insbesondere. *Der Naturforscher,* 9: 145-168.
	1782	Berschreibung zwoer seltenen neuen Gattungen Südländicher Conchylien mit eleuchterten Abbildungen. *Der Naturforscher,* 17: 24-31.
Spokes, S.	1927	*Gideon Algernon Mantell, LL.D., F.R.C.S., F.R.S. surgeon and geologist.* Bale and Danielsson, London.
Stancombe, G. H.	1966	*Barclay, Andrew (1759 – 1839)* in *Australian Dictionary of Biography,* A. G. L. Shaw and C. M. H. Clark, eds. Melbourne University Press, Melbourne.
Stearn, W. T.	1968	The botanical results of the Endeavour Voyage. *Endeavour,* 27: 3-10.
	1981	*The Natural History Museum at South Kensington.* Heinemann and the British Museum (Natural History), London.
Stenhouse, J.	1984	'The Wretched Gorilla Damnification of Humanity': the 'battle' between science and religion over evolution in nineteenth-century New Zealand. *The New Zealand Journal of History,* 18 (2): 143-162.
Stephen, L., & Lee, S., (eds.)	1908–1912	*Dictionary of National Biography,* 21 v. + supplements. Smith, Elder and Co., London.
Strange, F.	1847	Notes on some Rare Birds of New Zealand and Australia. [letter from F. Strange to J. Gould]. *Proceedings of the Zoological Society of London,* 15: 50-51.
Stresemann, E.	1975	*Ornithology from Aristotle to the Present.* Translated by H. J. and C. Epstein, Harvard University Press, Cambridge, Massachusetts.
Suter, H.	1913	*Manual of the New Zealand Mollusca with an Atlas of quarto plates.* John Mackay, Wellington.
Swainson, W.	1840	*Taxidermy; with the biographies of zoologists and notices of their works.* Longman, Swainson, W. London.
Taylor, N.	1959	*Early Travellers in New Zealand.* Oxford University Press, London.
Taylor, R.	1848	*A leaf from the natural history of New Zealand; or A vocabulary of its different*

productions, &c., &c., with their native names. Robert Stokes and J. Williamson, Wellington and Auckland.

1855   *Te Ika-a-Maui, or New Zealand and its inhabitants, Illustrating the Origin, manners, customs, mythology, religion, rites, Songs, proverbs, fables and language of the natives. Together with the Geology, natural history, productions, and climate of the country; etc.* Wertheim and Macintosh, London (2nd edition 1870).

1873   An Account of the First Discovery of Moa Remains. *Transactions and Proceedings of the New Zealand Institute,* 5: 97-101.

Thomson, A. S.   1853   On the Discovery of a Frog in New Zealand. *Edinburgh New Philosophical Journal,* 1853: 66-69.

1859   *The story of New Zealand: Past and present – savage and civilised.* 2 v. John Murray, London.

Thomson, C W.   1885   *Report on the scientific results of the Voyage of H.M.S. Challenger during the years 1873-76 under the command of Captain G. S. Nares, R.N., F.R.S., and the late Captain Frank Tourle Thomson, R.N.* 2 v. H.M.S.O., London.

Thomson, G. M.   1922   *The Naturalisation of Animals and Plants in New Zealand.* Cambridge University Press, Cambridge.

Tomes, R. F.   1857   On two Species of Bats inhabiting New Zealand. *Proceedings of the Zoological Society of London,* 25: 134-142

Travers, W. T. L.   1870   On the Changes effected in the Natural Features of a New Country by the Introduction of Civilized Races. *Transactions and Proceedings of the New Zealand Institute,* 2: 299-330.

1876   Notes on the Extinction of the Moa, with a review of the discussions of the subject, published in the 'Transactions of the New Zealand Institute'. *Transactions and Proceedings of the New Zealand Institute;* 8: 58-83.

Turbott, E. G. (ed.)   1967   *Buller's Birds of New Zealand.* Whitcoulls, Christchurch.

Tuxen, S. L.,   1967   The Entomologist, J. C. Fabricius. *Annual Review of Entomology,* 12: 1-14.

Vancouver, G.   1798   *A Voyage of discovery to the North Pacific Ocean, and round the World; ... performed in the years 1790, 1791, 1792, 1793, 1794, and 1795, in the Discovery Sloop of War, and Armed Tender Chatham under the command of Captain George Vancouver.* 3 v. G. G. and J. Robinson and J. Edwards, London.

1984   *A voyage of discovery to the North Pacific Ocean, and round the World, 1791 – 1795.* 4 v. W. K. Lamb, ed. The Hakluyt Society, London.

Venn, J. A.   1940   *Alumni Cantabrigienses, a biographical list of all known students, graduates and holders of office at the University of Cambridge, from the earliest times to 1900.* Pt II v.1. Cambridge University Press, Cambridge.

Von Haast   1948   The life and times of Sir Julius von Haast K.C.M.G., PH.D., D.Sc., F.R.S., explorer geologist, museum builder. Published by the author, Wellington.

Von Hochstetter, F.   1863   *Neu-Seeland.* Cotta'scher Verlag. Stuttgart.

1867   *New Zealand, its physical geography, geology and natural history with special reference to the results of government expeditions in the Provinces of Auckland and Nelson.* J. G. Cotta, Stuttgart.

Von Lendenfeld, R. J. E.   1889   *A monograph of the horny sponges.* Trübner and Co., London.

Wade, W. R.   1842   *A journey in the northern island of New Zealand: Interspersed with various information relative to the country and people.* George Rolwegan, Hobart.

Walch, J. E. I.   1774   Beischreibung einiger neuentdeckten Conchylien. *Der Naturforscher,* 4: 33-56.

1776   Beschreibung einiger neu entdeckten Conchylien. Drittes Stück. 2. von einer sehr seltenen träuselförmigen Mondschnecke. *Der Naturforscher,* 9: 203-204.

Wallace, A. R.	1876	*The geographical Distribution of animals With a study of the relations of living and extinct faunas as elucidating the past changes of the earth's surface.* 2 v. Macmillan and Co., London.
Walsh, P.	1893	The Effect of Deer on the New Zealand Bush: A Plea for the protection of our Forest Reserves. *Transactions and Proceedings of the New Zealand Institute,* 25: 435-439.
Waterhouse, F. H.	1885	*The dates of publication of some of the zoological works of the late John Gould.* R. H. Porter, London.
Waterhouse, G. A.	1938	Notes on Jones' Icones (Lepidoptera). *Proceedings of the Entomological Society of London* (series A), 13: 9-15.
Watt, J.	1979	*Medical Aspects and Consequences of Cook's Voyages* in *Captain James Cook and His Times,* R. Fisher and H. Johnston eds. Douglas and McIntyre, Vancouver; Croon Helm, London.
Watt, J. C.	1960	The New Zealand Onychophora. *Tane,* 8 : 95-105.
Weber, F.	1795	*Nomenclator Entomologicus* C. E. Bohn, Chilonii et Hamburgi.
Westwood, J.	1872	Description of some new Papilionidae. *Transactions of the Entomological Society of London,* 1872 (1): 85-110.
Wheeler, A.	1981	*The Forsters' Fishes in James Cook: The Journal of H.M.S. Resolution 1772 – 1775* Genesis Publications, Guildford.
White, G.	1789	*The Natural History and Antiquities of Selbourne, in the County of Southampton.* B. White and Son, London.
Whitehead, P. J. P.	1968	*Forty drawings of fishes made by the artists who accompanied Captain James Cook on his three voyages to the Pacific 1768-71, 1772-75, 1776-80, some being used by authors in the description of new species.* British Museum (Natural History), London.
	1969	Zoological Specimens from Captain Cook's Voyages. *Journal of the Society for the Bibliography of Natural History,* 5(3): 161-201.
	1973	Some further notes on Jacob Forster (1739 – 1806) mineral collector and dealer. *Mineralogical Magazine,* 39: 361-363.
	1978a	The Forster collection of Zoological drawings in the British Museum (Natural History). *Bulletin of the British Museum (Natural History), Historical series,* 6(2): 25-47.
	1978b	A Guide to the Dispersal of Zoological Material from Captain Cook's Voyages. *Pacific Studies,* 2 : 53-93.
	1984	'Should fate command me to the farthest verge': The Reverend Richard Laishley in New Zealand, 1860 – 1897. *Bulletin of the Royal Society of New Zealand,* 21: 101-112.
Whitley, G. P.	1959	Sidelights on New Zealand Ichthyology. *Australian Zoologist,* 12: 110-119.
	1968	A Check-list of the Fishes Recorded from the New Zealand Region. *Australian Zoologist,* 25: 1-102.
Whittell, H. M.	1947	Frederick Strange. *Australian Zoologist,* 11 : 96-114.
Wilkes, C.	1845	*Narrative of the United States Exploring Expedition, During the years 1838, 1839, 1840, 1841, 1842,* 5 v. Lea and Blanchard, Philadelphia. [The zoology published as follows: v. 7 Dana, 1848, zoophytes, Atlas 1849; v. 8 Cassin, 1858, mammology and ornithology, Atlas 1858; v. 12 Gould, 1852, Mollusca, Atlas 1856; v. 13, 14 Dana, 1852, Crustacea, Atlas, 1855; v. 20, Girard, 1858, herpetology, Atlas 1858.]
Wilkins, G. L.	1954	Captain Cook's Imperial Sun Trochus, *Journal of Conchology,* 24 (1): 7-12.
	1955	A Catalogue and Historical Account of the Banks Shell Collection. *Bulletin of the British Museum (Natural History) Historical Series* 1 (3): 71-119.
Willey, B.	1960	*Darwin and Butler, Two Versions of Evolution.* Chatto and Windus, London.

Williams, W. L.   1872   On the Occurrence of Footprints of a Large Bird, found at Turanganui, Poverty Bay. *Transactions and Proceedings of the New Zealand Institute*, 4: 124-127.

Wodzicki, K. A.   1950   Introduced Mammals of New Zealand. *Department of Scientific and Industrial Research Bulletin No. 98*: 1-250.

Wright, O. (trl.)   1950   *New Zealand 1826 – 1827; An English translation of the* Voyage de l'Astrolabe *in New Zealand waters, with an introductory essay by Olive Wright.* Wingfield Press, Wellington.

   1955   *The voyage of the Astrolabe – 1840; an English rendering of the journals of Dumont d'Urville and his officers of their visit to New Zealand in 1840, together with some account of Bishop Pompallier and Charles, Baron de Thierry.* A. H. and A. W. Reed, Wellington.

Wynn, G.   1977   Conservation and Society in late ninteenth-century New Zealand. *New Zealand Journal of History*, 2(1): 124-136.

Yarrell, W.   1833a   [A 'Description, with Additional Particulars, of the *Apteryx Australis* of Shaw'.] *Proceedings of the Zoological Society of London*, 1 : 80.

   1833b   Description, with some additional Particulars, of the *Apteryx Australis* of Shaw. *Transactions of the Zoological Society of London*, 1: 71-76.

Yate, W.   1835   *An Account of New Zealand; and of the Formation and Progress of the Church Missionary Society's mission in the Northern Island.* Seely and Burnside, London.

Zimsen, E.   1964   *The Type Material of I. C. Fabricius.* Munksgaard, Copenhagen.

Zorn von Plobsheim, F. A.   1775   Beschreibung einiger seltnen Conchylien aus der Sammlung der naturforschenden Gesellschaft zu Danzig. *Der Naturforscher*, 7: 151-168. (although his name is not mentioned, this work is attributed to Zorn)

   1778   *Neue Sammlung, von Versuchen und Abhandlungen der Naturforschenden Gesellschaft in Danzig*, 1: 247-288.

GENERAL INDEX

229

ZOOLOGICAL INDEX